# Stochastic Optimization Methods

Kurt Marti

# Stochastic Optimization Methods

With 14 Figures

Springer

Univ. Professor Dr. sc. math. Kurt Marti
Federal Armed Forces University Munich
Aero-Space Engineering and Technology
85577 Neubiberg/Munich
Germany
kurt.marti@unibw-muenchen.de

Cataloging-in-Publication Data
Library of Congress Control Number: 2004112585

ISBN 3-540-22272-3 Springer Berlin Heidelberg New York



Springer is a part of Springer Science+Business Media

springeronline.com


Printed in Germany



Hardcover-Design: Erich Kirchner, Heidelberg

SPIN 11015260 42/3130-5 4 3 2 1 0 – Printed on acid-free paper

# Preface

Optimization problems in practice depend mostly on several model parameters, noise factors, uncontrollable parameters, etc., which are not given fixed quantities at the planning stage. Typical examples from engineering and economics/operations research are: Material parameters (e.g. elasticity moduli, yield stresses, allowable stresses, moment capacities, specific gravity), external loadings, friction coefficients, moments of inertia, length of links, mass of links, location of center of gravity of links, manufacturing errors, tolerances, noise terms, demand parameters, technological coefficients in input-output functions, cost factors, etc.. Due to several types of stochastic uncertainties (physical uncertainty, economic uncertainty, statistical uncertainty, model uncertainty) these parameters must be modelled by random variables having a certain probability distribution. In most cases at least certain moments of this distribution are known.

In order to cope with these uncertainties, a basic procedure in the engineering/economic practice is to replace first the unknown parameters by some chosen nominal values, e.g. estimates, guesses, of the parameters. Then, the resulting and mostly increasing deviation of the performance (output, behavior) of the structure/system from the prescribed performance (output, behavior), i.e., the "tracking error", is compensated by (online) input corrections. However, the online correction of a system/structure is often time-consuming and causes mostly increasing expenses (correction or recourse costs). Very large recourse costs may arise in case of damages or failures of the plant. This can be omitted to a large extent by taking into account already at the planning stage the possible consequences of the tracking errors and the known prior and sample information about the random data of the problem. Hence, instead of relying on ordinary deterministic parameter optimization methods - based on some nominal parameter values - and applying then just some correction actions, stochastic optimization methods should be applied: Incorporating the stochastic parameter variations into the optimization process, expensive and increasing online correction expenses can be omitted or at least reduced to a large extent.

Consequently, for the computation of robust optimal decisions/designs, i.e., optimal decisions which are insensitive with respect to random parameter variations, appropriate deterministic substitute problems must be formulated first. Based on decision theoretical principles, these substitute problems depend on probabilities of failure/success and/or on more general expected cost/loss terms. Since probabilities and expectations are defined by multiple integrals in general, the resulting often nonlinear and also nonconvex deterministic substitute problems can be solved by approximative methods only. Hence, the analytical properties of the substitute problems are examined, and appropriate deterministic and stochastic solution procedures are presented. The aim of the present book is therefore to provide analytical and numerical tools – including their mathematical foundations – for the approximate computation of robust optimal decisions/designs as needed in concrete engineering/economic applications.

Two basic types of deterministic substitute problems occur mostly in practice:

* Reliability-Based Optimization Problems:
  Minimization of the expected primary costs subject to expected recourse cost constraints (reliability constraints) and remaining deterministic constraints, e.g. box constraints. In case of piecewise constant cost functions, probabilistic objective functions and/or probabilistic constraints occur;
* Expected Total Cost Minimization Problems:
  Minimization of the expected total costs (costs of construction, design, recourse costs, etc.) subject to the remaining deterministic constraints.

Basic methods and tools for the construction of appropriate deterministic substitute problems of different complexity and for various practical applications are presented in Chapter 1 and part of Chapter 2 together with fundamental analytical properties of reliability-based optimization/expected cost minimization problems. Here, a main problem is the analytical treatment and numerical computation of the occurring expectations and probabilities depending on certain input variables.

For this purpose deterministic and stochastic approximation methods are provided which are based on Taylor expansion methods, regression and response surface methods (RSM), probability inequalities, differentiation formulas for probability and expectation value functions. Moreover, first order reliability methods (FORM), Laplace expansions of integrals, convex approximation/deterministic descent directions/efficient points, (hybrid) stochastic gradient methods, stochastic approximation procedures are considered. The mathematical properties of the approximative problems and iterative solution procedures are described.

After the presentation of the basic approximation techniques in Chapter 2, in Chapter 3 derivatives of probability and mean value functions are obtained by the following methods: Transformation method, stochastic completion and transformation, orthogonal function series expansion. Chapter 4 shows how nonconvex deterministic substitute problems can be approximated by convex mean value minimization problems. Depending on the type of loss functions and the parameter distribution, feasible descent directions can be constructed at non-efficient points. Here, efficient points, determined often by linear/quadratic conditions involving first and second order moments of the parameters, are candidates for optimal solutions to be obtained which much larger effort. The numerical/iterative solution techniques presented in Chapter 5 are based on hybrid stochastic approximation, hence, methods using not only simple stochastic (sub) gradients, but also deterministic descent directions and/or more exact gradient estimators. More exact gradient estimators based on Response Surface Methods (RSM) are described in the first part of Chapter 5. The convergence in the mean square sense is considered, and results on the speed up of the convergence by using improved step directions at certain iteration points are given. Various extensions of hybrid stochastic approximation methods are treated in Chapter 6 on stochastic approximation methods with changing variance of the estimation error. Here, second order search directions, based e.g. on the approximation of the Hessian, and asymptotic optimal scalar and matrix valued step sizes are constructed. Moreover, using different types of probabilistic convergence concepts, related convergence properties and convergence rates are given. Mathematical definitions and theorems needed here are given in the Appendix. Applications to the approximate computation of survival/failure probabilities for elastoplastic mechanical structures and its application to stochastic structural optimization are given in the last Chapter 7. The reader of this book needs some basic knowledge in linear algebra, multivariate analysis and stochastics.

Realizing this monograph, the author was supported by several collaborators: First of all I would like to thank Ms Elisabeth Lössl from the Institute for Mathematics and Computer Sciences for the excellent LaTeXtypesetting of the whole manuscript. Without her very precise and careful work, this project could not have been realized. I also owe thanks to my former collaborators, Dr. Andreas Aurnhammer, for further LaTeXsupport and for providing English translation of some German parts of the original manuscript, and to Thomas Platzer for proof reading of some parts of the book. Last but not least I am indebted to Springer-Verlag for inclusion the book into the Springer-program. I would like to thanks especially to the Publishing Director Economics and Management Science of Springer-Verlag, Dr. Werner A. Müller, for his really very long patience until the completion of this book.

Munich, August 2004
Kurt Marti

# Contents

## Part I Basic Stochastic Optimization Methods

## Part II Differentiation Methods



## Part III Deterministic Descent Directions

# Part I

# Basic Stochastic Optimization Methods

# 1

# Decision/Control Under Stochastic Uncertainty

## 1.1 Introduction

Many concrete problems from engineering, economics, operations research, etc., can be formulated by an optimization problem of the type

$$\min f_0(a, x) \tag{1.1a}$$

s.t.

$$f_i(a, x) \leq 0, i = 1, \ldots, m_f \tag{1.1b}$$

$$g_i(a, x) = 0, i = 1, \ldots, m_g \tag{1.1c}$$

$$x \in D_0. \tag{1.1d}$$

Here, the objective (goal) function $f_0 = f_0(a, x)$ and the constraint functions$f_i = f_i(a, x), i = 1, \ldots, m_f, g_i = g_i(a, x), i = 1, \ldots, m_g$, defined on a joint subset of $\mathbb{R}^\nu \times \mathbb{R}^r$, depend on a decision, control or input vector $x = (x_1, x_2, \ldots, x_r)^T$ and a vector $a = (a_1, a_2, \ldots, a_\nu)^T$ of model parameters. Typical model parameters in technical applications, operations research and economics are material parameters, external load parameters, cost factors, technological parameters in input–output operators, demand factors. Furthermore, manufacturing and modelling errors, disturbances or noise factors, etc., may occur. Frequent decision, control or input variables are material, topological, geometrical and cross-sectional design variables in structural optimization [55], forces and moments in optimal control of dynamic systems and factors of production in operations research and economic design.

The objective function (1.1a) to be optimized describes the aim, the goal of the modelled optimal decision/design problem or the performance of a technical, economic system or process to be controlled optimally. Furthermore, the constraints (1.1b-d) represent the operating conditions guaranteeing a safe structure, a correct functioning of the underlying system, process, etc.. Note that the constraint (1.1d) with a given, fixed convex subset $D_0 \subset \mathbb{R}^r$ summarizes all (deterministic) constraints being independent of unknown model parameters $a$, as e.g. box constraints:

$$x^L \leq x \leq x^U \tag{1.1d'}$$

with given bounds $x^L, x^U$.

Important concrete optimization problems, which may be formulated, at least approximatively, this way are problems from optimal design of mechanical structures and structural systems [2, 55, 125, 135], adaptive trajectory planning for robots [3, 6, 25, 88, 100, 130], adaptive control of dynamic system [131,134], optimal design of economic systems [112], production planning, manufacturing [65, 107] and sequential decision processes [91] , etc..

A basic problem in practice is that the vector of model parameters $a = (a_1, \ldots, a_\nu)^T$ is not a given, fixed quantity. Model parameters are often unknown, only partly known and/or may vary randomly to some extent.

Several techniques have been developed in the recent years in order to cope with uncertainty with respect to model parameters $a$. A well known basic method, often used in engineering practice, is the following two-step procedure [6, 25, 100, 130, 131]:

I) *Approximation*: Replace first the $\nu$-vector $a$ of the unknown or stochastic varying model parameters $a_1, \ldots, a_\nu$ by some fixed vector $a_0$ of so-called *nominal* values $a_{0l}, l = 1, \ldots, \nu$. Apply then an optimal decision (control) $x^* = x^*(a_0)$ with respect to the resulting approximate optimization problem

$$\min f_0(a_0, x) \tag{1.2a}$$

s.t.

$$f_i(a_0, x) \leq 0, i = 1, \ldots, m_f \tag{1.2b}$$
$$g_i(a_0, x) = 0, i = 1, \ldots, m_g \tag{1.2c}$$
$$x \in D_0. \tag{1.2d}$$

Due to the deviation of the actual parameter vector $a$ from the nominal vector $a_0$ of model parameters, deviations of the actual state, trajectory or performance of the system from the prescribed state, trajectory, goal values occur.

II) *Compensation*: The deviation of the actual state, trajectory or performance of the system from the prescribed values/functions is compensated then by online measurement and correction actions (decisions or controls). Consequently, in general, increasing measurement and correction expenses result in course of time.

Considerable improvements of this standard procedure can be obtained by taking into account already at the planning stage, i.e. offline, the mostly available a priori (e.g. the type of random variability) and sample information about the parameter vector $a$. Indeed, based e.g. on some structural insight, or by parameter identification methods, regression techniques, calibration methods, etc., in most cases information about the vector $a$ of model/structural parameters can be extracted. Repeating this information gathering procedure at some later time points $t_j > t_0$ (= initial time point), $j = 1, 2, \ldots$, adaptive decision/control procedures occur [91].

Based on the inherent random nature of the parameter vector $a$, the observation or measurement mechanism, resp., or adopting a Bayesian approach concerning unknown parameter values [10], here we make the following basic assumption:

Stochastic Uncertainty : The unknown parameter vector $a$ is a realization

$$a = a(\omega), \omega \in \Omega, \tag{1.3}$$

of a random $\nu$-vector $a(\omega)$ on a certain probability space $(\Omega, \mathcal{A}_0, \mathcal{P})$, where the probability distribution $\mathcal{P}_{a(\cdot)}$ of $a(\omega)$ is known, or it is known that $\mathcal{P}_{a(\cdot)}$ lies within a given range $W$ of probability measures on $\mathbb{R}^\nu$. Using a Bayesian approach, the probability distribution $\mathcal{P}_{a(\cdot)}$ of $a(\omega)$ may also describe the subjective or personal probability of the decision maker, the designer.

Hence, in order to take into account the stochastic variations of the parameter vector $a$, to incorporate the a priori and/or sample information about the unknown vector $a$, resp., the standard approach "insert a certain nominal parameter vector $a_0$, and correct then the resulting error" must be replaced by a more appropriate deterministic substitute problem for the basic optimization problem (1.1a-d) under stochastic uncertainty.

## 1.2 Deterministic Substitute Problems: Basic Formulation

The proper selection of a deterministic substitute problem is a decision theoretical task, see [69]. Hence, for (1.1a-d) we have first to consider the *outcome map*

$$e = e(a, x) := \Big(f_0(a, x), f_1(a, x), \ldots, f_{m_f}(a, x), g_1(a, x), \ldots, g_{m_g}(a, x)\Big)^T, \tag{1.4a}$$
$$a \in \mathbb{R}^\nu, x \in \mathbb{R}^r (x \in D_0),$$

and to evaluate then the outcomes $e \in \mathcal{E} \subset \mathbb{R}^{1+m_f+m_g}$ by means of certain loss or cost functions

$$\gamma_i : \mathcal{E} \to \mathbb{R}, i = 0, 1, \ldots, m, \tag{1.4b}$$

with an integer $m \geq 0$. For the processing of the numerical outcomes $\gamma_i\big(e(a, x)\big), i = 0, 1, \ldots, m$, there are two basic concepts:

### 1.2.1 Minimum or Bounded Expected Costs

Consider the vector of (conditional) expected losses or costs

$$\mathbf{F}(x) = \begin{pmatrix} F_0(x) \\ F_1(x) \\ \vdots \\ F_m(x) \end{pmatrix} := \begin{pmatrix} E\gamma_0(e(a(\omega), x)) \\ E\gamma_1(e(a(\omega), x)) \\ \vdots \\ E\gamma_m(e(a(\omega), x)) \end{pmatrix}, x \in \mathbb{R}^r, \tag{1.5}$$

where the (conditional) expectation "$E$" is taken with respect to the time history $\mathcal{A} = \mathcal{A}_t, (\mathcal{A}_j) \subset \mathcal{A}_0$ up to a certain time point $t$ or stage $j$. A short definition of expectations is given in Section 2.1, for more details, see e.g. [7,38,113].

Having different expected cost or performance functions $F_0, F_1, \ldots, F_m$ to be minimized or bounded, as a basic deterministic substitute problem for (1.1a-d) with a random parameter vector $a = a(\omega)$ we may consider the multi–objective expected cost minimization problem

$$\text{" min "} \mathbf{F}(x) \tag{1.6a}$$

s.t.

$$x \in D_0. \tag{1.6b}$$

Obviously, a good compromise solution $x^*$ of this vector optimization problem should have at least one of the following properties [23,121]:

**Definition 1.1 a)** *A vector $x^0 \in D_0$ is called a* **functional–efficient** *or* **Pareto optimal** *solution of the vector optimization problem (1.6a,b) if there is no $x \in D_0$ such that*

$$F_i(x) \leq F_i(x^0), i = 0, 1, \ldots, m \tag{1.7a}$$

*and*

$$F_{i_0}(x) < F_{i_0}(x^0) \text{ for at least one } i_0, 0 \leq i_0 \leq m. \tag{1.7b}$$

**b)** *A vector $x^0 \in D_0$ is called a* **weak functional–efficient** *or* **weak Pareto optimal** *solution of (1.6a,b) if there is no $x \in D_0$ such that*

$$F_i(x) < F_i(x^0), i = 0, 1, \ldots, m \tag{1.8}$$

(Weak) Pareto optimal solutions of (1.6a,b) may be obtained now by means of scalarizations of the vector optimization problem (1.6a,b). Three main versions are stated in the following.

I) *Minimization of primary expected cost/loss under expected cost constraints*

$$\min F_0(x) \tag{1.9a}$$

s.t.

$$F_i(x) \le F_i^{\max}, i = 1, \dots, m \tag{1.9b}$$
$$x \in D_0. \tag{1.9c}$$

Here, $F_0 = F_0(x)$ is assumed to describe the primary goal of the design/decision making problem, while $F_i = F_i(x), i = 1, \dots, m$, describe secondary goals. Moreover, $F_i^{\max}, i = 1, \dots, m$, denote given upper cost/loss bounds.

**Remark 1.1** *An optimal solution $x^*$ of (1.9a-c) is a weak Pareto optimal solution of (1.6a,b).*

II) *Minimization of the total weighted expected costs*
Selecting certain positive weight factors $c_0, c_1, \dots, c_m$, the expected weighted total costs are defined by

$$\tilde{F}(x) := \sum_{i=0}^{m} c_i F_i(x) = Ef\Big(a(\omega), x\Big), \tag{1.10a}$$

where

$$f(a, x) := \sum_{i=0}^{m} c_i \gamma_i \Big(e(a, x)\Big). \tag{1.10b}$$

Consequently, minimizing the expected weighted total costs $\tilde{F} = \tilde{F}(x)$ subject to the remaining deterministic constraint (1.1d), the following deterministic substitute problem for (1.1a-d) occurs:

$$\min \sum_{i=0}^{m} c_i F_i(x) \tag{1.11a}$$

s.t.

$$x \in D_0. \tag{1.11b}$$

**Remark 1.2** *Let $c_i > 0, i = 1, 1, \dots, m$, be any positive weight factors. Then an optimal solution $x^*$ of (1.11a,b) is a Pareto optimal solution of (1.6a,b).*

III) *Minimization of the maximum weighted expected costs*
Instead of adding weighted expected costs, we may consider the maximum of the weighted expected costs:

$$\tilde{F}(x) := \max_{0 \le i \le m} c_i F_i(x) = \max_{0 \le i \le m} c_i E \gamma_i \Big(e\big(a(\omega), x\big)\Big). \tag{1.12}$$

Here $c_0, c_1, \dots, m$, are again positive weight factors. Thus, minimizing $\tilde{F} = \tilde{F}(x)$ we have the deterministic substitute problem

$$\min \max_{0 \le i \le m} c_i F_i(x) \tag{1.13a}$$

s.t.

$$x \in D_0. \tag{1.13b}$$

**Remark 1.3** *Let $c_i, i = 0, 1, \ldots, m$, be any positive weight factors. An optimal solution of $x^*$ of (1.13a,b) is a weak Pareto optimal solution of (1.6a,b).*

### 1.2.2 Minimum or Bounded Maximum Costs (Worst Case)

Instead of taking expectations, we may consider the worst case with respect to the cost variations caused by the random parameter vector $a = a(\omega)$. Hence, the random cost function

$$\omega \to \gamma_i\Big(e\big(a(\omega), x\big)\Big) \tag{1.14a}$$

is evaluated by means of

$$F_i^{\text{sup}}(x) := \text{ ess sup } \gamma_i\Big(e\big(a(\omega), x\big)\Big), i = 0, 1, \ldots, m. \tag{1.14b}$$

Here, ess sup $(\ldots)$ denotes the (conditional) essential supremum with respect to the random vector $a = a(\omega)$, given information $\mathcal{A}$, i.e. the infimum of the supremum of (1.14a) on sets $A \in \mathcal{A}_0$ of (conditional) probability one, see e.g. [113].

Consequently, the vector function $\mathbf{F} = \mathbf{F}^{\text{sup}}(x)$ is then defined by

$$\mathbf{F}^{\text{sup}}(x) = \begin{pmatrix} F_0(x) \\ F_1(x) \\ \vdots \\ F_m(x) \end{pmatrix} := \begin{pmatrix} \text{ess sup}\, \gamma_0\Big(e\big(a(\omega), x\big)\Big) \\ \text{ess sup}\, \gamma_1\Big(e\big(a(\omega), x\big)\Big) \\ \vdots \\ \text{ess sup}\, \gamma_m\Big(e\big(a(\omega), x\big)\Big) \end{pmatrix}. \tag{1.15}$$

Working with the vector function $\mathbf{F} = \mathbf{F}^{\text{sup}}(x)$, we have then the vector minimization problem

$$\text{"}\min\text{"}\, \mathbf{F}^{\text{sup}}(x) \tag{1.16a}$$

s.t.

$$x \in D_0. \tag{1.16b}$$

By scalarization of (1.16a,b) we obtain then again deterministic substitute problems for (1.1a-d) related to the substitute problem (1.6a,b) introduced in Section 1.2.1.

More details for the selection and solution of appropriate deterministic substitute problems for (1.1a-d) are given in the next sections.

# 2

# Deterministic Substitute Problems in Optimal Decision Under Stochastic Uncertainty

## 2.1 Optimum Design Problems with Random Parameters

In the optimum design of technical or economic structures/systems two basic classes of criterions appear:

First there is a primary cost function

$$G_0 = G_0(a, x). \tag{2.1a}$$

Important examples are the total weight or volume of a mechanical structure, the costs of construction, design of a certain technical or economic structure/system, or the negative utility or reward in a general decision situation.

For the representation of the structural/system safety or failure, for the representation of the admissibility of the state, or for the formulation of the basic operating conditions of the underlying structure/system certain **state functions**

$$y_i = y_i(a, x), \ i = 1, \ldots, m_y, \tag{2.1b}$$

are chosen. In structural design these functions are also called "limit state functions" or "safety margins". Frequent examples are some displacement, stress, load (force and moment) components in structural design, or production and cost functions in production planning problems, optimal mix problems, transportation problems, allocation problems and other problems of economic design.

In (2.1a,b), the design or input vector $x$ denotes the $r$-vector of design or input variables, $x_1, x_2, \ldots, x_r$, as e.g. structural dimensions, sizing variables, such as cross-sectional areas, thickness in structural design, or factors of production, actions in economic design. For the design or input vector $x$ one has mostly some basic deterministic constraints, e.g. nonnegativity constraints, box constraints, represented by

$$x \in D, \tag{2.1c}$$

where $D$ is a given convex subset of $\mathbb{R}^r$. Moreover, $a$ is the $\nu$-vector of model parameters. In optimal structural design,

$$a = \begin{pmatrix} p \\ P \end{pmatrix} \tag{2.1d}$$

is composed of the following two subvectors: $P$ is the $m$-vector of the acting external loads, e.g. wave, wind loads, payload, etc.. Moreover, $p$ denotes the $(\nu - m)$–vector of the further model parameters, as e.g. material parameters (like strength parameters, yield/allowable stresses, elastic moduli, plastic capacities, etc.) of each member of the structure, the manufacturing tolerances and the weight or more general cost coefficients.

In production planning problems,

$$a = (A, b, c) \tag{2.1e}$$

is composed of the $m \times r$ matrix $A$ of technological coefficients, the demand $m$-vector $b$ and the $r$-vector $c$ of unit costs.

Based on the $m_y$-vector of state functions

$$y(a,x) := \Big( y_1(a,x), y_2(a,x), \dots, y_{m_y}(a,x) \Big)^T, \tag{2.1f}$$

the admissible or safe states of the structure/system can be characterized by the condition

$$y(a,x) \in B, \tag{2.1g}$$

where $B$ is a certain subset of $\mathbb{R}^{m_y}$; $B = B(a)$ may depend also on some model parameters.

In production planning problems, typical operating conditions are given, cf. (2.1e), by

$$y(a,x) := Ax - b \geq 0 \text{ or } y(a,x) = 0,\ x \geq 0. \tag{2.2a}$$

In mechanical structures/structural systems, the safety (survival) of the structure/system is described by the operating conditions

$$y_i(a,x) > 0 \text{ for all } i = 1, \dots, m_y \tag{2.2b}$$

with state functions $y_i = y_i(a,x), i = 1, \dots, m_y$, depending on certain response components of the structure/system, such as displacement, stress, force, moment components. Hence, a failure occurs if and only of the structure/system is in the $i$-th failure mode (failure domain)

$$y_i(a,x) \leq 0 \tag{2.2c}$$

for at least one index $i, 1 \leq i \leq m_y$.

*Note.* The number $m_y$ of safety margins or limit state functions $y_i = y_i(a,x), i = 1, \dots, m_y$, may be very large. E.g., in optimal plastic design the limit state functions are determined by the extreme points of the admissible domain of the dual pair of static/kinematic LPs related to the equilibrium and linearized convex yield condition, see Section 7.

Basic problems in optimal decision/design are:

I) *Primary (construction, planning, investment, etc.) cost minimization under operating or safety conditions*

$$\min G_0(a, x) \tag{2.3a}$$

s.t.

$$y(a, x) \in B \tag{2.3b}$$
$$x \in D. \tag{2.3c}$$

Obviously we have $B = (0, +\infty)^{m_y}$ in (2.2b) and $B = [0, +\infty)^{m_y}$ or $B = \{0\}$ in (2.2a).

II) *Failure or recourse cost minimization under primary cost constraints*

$$\text{"}\min\text{"}\,\gamma\Big(y(a, x)\Big) \tag{2.4a}$$

s.t.

$$G_0(a, x) \leq G^{\max} \tag{2.4b}$$
$$x \in D. \tag{2.4c}$$

In (2.4a) $\gamma = \gamma(y)$ is a scalar or vector valued cost/loss function evaluating violations of the operating conditions (2.3b). Depending on the application, these costs are called "failure" or "recourse" costs [51,52,89,109,124,125]. As already discussed in Section 1, solving problems of the above type, a basic difficulty is the uncertainty about the true value of the vector $a$ of model parameters or the (random) variability of $a$:

In practice, due to several types of uncertainties such as, see [140],

* physical uncertainty (variability of physical quantities, like material, loads, dimensions, etc.)
* economic uncertainty (trade, demand, costs, etc.)
* statistical uncertainty (e.g. estimation errors of parameters due to limited sample data)
* model uncertainty (model errors),

the $\nu$-vector $a$ of model parameters must be modelled by a random vector

$$a = a(\omega), \omega \in \Omega, \tag{2.5a}$$

on a certain probability space $(\Omega, \mathcal{A}_0, \mathcal{P})$ with sample space $\Omega$ having elements $\omega$, see (1.3). For the mathematical representation of the corresponding (conditional) probability distribution $\mathcal{P}_{a(\cdot)} = \mathcal{P}^{\mathcal{A}}_{a(\cdot)}$ of the random vector $a = a(\omega)$ (given the time history or information $\mathcal{A} \subset \mathcal{A}_0$), two main distribution models are taken into account in practice:

i) Discrete probability distributions,
ii) Continuous probability distributions.

In the first case there is a finite or countably infinite number $l_0 \in \mathbb{N} \cup \{\infty\}$ of realizations or scenarios $a^l \in \mathbb{R}^\nu, l = 1, \ldots, l_0,$

$$\mathcal{P}\Big(a(\omega) = a^l\Big) = \alpha_l, l = 1, \ldots, l_0, \tag{2.5b}$$

taken with probabilities $\alpha_l, l = 1, \ldots, l_0$. In the second case, the probability that the realization $a(\omega) = a$ lies in a certain (measurable) subset $B \subset \mathbb{R}^\nu$ is described by the multiple integral

$$\mathcal{P}\Big(a(\omega) \in B\Big) = \int_B \varphi(a)\, da \tag{2.5c}$$

with a certain probability density function $\varphi = \varphi(a) \geq 0, a \in \mathbb{R}^\nu, \int \varphi(a) da = 1$.

The properties of the probability distribution $\mathcal{P}_{a(\cdot)}$ may be described - fully or in part - by certain numerical characteristics, called parameters of $\mathcal{P}_{a(\cdot)}$. These distribution parameters $\theta = \theta_h$ are obtained by considering expectations

$$\theta_h := Eh\Big(a(\omega)\Big) \tag{2.6a}$$

of (measurable) functions

$$(h \circ a)(\omega) := h\Big(a(\omega)\Big) \tag{2.6b}$$

composed of the random vector $a = a(\omega)$ with certain (measurable) mappings

$$h : \mathbb{R}^\nu \longrightarrow \mathbb{R}^{s_h}, s_h \geq 1. \tag{2.6c}$$

According to the type of the probability distribution $\mathcal{P}_{a(\cdot)}$ of $a = a(\omega)$, the expectation $Eh\Big(a(\omega)\Big)$ is defined by

$$Eh\Big(a(\omega)\Big) = \begin{cases} \sum\limits_{l=1}^{l_0} h\Big(a^l\Big)\ \alpha_l, & \text{in the discrete case (2.5b)} \\ \int\limits_{\mathbb{R}^\nu} h(a)\varphi(a)\, da, & \text{in the continuous case (2.5c).} \end{cases} \tag{2.6d}$$

Further distribution parameters $\theta$ are functions

$$\theta = \Psi(\theta_{h_1}, \dots, \theta_{h_s}) \tag{2.7}$$

of certain "$h$-moments" $\theta_{h_1}, \dots, \theta_{h_s}$ of the type (2.6a). Important examples of the type (2.6a), (2.7), resp., are the expectation

$$\overline{a} = Ea(\omega) \text{ (for } h(a) := a, a \in \mathbb{R}^\nu) \tag{2.8a}$$

and the covariance matrix

$$Q := E\Big(a(\omega) - \overline{a}\Big)\Big(a(\omega) - \overline{a}\Big)^T = Ea(\omega)a(\omega)^T - \overline{a}\,\overline{a}^T \tag{2.8b}$$

of the random vector $a = a(\omega)$.

Due to the stochastic variability of the random vector $a(\cdot)$ of model parameters, and since the realization $a(\omega) = a$ is not available at the decision making stage, the optimal design problem (2.3a-c) or (2.4a-c) under stochastic uncertainty cannot be solved directly.

Hence, appropriate deterministic substitute problems must be chosen taking into account the randomness of $a = a(\omega)$, cf. Section 1.2.

### 2.1.1 Deterministic Substitute Problems in Optimal Design

According to Section 1.2, a basic deterministic substitute problem in optimal design under stochastic uncertainty is the minimization of the total expected costs including the expected costs of failure

$$\min c_G E G_0\Big(a(\omega), x\Big) + c_f p_f(x) \tag{2.9a}$$

s.t.

$$x \in D. \tag{2.9b}$$

Here,

$$p_f - p_f(x) := \mathcal{P}\left(y\Big(a(\omega), x\Big) \not\in B\right) \tag{2.9c}$$

is the probability of failure or the probability that a safe function of the structure, the system is not guaranteed. Furthermore, $c_G$ is a certain weight factor, and $c_f > 0$ describes the failure or recourse costs. In the present definition of expected failure costs, constant costs for each realization $a = a(\omega)$ of $a(\cdot)$ are assumed. Obviously, it is

$$p_f(x) = 1 - p_s(x) \tag{2.9d}$$

with the probability of safety or survival

$$p_s(x) := \mathcal{P}\left(y\Big(a(\omega), x\Big) \in B\right). \tag{2.9e}$$

In case (2.2b) we have

$$p_f(x) = \mathcal{P}\left(y_i\left(a(\omega), x\right) \le 0 \text{ for at least one index } i, 1 \le i \le m_y\right). \quad (2.9f)$$

The objective function (2.9a) may be interpreted as the Lagrangian (with given cost multiplier $c_f$) of the following reliability based optimization (RBO) problem, cf. [2, 26, 79, 109, 125, 140]:

$$\min EG_0\left(a(\omega), x\right) \quad (2.10a)$$

s.t.

$$p_f(x) \le \alpha^{\max} \quad (2.10b)$$
$$x \in D, \quad (2.10c)$$

where $\alpha^{\max} > 0$ is a prescribed maximum failure probability, e.g. $\alpha^{\max} = 0.001$, cf. (2.3a-c).

The "dual" version of (2.10a-c) reads

$$\min p_f(x) \quad (2.11a)$$

s.t.

$$EG_0\left(a(\omega), x\right) \le G^{\max} \quad (2.11b)$$
$$x \in D \quad (2.11c)$$

with a maximum (upper) cost bound $G^{\max}$, see (2.4a-c).

Further substitute problems are obtained by considering more general expected failure or recourse cost functions

$$\Gamma(x) = E\gamma\left(y\left(a(\omega), x\right)\right) \quad (2.12a)$$

arising from structural systems weakness or failure, or because of false operation. Here,

$$y\left(a(\omega), x\right) := \left(y_1\left(a(\omega), x\right), \ldots, y_{m_y}\left(a(\omega), x\right)\right)^T \quad (2.12b)$$

is again the random vector of state functions, and

$$\gamma : \mathbb{R}^{m_y} \to \mathbb{R}^{m_\gamma} \quad (2.12c)$$

is a scalar or vector valued cost or loss function. In case $B = (0, +\infty)^{m_y}$ or $B = [0, +\infty)^{m_y}$ it is often assumed that $\gamma = \gamma(y)$ is a nonincreasing function, hence,

$$\gamma(y) \ge \gamma(z), \quad \text{if } y \le z, \quad (2.12d)$$

where inequalities between vectors are defined componentwise.

*Example 2.1.* If $\gamma(y) = 1$ for $y \in B^c$ (complement of $B$) and $\gamma(y) = 0$ for $y \in B$, then $\Gamma(x) = p_f(x)$.

*Example 2.2.* Suppose that $\gamma = \gamma(y)$ is a nonnegative scalar function on $\mathbb{R}^{m_y}$ such that

$$\gamma(y) \geq \gamma_0 > 0 \text{ for all } y \notin B \tag{2.13a}$$

with a constant $\gamma_0 > 0$. Then for the probability of failure we find the following upper bound

$$p_f(x) = \mathcal{P}\left(y\Big(a(\omega), x\Big) \notin B\right) \leq \frac{1}{\gamma_0} E\gamma\left(y\Big(a(\omega), x\Big)\right), \tag{2.13b}$$

where the right hand side of (2.13b) is obviously an expected cost function of type (2.12a-c). Hence, the condition (2.10b) can be guaranteed by the cost constraint

$$E\gamma\Big(y\big(a(\omega), x\big)\Big) \leq \gamma_0 \alpha^{\max}. \tag{2.13c}$$

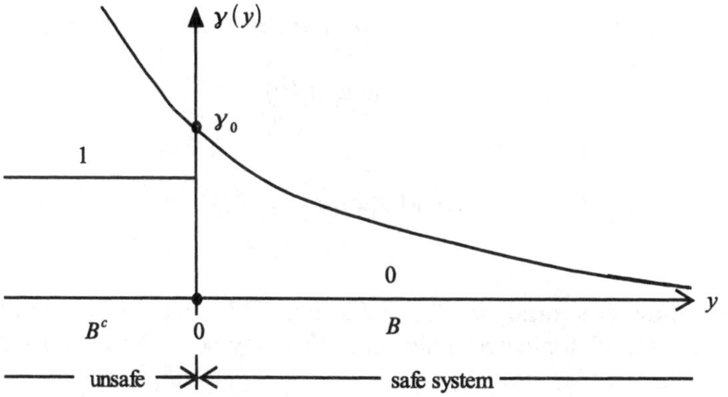


**Fig. 2.1.** Loss function $\gamma$

*Example 2.3.* If the loss function $\gamma(y)$ is defined by a vector of individual loss functions $\gamma_i$ for each state function $y_i = y_i(a, x), i = 1, \ldots, m_y$, hence,

$$\gamma(y) = \Big(\gamma_1(y_1), \ldots, \gamma_{m_y}(y_{m_y})\Big)^T, \tag{2.14a}$$

then

$$\Gamma(x) = \big(\Gamma_1(x), \ldots, \Gamma_{m_y}(x)\big)^T, \; \Gamma_i(x) := E\gamma_i\left(y_i\Big(a(\omega), x\Big)\right), 1 \leq i \leq m_y, \tag{2.14b}$$

i.e. the $m_y$ state functions $y_i, i = 1, \ldots, m_y$, will be treated separately.

Working with the more general expected failure or recourse cost functions $\Gamma = \Gamma(x)$, instead of (2.9a-c), (2.10a-c) and (2.11a-c) we have the related substitute problems

I) *Expected total cost minimization*

$$\min c_G EG_0\Big(a(\omega), x\Big) + c_f^T \Gamma(x), \tag{2.15a}$$

s.t.

$$x \in D \tag{2.15b}$$

II) *Expected primary cost minimization under expected failure or recourse cost constraints*

$$\min EG_0\Big(a(\omega), x\Big) \tag{2.16a}$$

s.t.

$$\Gamma(x) \le \Gamma^{\max} \tag{2.16b}$$

$$x \in D, \tag{2.16c}$$

III) *Expected failure or recourse cost minimization under expected primary cost constraints*

$$\text{"} \min \text{"}\, \Gamma(x) \tag{2.17a}$$

s.t.

$$EG_0\Big(a(\omega), x\Big) \le G^{\max} \tag{2.17b}$$

$$x \in D. \tag{2.17c}$$

Here, $c_G, c_f$ are (vectorial) weight coefficients, $\Gamma^{\max}$ is the vector of upper loss bounds, and "min" indicates again that $\Gamma(x)$ may be a vector valued function.

### 2.1.2 Deterministic Substitute Problems in Quality Engineering

In quality engineering [99,102,103] the aim is to optimize the performance of a certain product or a certain process, and to make the performance insensitive (robust) to any noise to which the product, the process may be subjected. Moreover, the production costs (development, manufacturing, etc.) should be bounded or minimized.

Hence, the control (input, design) factors $x$ are then to set as to provide

* an optimal mean performance (response)
* a high degree of immunity (robustness) of the performance (response) to noise factors
* low or minimum production costs.

Of course, a certain trade-off between these criterions may be taken into account.

For this aim, the quality characteristics or performance (e.g. dimensions, weight, contamination, structural strength, yield, etc.) of the product/process is described [99] first by a function

$$y = y\Big(a(\omega), x\Big) \tag{2.18}$$

depending on a $\nu$-vector $a = a(\omega)$ of random varying parameters (noise factors) $a_j = a_j(\omega), j = 1, \ldots, \nu$, causing quality variations, and the $r$-vector $x$ of control/input factors $x_k, k = 1, \ldots, r$.

The **quality loss** of the product/process under consideration is measured then by a certain loss function $\gamma = \gamma(y)$. Simple loss functions $\gamma$ are often chosen [102, 103] in the following way:

i) *Nominal-is-best*
Having a given, finite target or nominal value $y^*$, e.g. a nominal dimension of a product, the quality loss $\gamma(y)$ reads

$$\gamma(y) := (y - y^*)^2. \tag{2.19a}$$

ii) *Smallest-is-best*
If the target or nominal value $y^*$ is zero, i.e., if the absolute value $|y|$ of the product quality function $y = y(a, x)$ should be as small as possible, e.g. a certain product contamination, then the corresponding quality loss is defined by

$$\gamma(y) := y^2. \tag{2.19b}$$

iii) *Biggest-is-best*
If the absolute value $|y|$ of the product quality function $y = y(a, x)$ should be as large as possible, e.g. the strength of a bar or the yield of a process, then a possible quality loss is given by

$$\gamma(y) := \left(\frac{1}{y}\right)^2. \tag{2.19c}$$

Obviously, (2.19a) and (2.19b) are convex loss functions. Moreover, since the quality or response function $y = y(a, x)$ takes only positive or only negative values $y$ in many practical applications, also (2.19c) yields a convex loss function in many cases.

Further decision situations may be modelled by choosing more general convex loss functions $\gamma = \gamma(y)$.

A high mean product/process level or performance at minimum random quality variations and low or bounded production/manufacturing costs is achieved then [99, 102, 103] by minimization of the expected quality loss function

$$\Gamma(x) := E\gamma\left(y\Big(a(\omega), x\Big)\right), \tag{2.19d}$$

subject to certain constraints for the input $x$, as e.g. constraints for the production/manufacturing costs. Obviously, $\Gamma(x)$ can also be minimized by maximizing the negative logarithmic mean loss (sometimes called "signal-to-noise-ratio" [99]):

$$SN := -\log \Gamma(x). \tag{2.19e}$$

Since $\Gamma = \Gamma(x)$ is a mean value function as described in the previous sections, for the computation of a robust optimal input vector $x^*$ one has the following stochastic optimization problem

$$\min \Gamma(x) \tag{2.20a}$$

s.t.

$$EG_0\Big(a(\omega), x\Big) \le G^{\max} \tag{2.20b}$$

$$x \in D. \tag{2.20c}$$

## 2.2 Basic Properties of Substitute Problems

As can be seen from the conversion of an optimization problem with random parameters into a deterministic substitute problem, cf. Sections 2.1.1 and 2.1.2, a central role is played by expectation or mean value functions of the type

$$\Gamma(x) = E\gamma\left(y\Big(a(\omega), x\Big)\right), x \in D_0, \tag{2.21a}$$

or more general

$$\Gamma(x) = Eg\Big(a(\omega), x\Big), x \in D_0. \tag{2.21b}$$

Here, $a = a(\omega)$ is a random $\nu$-vector, $y = y(a,x)$ is an $m_y$-vector valued function on a certain subset of $\mathbb{R}^\nu \times \mathbb{R}^r$, and $\gamma = \gamma(z)$ is a real valued function on a certain subset of $\mathbb{R}^{m_y}$.

Furthermore, $g = g(a,x)$ denotes a real valued function on a certain subset of $\mathbb{R}^\nu \times \mathbb{R}^r$. In the following we suppose that the expectation in (2.21a,b) exists and is finite for all input vectors $x$ lying in an appropriate set $D_0 \subset \mathbb{R}^r$, cf. [11].

The following basic properties of the mean value functions $\Gamma$ are needed in the following again and again.

**Lemma 2.1. Convexity.** *Suppose that $x \to g\Big(a(\omega), x\Big)$ is convex a.s. (almost sure) on a fixed convex domain $D_0 \subset \mathbb{R}^r$. If $Eg\Big(a(\omega), x\Big)$ exists and is finite for each $x \in D_0$, then $\Gamma = \Gamma(x)$ is convex on $D_0$.*

*Proof.* This property follows [51, 52, 69] directly from the linearity of the expectation operator.

If $g = g(a,x)$ is defined by $g(a,x) := \gamma\Big(y(a,x)\Big)$, see (2.21a), then the above theorem yields the following result:

**Corollary 2.1.** *Suppose that $\gamma$ is convex and $E\gamma\Big(y\big(a(\omega),x\big)\Big)$ exists and is finite for each $x \in D_0$. a) If $x \to y\big(a(\omega),x\big)$ is linear a.s., then $\Gamma = \Gamma(x)$ is convex. b) If $x \to y\big(a(\omega),x\big)$ is convex a.s., and $\gamma$ is a convex, monotoneous nondecreasing function, then $\Gamma = \Gamma(x)$ is convex.*

It is well known [64] that a convex function is continuous on each open subset of its domain. A general sufficient condition for the continuity of $\Gamma$ is given next.

**Lemma 2.2. Continuity.** *Suppose that $Eg\big(a(\omega),x\big)$ exists and is finite for each $x \in D_0$, and assume that $x \to g\big(a(\omega),x\big)$ is continuous at $x_0 \in D_0$ a.s.. If there is a function $\psi = \psi\big(a(\omega)\big)$ having finite expectation such that*

$$\Big|g\big(a(\omega),x\big)\Big| \leq \psi\big(a(\omega)\big) \text{ a.s. for all } x \in U(x_0) \cap D_0, \tag{2.22}$$

*where $U(x_0)$ is a neighborhood of $x_0$, then $\Gamma = \Gamma(x)$ is continuous at $x_0$.*

*Proof.* The assertion can be shown by using Lebesgue's dominated convergence theorem, see e.g. [69].

For the consideration of the differentiability of $\Gamma = \Gamma(x)$, let $D$ denote an open subset of the domain $D_0$ of $\Gamma$.

**Lemma 2.3. Differentiability.** *Suppose that*

*i) $Eg\big(a(\omega),x\big)$ exists and is finite for each $x \in D_0$,*

*ii) $x \to g\big(a(\omega),x\big)$ is differentiable on the open subset $D$ of $D_0$ a.s. and*

*iii)*

$$\Big\|\nabla_x g\big(a(\omega),x\big)\Big\| \leq \psi\big(a(\omega)\big), x \in D, \text{ a.s.}, \tag{2.23a}$$

*where $\psi = \psi\big(a(\omega)\big)$ is a function having finite expectation. Then the expectation of $\nabla_x g\big(a(\omega),x\big)$ exists and is finite, $\Gamma = \Gamma(x)$ is differentiable on $D$ and*

$$\nabla\Gamma(x) = \nabla_x Eg\big(a(\omega),x\big) = E\nabla_x g\big(a(\omega),x\big), x \in D. \tag{2.23b}$$

*Proof.* Considering the difference quotients $\frac{\Delta\Gamma}{\Delta x_k}, k = 1,\dots,r$, of $\Gamma$ at a fixed point $x_0 \in D$, the assertion follows by means of the mean value theorem, inequality (2.23a) and Lebesque's dominated convergence theorem, cf. [51,52, 69].

*Example 2.4.* In case (2.21a), under obvious differentiability assumptions concerning $\gamma$ and $y$ we have $\nabla_x g(a,x) = \nabla_x y(a,x)^T \nabla\gamma\Big(y(a,x)\Big)$, where $\nabla_x y(a,x)$ denotes the Jacobian of $y = y(a,x)$ with respect to $a$. Hence, if (2.23b) holds, then

$$\nabla\Gamma(x) = E\nabla_x y\Big(a(\omega),x\Big)^T \nabla\gamma\left(y\Big(a(\omega),x\Big)\right). \tag{2.23c}$$

## 2.3 Approximations of Deterministic Substitute Problems in Optimal Design

The main problem in solving the deterministic substitute problems defined above is that the arising probability and expected cost functions $p_f = p_f(x), \Gamma = \Gamma(x), x \in \mathbb{R}^r$, are defined by means of multiple integrals over a $\nu$-dimensional space.

Thus, the substitute problems may be solved, in practice, only by some approximative analytical and numerical methods [32, 150]. In the following we consider possible approximations for substitute problems based on general expected recourse cost functions $\Gamma = \Gamma(x)$ according to (2.21a) having a real valued convex loss function $\gamma(z)$. Note that the probability of failure function $p_f = p_f(x)$ may be approximated from above, see (2.13a,b), by expected cost functions $\Gamma = \Gamma(x)$ having a nonnegative function $\gamma = \gamma(z)$ being bounded from below on the failure domain $B^c$. In the following several basic approximation methods are presented.

### 2.3.1 Approximation of the Loss Function

Suppose here that $\gamma = \gamma(y)$ is a continuously differentiable, convex loss function on $\mathbb{R}^{m_y}$. Let then denote

$$\overline{y}(x) := Ey\Big(a(\omega),x\Big) = \left(Ey_1\Big(a(\omega),x\Big),\dots,Ey_{m_y}\Big(a(\omega),x\Big)\right)^T \tag{2.24}$$

the expectation of the vector $y = y\Big(a(\omega),x\Big)$ of state functions $y_i = y_i\Big(a(\omega),x\Big)$, $i = 1,\dots,m_y$.

For an arbitrary continuously differentiable, convex loss function $\gamma$ we have

$$\gamma\left(y\Big(a(\omega),x\Big)\right) \geq \gamma\Big(\overline{y}(x)\Big) + \nabla\gamma\Big(\overline{y}(x)\Big)^T \left(y\Big(a(\omega),x\Big) - \overline{y}(x)\right). \tag{2.25a}$$

Thus, taking expectations in (2.25a), we find Jensen's inequality

$$\Gamma(x) = E\gamma\left(y\Big(a(\omega),x\Big)\right) \geq \gamma\Big(\overline{y}(x)\Big) \tag{2.25b}$$

which holds for any convex function $\gamma$. Using the mean value theorem, we have

$$\gamma(y) = \gamma(\overline{y}) + \nabla\gamma(\hat{y})^T(y - \overline{y}), \tag{2.25c}$$

where $\hat{y}$ is a point on the line segment $\overline{y}y$ between $\overline{y}$ and $y$. By means of (2.25b,c) we get

$$0 \leq \Gamma(x) - \gamma\Big(\overline{y}(x)\Big) \leq E\left\|\nabla\gamma\left(\hat{y}\Big(a(\omega), x\Big)\right)\right\| \cdot \left\|y\Big(a(\omega), x\Big) - \overline{y}(x)\right\|. \tag{2.25d}$$

a) Bounded gradient

If the gradient $\nabla\gamma$ is bounded on the range of $y = y\Big(a(\omega), x\Big), x \in D$, i.e., if

$$\left\|\nabla\gamma\left(y\Big(a(\omega), x\Big)\right)\right\| \leq \vartheta^{\max} \text{ a.s. for each } x \in D, \tag{2.26a}$$

with a constant $\vartheta^{\max} > 0$, then

$$0 \leq \Gamma(x) - \gamma\Big(\overline{y}(x)\Big) \leq \vartheta^{\max} E\left\|y\Big(a(\omega), x\Big) - \overline{y}(x)\right\|, x \in D. \tag{2.26b}$$

Since $t \to \sqrt{t}, t \geq 0$, is a concave function, we get

$$0 \leq \Gamma(x) - \gamma\Big(\overline{y}(x)\Big) \leq \vartheta^{\max}\sqrt{q(x)}, \tag{2.26c}$$

where

$$q(x) := E\left\|y\Big(a(\omega), x\Big) - \overline{y}(x)\right\|^2 = trQ(x) \tag{2.26d}$$

is the generalized variance, and

$$Q(x) := \mathrm{cov}\left(y\Big(a(\cdot), x\Big)\right) \tag{2.26e}$$

denotes the covariance matrix of the random vector $y = y\Big(a(\omega), x\Big)$. Consequently, the expected loss function $\Gamma(x)$ can be approximated from above by

$$\Gamma(x) \leq \gamma\Big(\overline{y}(x)\Big) + \vartheta^{\max}\sqrt{q(x)} \text{ for } x \in D. \tag{2.26f}$$

b) Bounded eigenvalues of the Hessian

Considering second order expansions of $\gamma$, with a vector $\tilde{y} \in \bar{y}y$ we find

$$\gamma(y) - \gamma(\overline{y}) = \nabla\gamma(\overline{y})^T(y - \overline{y}) + \frac{1}{2}(y - \overline{y})^T\nabla^2\gamma(\tilde{y})(y - \overline{y}). \tag{2.27a}$$

Suppose that the eigenvalues $\lambda$ of $\nabla^2\gamma(y)$ are bounded from below and above on the range of $y = y\Big(a(\omega), x\Big)$ for each $x \in D$, i.e.

$$0 < \lambda^{\min} \leq \lambda\left(\nabla^2\gamma\left(y\Big(a(\omega), x\Big)\right)\right) \leq \lambda^{\max} < +\infty, \text{ a.s.}, x \in D \tag{2.27b}$$

with constants $0 < \lambda^{\min} \leq \lambda^{\max}$. Taking expectations in (2.27a), we get

$$\gamma\Big(\overline{y}(x)\Big) + \frac{\lambda^{\min}}{2} q(x) \leq \Gamma(x) \leq \gamma\Big(\overline{y}(x)\Big) + \frac{\lambda^{\max}}{2} q(x), x \in D. \tag{2.27c}$$

Consequently, using (2.26f) or (2.27c), various approximations for the deterministic substitute problems (2.15a,b), (2.16a-c), (2.17a-c) may be obtained.

Based on the above approximations of expected cost functions, we state the following two approximates to (2.16a-c), (2.17a-c), resp., which are well known in *robust optimal design:*

i) *Expected primary cost minimization under approximate expected failure or recourse cost constraints*

$$\min EG_0\Big(a(\omega), x\Big) \tag{2.28a}$$

s.t.

$$\gamma\Big(\overline{y}(x)\Big) + c_0 q(x) \leq \Gamma^{\max} \tag{2.28b}$$

$$x \in D, \tag{2.28c}$$

where $c_0$ is a scale factor, cf. (2.26f) and (2.27c);

ii) *Approximate expected failure or recourse cost minimization under expected primary cost constraints*

$$\min \gamma\Big(\overline{y}(x)\Big) + c_0 q(x) \tag{2.29a}$$

s.t.

$$EG_0\Big(a(\omega), x\Big) \leq G^{\max} \tag{2.29b}$$

$$x \in D. \tag{2.29c}$$

Obviously, by means of (2.28a-c) or (2.29a-c) optimal designs $x^*$ are achieved which

* yield a high mean performance of the structure/structural system
* are minimally sensitive or have a limited sensitivity with respect to random parameter variations (material, load, manufacturing, process, etc.) and
* cause only limited costs for design, construction, maintenance, etc..

### 2.3.2 Regression Techniques, RSM

a) *Approximation of state functions.* The numerical solution is simplified considerably if one can work with one single state function $y = y(a, x)$. Formally, this is possible by defining the function

$$y^{\min}(a,x) := \min_{1 \le i \le m_y} y_i(a,x). \tag{2.30a}$$

Indeed, according to (2.2b,c) the failure of the structure, the system can be represented by the condition

$$y^{\min}(a,x) \le 0. \tag{2.30b}$$

Thus, the weakness or failure of the technical or economic device can be evaluated numerically by the function

$$\Gamma(x) := E\gamma\left(y^{\min}\left(a(\omega),x\right)\right) \tag{2.30c}$$

with a nonincreasing loss function $\gamma : \mathbb{R} \to \mathbb{R}_+$, see Fig. 2.1.
However, the "min"-operator in (2.30a) yields a nonsmooth function $y^{\min} = y^{\min}(a,x)$ in general, and the computation of the mean and covariance function

$$\overline{y^{\min}}(x) := Ey^{\min}\left(a(\omega),x\right) \tag{2.30d}$$

$$Q_{y^{\min}}(x) := \text{cov}\left(y^{\min}\left(a(\cdot),x\right)\right) \tag{2.30e}$$

by means of Taylor expansion with respect to the model parameter vector $a$ at $\overline{a} = Ea(\omega)$ is not possible directly, cf. Section 2.3.3.

Using regression techniques, Response Surface Methods (RSM), etc., for given vector $x$, the function $a \to y^{\min}(a,x)$ can be approximated [13, 22, 49, 56, 123] by functions $\tilde{y} = \tilde{y}(a,x)$ being sufficiently smooth with respect to the parameter vector $a$.

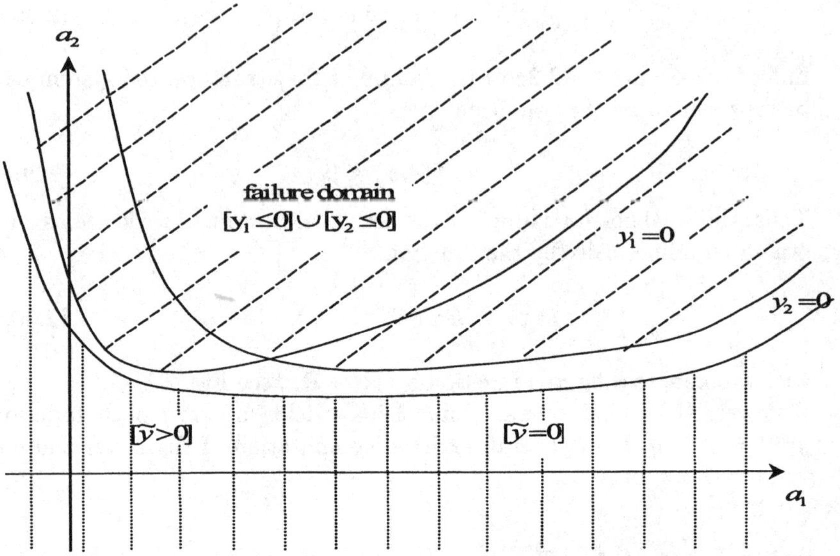


**Fig. 2.2.** Approximate smooth failure domain $[\widetilde{y}(a,x) \leq 0]$ for given $x$

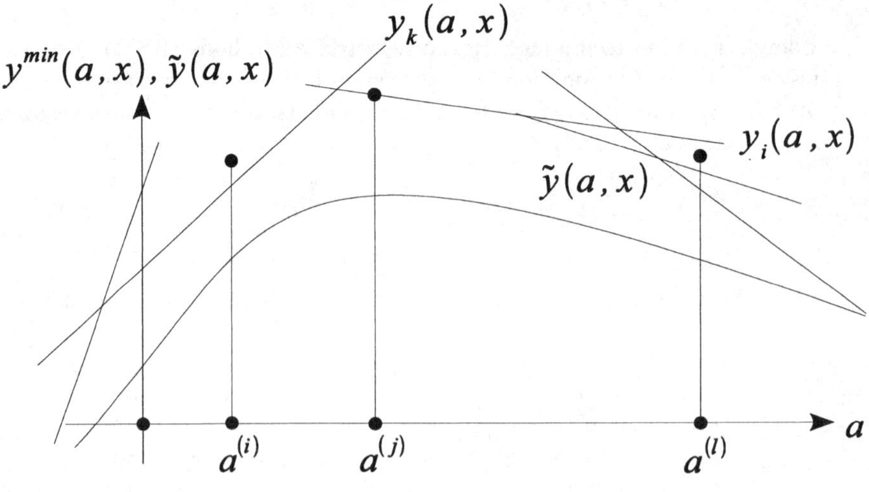


**Fig. 2.3.** Approximations of $y^{\min}(a,x)$ for given $x$

In many important cases, for each $i = 1, \ldots, m_y$, the state functions

$$(a, x) \longrightarrow y_i(a, x)$$

are bilinear functions. Thus, in this case $y^{\min} = y^{\min}(a, x)$ is a piecewise linear function with respect to $a$. Fitting a linear or quadratic Response Surface Model [16,17,95,96]

$$\widetilde{y}(a, x) = c(x) + q(x)^T(a - \overline{a}) + (a - \overline{a})^T Q(x)(a - \overline{a}) \tag{2.30f}$$

to $a \to y^{\min}(a, x)$, after the selection of appropriate reference points

$$a^{(j)} := \overline{a} + d_a^{(j)}, j = 1, \ldots, p, \tag{2.30g}$$

with "design" points $d_a^{(j)} \in \mathbb{R}^\nu, j = 1, \ldots, p$, the unknown coefficients $c = c(x), q = q(x)$ and $Q = Q(x)$ are obtained by minimizing the mean square error

$$\rho(c, q, Q) := \sum_{j=1}^{p} \left( \widetilde{y}(a^{(j)}, x) - y^{\min}(a^{(j)}, x) \right)^2 \tag{2.30h}$$

with respect to $(c, q, Q)$. Since the model (2.30f) depends linearly on the function parameters $(c, q, Q)$, explicit formulas for the optimal coefficients

$$c^* = c^*(x), q^* = q^*(x), Q^* = Q^*(x), \tag{2.30i}$$

are obtained from this least squares estimation method, see Chapter 5.

b) *Approximation of expected loss functions.* Corresponding to the approximation (2.30f) of $y^{\min} = y^{\min}(a, x)$, using again least squares techniques, a mean value function $\Gamma(x) = E\gamma\left(y\left(a(\omega), x\right)\right)$, cf. (2.12a), can be approximated at a given point $x_0 \in \mathbb{R}^\nu$ by a linear or quadratic Response Surface Function

$$\widetilde{\Gamma}(x) := \beta_0 + \beta_I^T(x - x_0) + (x - x_0)^T B(x - x_0), \tag{2.30j}$$

with scalar, vector and matrix parameters $\beta_0, \beta_I, B$. In this case estimates $\hat{\Gamma}^{(i)}$ of $\Gamma(x)$ are needed at some reference points $x^{(i)} = x_0 + d^{(i)}, i = 1, \ldots, p$. Details are given in Chapter 5.

### 2.3.3 Taylor Expansion Methods

As can be seen above, cf. (2.21a,b), in the objective and/or in the constraints of substitute problems for optimization problems with random data mean value functions of the type

$$\Gamma(x) := Eg\left(a(\omega), x\right)$$

occur. Here, $g = g(a, x)$ is a real valued function on a subset of $\mathbb{R}^\nu \times \mathbb{R}^r$, and $a = a(\omega)$ is a $\nu$-random vector.

a) *Expansion with respect to a*: Suppose that on its domain the function $g = g(a, x)$ has partial derivatives $\nabla_a^l g(a, x), l = 0, 1, \ldots, l_g+1$, up to order $l_g+1$. Note that the gradient $\nabla_a g(a, x)$ contains the so-called *sensitivities* $\frac{\partial g}{\partial a_j}(a, x), j = 1, \ldots, \nu$, of $g$ with respect to the parameter vector $a$ at $(a, x)$. In the same way, the higher order partial derivatives $\nabla_a^l g(a, x), l > 1$, represent the *higher order sensitivities* of $g$ with respect to $a$ at $(a, x)$. Taylor expansion of $g = g(a, x)$ with respect to $a$ at $\bar{a} := Ea(\omega)$ yields

$$g(a,x) = \sum_{l=0}^{l_g} \frac{1}{l!} \nabla_a^l g(\bar{a}, x) \cdot (a - \bar{a})^l + \frac{1}{(l_g+1)!} \nabla_a^{lg+1} g(\hat{a}, x) \cdot (a - \bar{a})^{l_g+1} \tag{2.31a}$$

where $\hat{a} := \bar{a} + \vartheta(a - \bar{a}), 0 < \vartheta < 1$. If $g = g(a, x)$ is defined by

$$g(a,x) := \gamma\Big(y(a,x)\Big),$$

see (2.21a), then the partial derivatives $\nabla_a^l g$ of $g$ up to the second order read:

$$\nabla_a g(a,x) = \Big(\nabla_a y(a,x)\Big)^T \nabla\gamma\Big(y(a,x)\Big) \tag{2.31b}$$

$$\nabla_a^2 g(a,x) = \Big(\nabla_a y(a,x)\Big)^T \nabla^2\gamma\Big(y(a,x)\Big)\nabla_a y(a,x) + \nabla\gamma\Big(y(a,x)\Big) \cdot \nabla_a^2 y(a,x), \tag{2.31c}$$

where

$$(\nabla\gamma) \cdot \nabla_a^2 y := \left((\nabla y)^T \frac{\partial^2 y}{\partial a_k \partial a_l}\right)_{k,l=1,\ldots,\nu}. \tag{2.31d}$$

Taking expectations in (2.31a), $\Gamma(x)$ can be approximated, cf. Section 2.3.1, by

$$\widetilde{\Gamma}(x) := g(\bar{a}, x) + \sum_{l=2}^{l_g} \nabla_a^l g(\bar{a}, x) \cdot E\Big(a(\omega) - \bar{a}\Big)^l, \tag{2.32a}$$

where $E\Big(a(\omega) - \bar{a}\Big)^l$ denotes the system of mixed $l$th central moments of the random vector $a(\omega) = \Big(a_1(\omega), \ldots, a_\nu(\omega)\Big)^T$. Assuming that the domain of $g = g(a, x)$ is convex with respect to $a$, we get the error estimate

$$\Big|\Gamma(x) - \widetilde{\Gamma}(x)\Big| \leq \frac{1}{(l_g+1)!} E \sup_{0 \leq \vartheta \leq 1} \Big\|\nabla_a^{l_g+1} g\Big(\bar{a} + \vartheta\big(a(\omega) - \bar{a}\big), x\Big)\Big\| \times \Big\|a(\omega) - \bar{a}\Big\|^{l_g+1}. \tag{2.32b}$$

In many practical cases the random parameter $\nu$–vector $a = a(\omega)$ has a convex, bounded support, and $\nabla_a^{l_g+1} g$ is continuous. Then the $L_\infty$–norm

$$r(x) := \frac{1}{(l_g+1)!} \operatorname{ess\,sup} \left\| \nabla_a^{l_g+1} g\Big(a(\omega), x\Big) \right\| \tag{2.32c}$$

is finite for all $x$ under consideration, and (2.32b,c) yield the error bound

$$\left| \Gamma(x) - \widetilde{\Gamma}(x) \right| \le r(x) E \left\| a(\omega) - \bar{a} \right\|^{l_g+1}. \tag{2.32d}$$

*Remark 2.1.* The above described method can be extended to the case of vector valued loss functions $\gamma(z) = \Big(\gamma_1(z), \ldots, \gamma_{m_\gamma}(z)\Big)^T$.

b) *Inner expansions:* Suppose that $\Gamma(x)$ is defined by (2.21a), hence,

$$\Gamma(x) = E\gamma\left(y\Big(a(\omega), x\Big)\right).$$

Linearizing the vector function $y = y(a, x)$ with respect to $a$ at $\bar{a}$, thus,

$$y(a, x) \approx y_{(1)}(a, x) := y(\bar{a}, x) + \nabla_a y(\bar{a}, x)(a - \bar{a}), \tag{2.33a}$$

the mean value function $\Gamma(x)$ is approximated by

$$\widetilde{\Gamma}(x) := E\gamma\left(y(\bar{a}, x) + \nabla_a y(\bar{a}, x)\Big(a(\omega) - \bar{a}\Big)\right). \tag{2.33b}$$

Corresponding to (2.24), (2.26e), define

$$\overline{y}_{(1)}(x) := E y_{(1)}\Big(a(\omega), x\Big) = y(\bar{a}, x) \tag{2.33c}$$

$$Q(x) := \operatorname{cov}\left(y_{(1)}\Big(a(\cdot), x\Big)\right) = \nabla_a y(\bar{a}, x) \operatorname{cov}\Big(a(\cdot)\Big) \nabla_a y(\bar{a}, x)^T. \tag{2.33d}$$

In case of convex loss functions $\gamma$, approximates of $\widetilde{\Gamma}$ and the corresponding substitute problems based on $\widetilde{\Gamma}$ may be obtained now by applying the methods described in Section 2.3.1. Explicit representations for $\widetilde{\Gamma}$ are obtained in case of quadratic loss functions $\gamma$.
Error estimates can be derived easily for Lipschitz-continuous or convex loss function $\gamma$. In case of a Lipschitz-continuous loss function $\gamma$ with Lipschitz constant $L > 0$, e.g. for sublinear [69, 72] loss functions, using (2.33c) we have

$$\left| \Gamma(x) - \widetilde{\Gamma}(x) \right| \le L \cdot E \left\| y\Big(a(\omega), x\Big) - y_{(1)}\Big(a(\omega), x\Big) \right\|. \tag{2.33e}$$

Applying the mean value theorem [27], under appropriate 2nd order differentiability assumptions, for the right hand side of (2.33e) we find the following stochastic version of the mean value theorem

$$\begin{aligned} & E \left\| y\Big(a(\omega), x\Big) - y_{(1)}\Big(a(\omega), x\Big) \right\| \\ & \le E \left\| a(\omega) - \bar{a} \right\|^2 \sup_{0 \le \vartheta \le 1} \left\| \nabla_a^2 y\left(\bar{a} + \vartheta\Big(a(\omega) - \bar{a}\Big), x\right) \right\|. \end{aligned} \tag{2.33f}$$

## 2.4 Applications to Problems in Quality Engineering

Since the deterministic substitute problems arising in quality engineering are of the same type as in optimal design, the approximation techniques described in Section 2.3 can be applied directly to the robust design problem (2.20a-c). This is shown in the following for the approximation method described in Section 2.3.1.

Hence, the approximation techniques developed in Section 2.3.1 are applied now to the objective function

$$\Gamma(x) = E\gamma\Big(y\big(a(\omega), x\big)\Big)$$

of the basic robust design problem (2.20a-c). If $\gamma = \gamma(y)$ in an arbitrary continuously differentiable convex loss function on $\mathbb{R}^1$, then we find the following approximations:

a) Bounded 1st order derivative
In case that $\gamma$ has a bounded derivative on the convex support of the quality function, i.e., if

$$\Big|\gamma'\Big(y\big(a(\omega), x\big)\Big)\Big| \leq \vartheta^{\max} < +\infty \text{ a.s. }, x \in D, \tag{2.34a}$$

then

$$\Gamma(x) \leq \gamma\big(\overline{y}(x)\big) + \vartheta^{\max}\sqrt{q(x)}, x \in D, \tag{2.34b}$$

where

$$\overline{y}(x) := Ey\big(a(\omega), x\big), \tag{2.34c}$$

$$q(x) := V\Big(y\big(a(\cdot), x\big)\Big) = E\Big(y\big(a(\omega), x\big) - \overline{y}(x)\Big)^2 \tag{2.34d}$$

denote the mean and variance of the quality function $y = y\big(a(\omega), x\big)$.

b) Bounded 2nd order derivative
If the second order derivative $\gamma''$ of $\gamma$ is bounded from above and below, i.e., if

$$0 < \lambda^{\min} \leq \gamma''\Big(y\big(a(\omega), x\big)\Big) \leq \lambda^{\max} < +\infty \text{ a.s.}, x \in D, \tag{2.35a}$$

then

$$\gamma\big(\overline{y}(x)\big) + \frac{\lambda^{\min}}{2}q(x) \leq \Gamma(x) \leq \gamma\big(\overline{y}(x)\big) + \frac{\lambda^{\max}}{2}q(x), x \in D. \tag{2.35b}$$

Corresponding to (2.28a-c), (2.29a-c), we then have the following “dual” robust optimal design problems:

i) *Expected primary cost minimizationunder approximate expected failure or recourse cost constraints*

$$\min EG_0\Big(a(\omega), x\Big) \tag{2.36a}$$

s.t.

$$\gamma\Big(\overline{y}(x)\Big) + c_0 q(x) \leq \Gamma^{\max} \tag{2.36b}$$

$$x \in D, \tag{2.36c}$$

where $\Gamma^{\max}$ denotes an upper loss bound.

ii) *Approximate expected failure cost minimization under expected primary cost constraints*

$$\min \gamma\Big(\overline{y}(x)\Big) + c_0 q(x) \tag{2.37a}$$

s.t.

$$EG_0\Big(a(\omega), x\Big) \leq G^{\max} \tag{2.37b}$$

$$x \in D, \tag{2.37c}$$

where $c_0$ is again a certain scale factor, see (2.34a,b), (2.35a,b), and $G_0 = G_0\Big(a(\omega), x\Big)$ represents the production costs having an upper bound $G^{\max}$.

## 2.5 Approximation of Probabilities - Probability Inequalities

In reliability analysis of engineering/economic structures or systems, a main problem is the computation of probabilities

$$\mathcal{P}\left(\bigcup_{i=1}^{N} V_i\right) := \mathcal{P}\left(a(\omega) \in \bigcup_{i=1}^{N} V_i\right) \tag{2.38a}$$

or

$$\mathcal{P}\left(\bigcap_{j=1}^{N} S_j\right) := \mathcal{P}\left(a(\omega) \in \bigcap_{j=1}^{N} S_j\right) \tag{2.38b}$$

of unions and intersections of certain failure/survival domains (events) $V_j, S_j$, $j = 1, \ldots, N$. These domains (events) arise from the representation of the structure or system by a combination of certain series and/or parallel substructures/systems. Due to the high complexity of the basic physical relations, several approximation techniques are needed for the evaluation of (2.38a,b).

### 2.5.1 Bonferroni-Type Inequalities

In the following $V_1, V_2, \ldots, V_N$ denote arbitrary (Borel-)measurable subsets of the parameter space $\mathbb{R}^\nu$, and the abbreviation

$$\mathcal{P}(V) := \mathcal{P}\Big(a(\omega) \in V\Big) \tag{2.38c}$$

is used for any measurable subset $V$ of $\mathbb{R}^\nu$.

Starting from the representation of the probability of a union of $N$ events,

$$\mathcal{P}\left(\bigcup_{j=1}^{N} V_j\right) = \sum_{k=1}^{N} (-1)^{k-1} s_{k,N}, \tag{2.39a}$$

where

$$s_{k,N} := \sum_{1 \le i_1 < i_2 < \ldots < i_k \le N} \mathcal{P}\left(\bigcap_{l=1}^{k} V_{i_l}\right), \tag{2.39b}$$

we obtain [40] the well known basic Bonferroni bounds

$$\mathcal{P}\left(\bigcup_{j=1}^{N} V_j\right) \le \sum_{k=1}^{\rho} (-1)^{k-1} s_{k,N} \text{ for } \rho \ge 1, \rho \text{ odd} \tag{2.39c}$$

$$\mathcal{P}\left(\bigcup_{j=1}^{N} V_j\right) \ge \sum_{k=1}^{\rho} (-1)^{k-1} s_{k,N} \text{ for } \rho \ge 1, \rho \text{ even.} \tag{2.39d}$$

Besides (2.39c,d), a large amount of related bounds of different complexity are available, cf. [40, 141]. Important bounds of first and second degree are given below:

$$\max_{1 \le j \le N} q_j \le \mathcal{P}\left(\bigcup_{j=1}^{N} V_j\right) \le Q_1 \tag{2.40a}$$

$$Q_1 - Q_2 \le \mathcal{P}\left(\bigcup_{j=1}^{N} V_j\right) \le Q_1 - \max_{1 \le l \le N} \sum_{i \ne l} q_{il} \tag{2.40b}$$

$$\frac{Q_1^2}{Q_1 + 2Q_2} \le \mathcal{P}\left(\bigcup_{j=1}^{N} V_j\right) \le Q_1. \tag{2.40c}$$

The above quantities $q_j, q_{ij}, Q_1, Q_2$ are defined as follows:

$$Q_1 := \sum_{j=1}^{N} q_j \text{ with } q_j := \mathcal{P}(V_j) \tag{2.40d}$$

$$Q_2 := \sum_{j=2}^{N} \sum_{i=1}^{j-1} q_{ij} \text{ with } q_{ij} := \mathcal{P}(V_i \cap V_j). \tag{2.40e}$$

Moreover, defining

$$q := (q_1, \ldots, q_N), Q := (q_{ij})_{1 \le i,j \le N}, \tag{2.40f}$$

we have

$$\mathcal{P}\left(\bigcup_{j=1}^{N} V_j\right) \ge q^T Q^- q, \tag{2.40g}$$

where $Q^-$ denotes the generalized inverse of $Q$, cf. [141].

### 2.5.2 Tschebyscheff-Type Inequalities

In many cases the survival or feasible domain (event) $S = \bigcap_{i=1}^{m} S_i$, is represented by a certain number $m$ of inequality constraints of the type

$$y_{li} < (\le) y_i(a, x) < (\le) y_{ui}, i = 1, \ldots, m, \tag{2.41a}$$

as e.g. operating conditions, behavioral constraints. Hence, for a fixed input, design or control vector $x$, the event $S = S(x)$ is given by

$$S := \{a \in \mathbb{R}^\nu : y_{li} < (\le) y_i(a, x) < (\le) y_{ui}, i = 1, \ldots, m\}. \tag{2.41b}$$

Here,

$$y_i = y_i(a, x), i = 1, \ldots, m, \tag{2.41c}$$

are certain functions, e.g. response, output or performance functions of the structure, system, defined on (a subset of) $\mathbb{R}^\nu \times \mathbb{R}^r$.

Moreover, $y_{li} < y_{ui}, i = 1, \ldots, m$, are lower and upper bounds for the variables $y_i, i = 1, \ldots, m$. In case of one-sided constraints some bounds $y_{li}, y_{ui}$ are infinite.

#### Two-Sided Constraints

If $y_{li} < y_{ui}, i = 1, \ldots, m$, are finite bounds, (2.41a) can be represented by

$$|y_i(a, x) - y_{ic}| < (\le) \rho_i, i = 1, \ldots, m, \tag{2.41d}$$

where the quantities $y_{ic}, \rho_i, i = 1, \ldots, m$, are defined by

$$y_{ic} := \frac{y_{li} + y_{ui}}{2}, \rho_i := \frac{y_{ui} - y_{li}}{2}. \tag{2.41e}$$

Consequently, for the probability $\mathcal{P}(S)$ of the event $S$, defined by (2.41b), we have

$$\mathcal{P}(S) = \mathcal{P}\left(\left|y_i\left(a(\omega), x\right) - y_{ic}\right| < (\le) \rho_i, i = 1, \ldots, m\right). \tag{2.41f}$$

Introducing the random variables

$$\tilde{y}_i\Big(a(\omega),x\Big) := \frac{y_i\Big(a(\omega),x\Big) - y_{ic}}{\rho_i}, i = 1,\dots,m, \tag{2.42a}$$

and the set

$$B := \{y \in \mathbb{R}^m : |y_i| < (\leq) 1, i = 1,\dots,m\}, \tag{2.42b}$$

with $\tilde{y} = (\tilde{y}_i)_{1 \leq i \leq m}$, we get

$$\mathcal{P}(S) = \mathcal{P}\left(\tilde{y}\Big(a(\omega),x\Big) \in B\right). \tag{2.42c}$$

Considering any (measurable) function $\varphi : \mathbb{R}^m \to \mathbb{R}$ such that

$$i)\ \ \varphi(y) \geq 0, y \in \mathbb{R}^m \tag{2.42d}$$

$$ii)\ \ \varphi(y) \geq \varphi_0 > 0, \text{ if } y \notin B, \tag{2.42e}$$

with a positive constant $\varphi_0$, we find the following result:

**Theorem 2.1.** *For any (measurable) function $\varphi : \mathbb{R}^m \to \mathbb{R}$ fulfilling conditions (2.42d,e), the following* **Tschebyscheff-type inequality** *holds:*

$$\begin{aligned} \mathcal{P}\left(y_{li} < (\leq) y_i\Big(a(\omega),x\Big) < (\leq) y_{ui}, i = 1,\dots,m\right) \\ \geq 1 - \frac{1}{\varphi_0} E\varphi\left(\tilde{y}\Big(a(\omega),x\Big)\right), \end{aligned} \tag{2.43}$$

*provided that the expectation in (2.43) exists and is finite.*

*Proof.* If $\mathcal{P}_{\tilde{y}(a(\cdot),x)}$ denotes the probability distribution of the random $m$-vector $\tilde{y} = \tilde{y}\Big(a(\omega),x\Big)$, then

$$\begin{aligned} E\varphi\left(\tilde{y}\Big(a(\omega),x\Big)\right) &= \int\limits_{y \in B} \varphi(y)\mathcal{P}_{\tilde{y}(a(\cdot),x)}(dy) + \int\limits_{y \in B^c} \varphi(y)\mathcal{P}_{\tilde{y}(a(\cdot),x)}(dy) \\ &\geq \int\limits_{y \in B^c} \varphi(y)\mathcal{P}_{\tilde{y}(a(\cdot),x)}(dy) \geq \varphi_0 \int\limits_{y \in B^c} \mathcal{P}_{\tilde{y}(a(\cdot),x)}(dy) \\ &= \varphi_0 \mathcal{P}\left(\tilde{y}\Big(a(\omega),x\Big) \notin B\right) = \varphi_0\left(1 - \mathcal{P}\left(\tilde{y}\Big(a(\omega),x\Big) \in B\right)\right), \end{aligned}$$

which yields the assertion, cf. (2.42c).

*Remark 2.2.* Note that $\mathcal{P}(S) \geq \alpha_s$ with a given minimum reliability $\alpha_s \in (0,1]$ can be guaranteed by the expected cost constraint

$$E\varphi\left(\tilde{y}\Big(a(\omega),x\Big)\right) \leq (1-\alpha_s)\varphi_0.$$

*Example 2.5.* If $\varphi = 1_{B^c}$ is the indicator function of the complement $B^c$ of $B$, then $\varphi_0 = 1$ and (2.43) holds with the equality sign.

*Example 2.6.* For a given positive definite $m \times m$ matrix $C$, define $\varphi(y) := y^T C y, y \in \mathbb{R}^m$. Then, cf. (2.42b,d,e),

$$\min_{y \notin B} \varphi(y) = \min_{1 \le i \le m} \left\{ \min_{y_i \ge 1} y^T C y, \min_{y_i \le -1} y^T C y \right\}. \tag{2.44a}$$

Thus, the lower bound $\varphi_0$ follows by considering the convex optimization problems arising in the right hand side of (2.44a). Moreover, the expectation $E\varphi(\tilde{y})$ needed in (2.43) is given, see (2.42a), by

$$\begin{aligned} E\varphi(\tilde{y}) &= E\tilde{y}^T C \tilde{y} = E\mathrm{tr} C \tilde{y}\tilde{y}^T, \\ &= \mathrm{tr} C (\mathrm{diag}\ \rho)^{-1} \left( \mathrm{cov}\ y\Big(a(\cdot), x\Big) + \Big(\overline{y}(x) - y_c\Big)\Big(\overline{y}(x) - y_c\Big)^T \right) (\mathrm{diag}\ \rho)^{-1}, \end{aligned} \tag{2.44b}$$

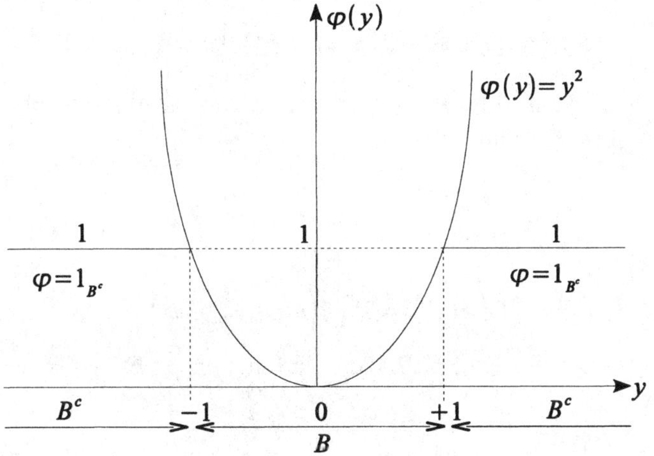


**Fig. 2.4.** Function $\varphi = \varphi(y)$

where "tr" denotes the trace of a matrix, diag $\rho$ is the diagonal matrix diag $\rho := (\rho_i \delta_{ij}), y_c := (y_{ic})$, see (2.41e), and $\overline{y} = \overline{y}(x) := \Big(Ey_i\big(a(\omega), x\big)\Big)$. Since $\|Q\| \le trQ \le m\|Q\|$ for any positive definite $m \times m$ matrix $Q$, an upper bound of $E\varphi(\tilde{y})$ reads

$$E\varphi(\tilde{y}) \le m\|C\| \, \|(\mathrm{diag}\ \rho)^{-1}\|^2 \|\mathrm{cov}\ y\Big(a(\cdot), x\Big) + \Big(\overline{y}(x) - y_c\Big)\Big(\overline{y}(x) - y_c\Big)^T\|. \tag{2.44c}$$

*Example 2.7.* Assuming in the above Example 2.6 that $C = \mathrm{diag}\,(c_{ii})$ is a diagonal matrix with positive elements $c_{ii} > 0, i = 1, \ldots, m$, then

$$\min_{y \notin B} \varphi(y) = \min_{1 \le i \le m} c_{ii} > 0, \tag{2.44d}$$

and $E\varphi(\tilde{y})$ is given by

$$\begin{aligned} E\varphi(\tilde{y}) &= \sum_{i=1}^{m} c_{ii} \frac{E\left(y_i\big(a(\omega), x\big) - y_{ic}\right)^2}{\rho_i^2} \\ &= \sum_{i=1}^{m} c_{ii} \frac{\sigma_{y_i}^2\big(a(\cdot), x\big) + \big(\overline{y}_i(x) - y_{ic}\big)^2}{\rho_i^2}. \end{aligned} \tag{2.44e}$$

## One-Sided Inequalities

Suppose that exactly one of the two bounds $y_{li} < y_{ui}$ is infinite for each $i = 1, \ldots, m$. Multiplying the corresponding constraints in (2.41a) by $-1$, the admissible domain $S = S(x)$, cf. (2.41b), can be represented always by

$$S(x) = \{a \in \mathbb{R}^{\nu} : \tilde{y}_i(a, x) < (\le)\, 0, i = 1, \ldots, m\}, \tag{2.45a}$$

where $\tilde{y}_i := y_i - y_{ui}$, if $y_{li} = -\infty$, and $\tilde{y}_i := y_{li} - y_i$, if $y_{ui} = +\infty$. If we set $\tilde{y}(a, x) := \big(\tilde{y}_i(a, x)\big)$ and

$$\tilde{B} := \Big\{y \in \mathbb{R}^m : y_i < (\le)\, 0, i = 1, \ldots, m\Big\}, \tag{2.45b}$$

then

$$P\big(S(x)\big) = P\Big(\tilde{y}\big(a(\omega), x\big) \in \tilde{B}\Big). \tag{2.45c}$$

Consider also in this case, cf. (2.42d,e), a function $\varphi : \mathbb{R}^m \to \mathbb{R}$ such that

$$i)\ \varphi(y) \ge 0, y \in \mathbb{R}^m \tag{2.46a}$$

$$ii)\varphi(y) \ge \varphi_0 > 0, \text{ if } y \notin \tilde{B}. \tag{2.46b}$$

Then, corresponding to Theorem 2.1, we have this result:

**Theorem 2.2 (Markov–type inequality).** *If $\varphi : \mathbb{R}^m \to \mathbb{R}$ is any (measurable) function fulfilling conditions (2.46a,b), then*

$$P\Big(\tilde{y}\big(a(\omega), x\big) < (\le)\, 0\Big) \ge 1 - \frac{1}{\varphi_0} E\varphi\Big(\tilde{y}\big(a(\omega), x\big)\Big), \tag{2.47}$$

*provided that the expectation in (2.47) exists and is finite.*

*Remark 2.3.* Note that a related inequality was used already in Example 2.2.

*Example 2.8.* If $\varphi(y) := \sum_{i=1}^{m} w_i e^{\alpha_i y_i}$ with positive constants $w_i, \alpha_i, i = 1, \ldots, m$, then

$$\inf_{y \notin \tilde{B}} \varphi(y) = \min_{1 \le i \le m} w_i > 0 \tag{2.48a}$$

and

$$E\varphi\left(\tilde{y}\left(a(\omega), x\right)\right) = \sum_{i=1}^{m} w_i E e^{\alpha_i \tilde{y}_i \left(a(\omega), x\right)}, \tag{2.48b}$$

where the expectation in (2.48b) can be computed approximatively by Taylor expansion:

$$\begin{aligned} Ee^{\alpha_i \tilde{y}_i} &= e^{\alpha_i \overline{\tilde{y}}_i(x)} E e^{\alpha_i \left(\tilde{y}_i - \overline{\tilde{y}}_i(x)\right)} \\ &\approx e^{\alpha_i \overline{\tilde{y}}_i(x)} \left(1 + \frac{\alpha_i^2}{2} E\left(y_i\left(a(\omega), x\right) - \overline{y}_i(x)\right)^2\right) \\ &= e^{\alpha_i \overline{\tilde{y}}_i(x)} \left(1 + \frac{\alpha_i^2}{2} \sigma^2_{y_i(a(\cdot), x)}\right). \end{aligned} \tag{2.48c}$$

Supposing that $y_i = y_i\left(a(\omega), x\right)$ is a normal distributed random variable, then

$$Ee^{\alpha_i \tilde{y}_i} = e^{\alpha_i \overline{\tilde{y}}_i(x)} e^{\frac{1}{2}\alpha_i^2 \sigma^2_{y_i(a(\cdot), x)}}. \tag{2.48d}$$

*Example 2.9.* Consider $\varphi(y) := (y - b)^T C (y - b)$, where, cf. Example 2.6, $C$ is a positive definite $m \times m$ matrix and $b < 0$ a fixed $m$-vector. In this case we again have

$$\min_{y \notin \tilde{B}} \varphi(y) = \min_{1 \le i \le m} \min_{y_i \ge 0} \varphi(y)$$

and

$$\begin{aligned} E\varphi(\tilde{y}) &= E\left(\tilde{y}\left(a(\omega), x\right) - b\right)^T C \left(\tilde{y}\left(a(\omega), x\right) - b\right) \\ &= \mathrm{tr} C E\left(\tilde{y}\left(a(\omega), x\right) - b\right)\left(\tilde{y}\left(a(\omega), x\right) - b\right)^T. \end{aligned}$$

Note that

$$\tilde{y}_i(a, x) - b_i = \begin{cases} y_i(a, x) - (y_{ui} + b_i), & \text{if } y_{li} = -\infty \\ y_{li} - b_i - y_i(a, x), & \text{if } y_{ui} = +\infty, \end{cases}$$

where $y_{ui} + b_i < y_{ui}$ and $y_{li} < y_{li} - b_i$.

*Remark 2.4.* The one-sided case can also be reduced approximatively to the two-sided case by selecting a sufficiently large, but finite upper bound $\tilde{y}_{ui} \in \mathbb{R}$, lower bound $\tilde{y}_{li} \in \mathbb{R}$, resp., if $y_{ui} = +\infty, y_{li} = -\infty$.

### 2.5.3 FORM (First Order Reliability Methods)

Special approximation methods for probabilities were developed in structural reliability analysis [18,20,48,94] for the probability of failure $p_f = p_f(x)$ given by (2.9c) with $B := (0, +\infty)^{m_y}$. Hence, see (2.2b),

$$p_f = 1 - P\Big(y\big(a(\omega), x\big) > 0\Big).$$

Applying certain probability inequalities, see Section 2.5.1 and 2.5.2, the case with $m(> 1)$ limit state functions is reduced first to the case with one limit state function only. Approximations of $p_f$ are obtained then by linearization or by quadratic approximation of a transformed state function $\tilde{y}_i = \tilde{y}_i(\tilde{a}, x)$ at a certain "design point" $\tilde{a}^* = \tilde{a}^{*i}(x)$ for each $i = 1, \ldots, m$.

Based on the above procedure, according to (2.9f) we have

$$\begin{aligned} p_f(x) &= P\left(y_i\big(a(\omega), x\big) \le 0 \text{ for at least one index } i, 1 \le i \le m_y\right) \\ &\le \sum_{i=1}^{m_y} P\left(y_i\big(a(\omega), x\big) \le 0\right) = \sum_{i=1}^{m_y} p_{fi}(x), \end{aligned} \tag{2.49a}$$

where

$$p_{fi}(x) := P\left(y_i\big(a(\omega), x\big) \le 0\right). \tag{2.49b}$$

Assume now that the random $\nu$-vector $a = a(\omega)$ can be represented [94] by

$$a(\omega) = U\Big(\tilde{a}(\omega)\Big), \tag{2.50a}$$

where $U : \mathbb{R}^\nu \to \mathbb{R}^\nu$ is a certain $1-1$–transformation in $\mathbb{R}^\nu$, and $\tilde{a} = \tilde{a}(\omega)$ is an $N(0, I)$-normal distributed random $\nu$-vector, i.e. with mean vector zero and covariance matrix equal to the identity matrix $I$. In the following, let $x$ be a given, fixed $r$-vector. Defining then

$$\tilde{y}_i(\tilde{a}, x) := y_i\Big(U(\tilde{a}), x\Big), \tag{2.50b}$$

we get

$$p_{fi}(x) = P\left(\tilde{y}_i\big(\tilde{a}(\omega), x\big) \le 0\right), i = 1, \ldots, m_y. \tag{2.50c}$$

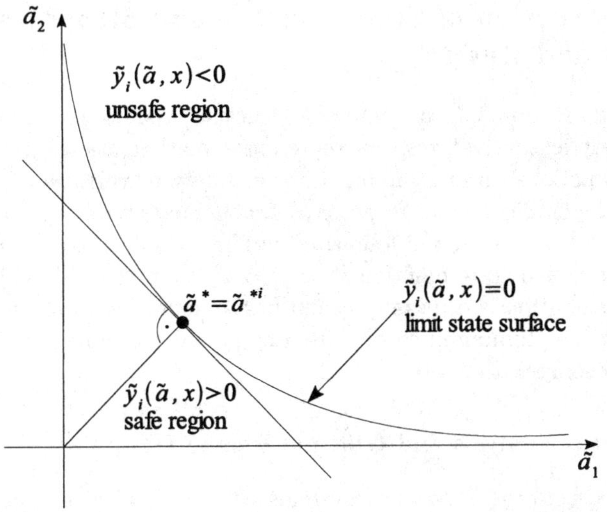


**Fig. 2.5.** FORM

In the First Order Reliability Method (FORM), cf. [48], the state function $\widetilde{y}_i = \widetilde{y}_i(\widetilde{a}, x)$ is linearized at a so-called design point $\widetilde{a}^* = \widetilde{a}^{*(i)}$. A design point $\widetilde{a}^* = \widetilde{a}^{*(i)}$ is obtained by projection of the origin $\widetilde{a} = 0$ onto the limit state surface $\widetilde{y}_i(\widetilde{a}, x) = 0$, see Fig. 2.5. Since $\widetilde{y}_i(\widetilde{a}^*, x) = 0$, we get

$$\begin{aligned} \widetilde{y}_i(\widetilde{a}, x) &\approx \widetilde{y}_i(\widetilde{a}^*, x) + \Big(\nabla_{\widetilde{a}}\widetilde{y}_i(\widetilde{a}^*, x)\Big)^T (\widetilde{a} - \widetilde{a}^*) \\ &= \Big(\nabla_{\widetilde{a}}\widetilde{y}_i(\widetilde{a}^*, x)\Big)^T (\widetilde{a} - \widetilde{a}^*) \end{aligned} \tag{2.50d}$$

and therefore

$$p_{fi}(x) \approx P\left(\Big(\nabla_{\widetilde{a}}\widetilde{y}_i(\widetilde{a}^*, x)\Big)^T (\widetilde{a}(\omega) - \widetilde{a}^*) \le 0\right) = \Phi\Big(-\beta_i(x)\Big). \tag{2.50e}$$

Here, $\Phi$ denotes the distribution function of the univariate $N(0,1)$-normal distribution, and the reliability index $\beta_i = \beta_i(x)$ is given by

$$\beta_i(x) := -\frac{\Big(\nabla_{\widetilde{a}}\widetilde{y}_i(\widetilde{a}^*, x)\Big)^T \widetilde{a}^*}{\|\nabla_{\widetilde{a}}\widetilde{y}_i(\widetilde{a}^*, x)\|}. \tag{2.50f}$$

Note that $\beta_i(x)$ is the minimum distance from $\widetilde{a} = 0$ to the limit state surface $\widetilde{y}_i(\widetilde{a}, x) = 0$ in the $\widetilde{a}$-domain.

For more details and Second Order Reliability Methods (SORM) based on second order Taylor expansions, see Chapter 7 and [48,94].

## 2.6 Construction of State Functions in Structural Analysis and Design

According to Section 2.1, the state of a structure/system, i.e., its weakness or failure, safety or survival, resp., is represented mathematically by one or several state functions (limit state functions or safety margins) $y_i = y_i(a, x), i = 1, \dots, m_y$, depending on the $\nu$–vector $a$ of model parameters and the $r$–vector $x$ of design variables. In the following, minimum value representations of a general real valued state function $s^* = s^*(a, x)$ are presented. This minimum value representation is especially useful in the definition of the minimum *recourse costs*, i.e., minimum costs caused by systems weakness, failure, or due to repair, reconstruction, etc..

### 2.6.1 Plastic Analysis and Optimal Plastic Design

In case of structures, structural systems consisting of elastoplastic materials, the survival or safety of the structure/system can be described [2,44,47, 58,89,117], after a possible Finite Element (FE) discretization [133], by the equilibrium equation and the yield condition:

$$C\sigma = P \tag{2.51a}$$
$$\pi(R_{id}^{-1}|K_i) \le 1, i = 1, \dots, n_G. \tag{2.51b}$$

Here, $C$ is the $m \times n$ equilibrium matrix having rank $C = m < n, P$ is the vector of external loads, $\sigma$ is the total stress (state) $n$–vector composed of the $n_0$–subvectors $\sigma_i$ of stress (state) components at the reference or nodal points $x_i, i = 1, \dots, n_G$, arising in the FE–discretization. Furthermore, $R_i = (R_{ij})$ is the $n_0$–vector of positive material resistance or strength parameters at point $x_i$, and $R_{id}$ denotes the diagonal matrix having the diagonal elements $R_{ij}, j = 1, \dots, n_0$. Finally, $K_i$ is the closed, convex admissible domain at point $x_i$, where $K_i$ contains the origin of $\mathbb{R}^{n_0}$ as an interior point. In the following, the function

$$\pi(z|K_i) = \inf\left\{\lambda > 0 : \frac{z}{\lambda} \in K_i\right\}, z \in \mathbb{R}^{n_0}, \tag{2.52}$$

denotes the distance or Minkowski functional of the convex set $K_i$.

According to [89,90], the state function $s^* = s^*(R, P)$ is then defined by the minimum value function of the convex optimization problem

$$\min s \tag{2.53a}$$

s.t.

$$C\sigma = P \tag{2.53b}$$
$$\pi(R_{id}^{-1}\sigma_i|K_i) - 1 \le s, i = 1, \dots, n_G. \tag{2.53c}$$

Note that $R = (R_i)$ is the $n$–vector consisting of the $n_0$–subvectors $R_i, i = 1, \dots, n_G$.

Structural survival, failure, i.e. the existence, lack of a feasible stress state, hence, the existence, lack, resp., of a vector $\sigma$ fulfilling the basic conditions (2.51a,b) can be described by the following criterion:

**Theorem 2.3.** *a) If the structure/structural system is in a safe state, i.e., if a stress (state) vector $\sigma$ exists fulfilling the safety conditions (2.51a,b), then*

$$-1 \leq s^*(R,P) \leq 0. \tag{2.54a}$$

*b) Conversely, if*

$$-1 \leq s^*(R,P) < 0, \tag{2.54b}$$

*then the structure/system is in a safe state.*

*Proof.* We first observe that $s \geq -1$ for any feasible point $(\sigma, s)$. Hence $s^*(R,P) \geq -1$. a) If the survival conditions (2.51a,b) can be fulfilled by a stress (state) vector $\sigma^0$, then $(\sigma, s) = (\sigma^0, s^0), s^0 := 0$, is feasible solution of the optimization problem (2.53a-c). Hence, $s^*(R,P) \leq s^0 = 0$. b) Suppose that $s^*(R,P) < 0$. Then there is a feasible point $(\sigma^0, s^0)$ in (2.53a-c) such that $s^0 < 0$. This yields now $C\sigma^0 = P$ and $\pi\left(R_{id}^{-1}\sigma_i|K_i\right) - 1 \leq s^0 < 0, i = 1,\ldots,n_G$. Hence, $\sigma^0$ fulfills the safety conditions (2.51a,b).

In many cases an optimal solution $(\sigma^*, s^*)$ is attained in (2.53a-c). Then Theorem 2.3 has the following consequence:

**Corollary 2.2.** *Suppose that a minimum point is attained in (2.53a-c) for all parameters $(R,P)$ under consideration. Then a safe stress (state) $\sigma$ exists if and only if condition (2.54a) holds.*

Introducing the new variable

$$\hat{s} := s + 1, \tag{2.55a}$$

the state function $s^* = s^*(R,P)$ can be represented by

$$s^*(R,P) = \hat{s}^*(R,P) - 1, \tag{2.55b}$$

where the modified state function $\hat{s}^* = \hat{s}^*(R,P)$ is given by the minimum value of the convex optimization problem

$$\min \hat{s} \tag{2.56a}$$

s.t.

$$C\sigma = P \tag{2.56b}$$

$$\pi(R_{id}^{-1}\sigma_i|K_i) \leq \hat{s}, i = 1,\ldots,n_G. \tag{2.56c}$$

However, this problem is equivalent to

$$\min \max_{1 \leq i \leq n_G} \pi(R_{id}^{-1}\sigma_i|K_i) \tag{2.57a}$$

s.t.

$$C\sigma = P. \tag{2.57b}$$

Obviously, (2.57a,b) can be interpreted as the weighted projection of the zero stress state vector $\sigma = 0$ onto linear manifold given by (2.57b).

A further state function $t^* = t^*(R, P)$ can be defined by the minimum value of the convex optimization problem

$$\min t \tag{2.58a}$$

s.t.

$$||C\sigma - P|| \leq t \tag{2.58b}$$

$$\pi(R_{id}^{-1}\sigma_i|K_i) \leq 1, i = 1, \ldots, n_G. \tag{2.58c}$$

Hence, in contrast to (2.53a-c), here the yield condition (2.51b) is required to hold, while variations of the equilibrium condition (2.51a) are allowed.

Corresponding to Theorem 2.3, for $t^* = t^*(R, P)$ we have the following criterion:

**Theorem 2.4.** *a) If a safe stress state exists, then*

$$t^*(R, P) = 0. \tag{2.59}$$

*b) Suppose that a minimum point is attained in (2.58a-c) for all parameters $(R, P)$ under consideration. Then a safe stress state exists if and only if (2.59) holds.*

*Proof.* This criterion can be shown by the same arguments as used in the above Theorem 2.3.

It is easy to see that the state function $t^* = t^*(R, P)$ can be described also by the minimum value of the convex program

$$\min ||C\sigma - P|| \tag{2.60a}$$

s.t.

$$\pi(R_{id}^{-1}\sigma_i|K_i) \leq 1, i = 1, \ldots, n_G. \tag{2.60b}$$

Hence, $t^* = t^*(R, P)$ is the distance between the external load vector $P$ and the convex set

$$\left\{C\sigma : \pi(R_{id}^{-1}\sigma_i|K_i) \leq 1, i = 1, \ldots, n_G\right\}$$

of feasible external loadings in case of the material resistance given by the vector $R$.

### 2.6.2 Optimal Elastic Design

In optimal elastic design [55], instead of (2.51a,b) we have the safety conditions:

$$K(R)u = P \tag{2.61a}$$

$$\eta(u) \geq 0. \tag{2.61b}$$

Here, $K = K(R)$ is the global $m \times m$ stiffness matrix of the structure depending on a vector $R$ of resistance parameters involving the elastic moduli of the structural elements. Moreover, $u$ denotes the $m$–vector of displacement components, and the vectorial constraint (2.61b) summarize different displacement, stress and force constraints:

$$\eta_j(u) \geq 0, j = 1, \ldots, m_\varphi. \tag{2.61b'}$$

Consequently, corresponding to (2.53a-c), a first state function $s^* = s^*(R, P)$ can be defined by the minimum value of the optimization problem:

$$\min s \tag{2.62a}$$

s.t.

$$K(R)u = P \tag{2.62b}$$

$$-\eta(u) \leq s 1_{m_\eta}, \tag{2.62c}$$

where $1_{m_\eta} := (1, 1, \ldots, 1)^T \in \mathbb{R}^{m_\eta}$.

In many cases $K(R)$ is positive definite, and the equation (2.62b) yields $u = K(R)^{-1}P$. Hence, in this case the minimum value of (2.62a-c) reads:

$$s^*(R, P) = -\min_{1 \leq j \leq m_\eta} \eta_j\left(K(R)^{-1}P\right). \tag{2.63}$$

In correspondence to Theorem 2.3, here we have immediately this criterion:

**Theorem 2.5.** *a) If a safe elastic stress state exists, then $s^*(R, P) \leq 0$. b) If $s^*(R, P) < 0$, then a safe stress state exists. Moreover, if (2.62a-c) has an optimal solution, then a safe elastic stress state exists in case of the parameter configuration $(R, P)$ if and only if $s^*(R, P) \leq 0$.*

Corresponding to Section 2.6.1, a further state function can be obtained by the minimum value of the convex optimization problem

$$\min t \tag{2.64a}$$

s.t.

$$\left\| K(R)u - P \right\| \leq t \tag{2.64b}$$

$$\varphi(u) \leq 0. \tag{2.64c}$$

Using the state function $t^* = t^*(R, P) \geq 0$, we have the following criterion:

**Theorem 2.6.** *a) If a safe elastic stress state exists, then* $t^*(R, P) = 0$. *b) If (2.64a-c) has an optimal solution, then a safe stress state exists if and only if* $t^*(R, P) = 0$.

# Part II

# Differentiation Methods

# 3

# Differentiation Methods for Probability and Risk Functions

## 3.1 Introduction

The reliability of a stochastic system from engineering or economics is often measured by probability functions of the type

$$P(x) = \mathcal{P}\left(y_{li} < (\leq) y_i\Big(a(\omega), x\Big) < (\leq) y_{ui}, i = 1, \ldots, m\right)$$

$$= \int\limits_{B(x)} f(a)da \tag{3.1a}$$

and/or by the separated probability functions

$$P_i(x) = \mathcal{P}\left(y_{li} < (\leq) y_i\Big(a(\omega), x\Big) < (\leq), y_{ui}\right), i = 1, \ldots, m,$$

$$= \int\limits_{B_i(x)} f(a)da. \tag{3.1b}$$

Here, cf. (2.2a-c),

$$y_{li} < (\leq) y_i(a, x) < (\leq) y_{ui}, i = 1, \ldots, m, \tag{3.2}$$

are the basic operating, safety conditions or behavioral constraints of the underlying system. Furthermore, $y = (y_1, \ldots, y_i, \ldots, y_m)'$ are certain response, output, performance variables, e.g. displacements, stresses, strains, forces, weight in structural design or the total costs, deviation between the demand vector and the production output in a production problem. The response variables $y_i$ are functions

$$y_i = y_i(a, x), i = 1, \ldots, m, \tag{3.3a}$$

of a decision, design or input $r$-vector $x = (x_1, \ldots, x_i, \ldots, x_r)'$ and a parameter $\nu$-vector $a = (a_1, \ldots, a_j, \ldots, a_\nu)'$.

Moreover,

$$B(x) := \{a \in \mathbb{R}^\nu : y_{li} < (\le) y_i(a,x) < (\le) y_{ui}, i = 1, \ldots, m\}, \quad (3.3b)$$
$$B_i(x) := \{a \in \mathbb{R}^\nu : y_{li} < (\le) y_i(a,x) < (\le) y_{ui}\}, i = 1, \ldots, m, \quad (3.3c)$$

denotes the region of parameter vectors $a \in \mathbb{R}^\nu$ fulfilling the basic constraints (3.2), one single constraint in (3.2), respectively.

Corresponding to Section 1.1 and 1.2, $x_k, k = 1, \ldots, r$, are the *deterministic (nominal)* design, input variables or deterministic system coefficients, as e.g. sizing variables, geometrical variables (e.g. nodal coordinates), degree of refinement of the material, manufacturing tolerances in structural design problems, or factors of production, production inputs in production problems. Moreover, $a_j = a_j(\omega), j = 1, \ldots, \nu$, are the random system parameters or coefficients, e.g. noise, material, tolerance, load parameters, manufacturing errors in technical design problems, or demands, technological coefficients, cost factors in production problems. We assume that the random $\nu$-vector

$$a(\omega) = (a_1(\omega), \ldots, a_\nu(\omega))', \quad (3.3d)$$

defined on a probability space $(\Omega, \mathcal{A}, P)$, has a given probability density function $f = f(a)$ on $\mathbb{R}^\nu$. Finally,

$$y_l = (y_{l1}, \ldots, y_{li}, \ldots, y_{lm})', y_u = (y_{u1}, \ldots, u_{ui}, \ldots, y_{um})' \quad (3.3e)$$

are the $m$-vectors of lower, upper bounds (margins) $y_{li} < y_{ui}, i = 1, \ldots, m$, for the response vector $y$. In some concrete applications [33, 55], the bounds $y_{li}, y_{ui}$ are also functions

$$y_{li} = y_{li}\Big(a(\omega)\Big), y_{ui} = y_{ui}\Big(a(\omega)\Big), i = 1, \ldots, m \quad (3.3f)$$

of the random parameter vector $a = a(\omega)$.

*Example 3.1.* Structural reliability and design [2, 26, 79, 94, 109, 124, 125, 140]. Here, $y_i = y_i(a,x), i = 1, \ldots, m$, are the relevant stress, strain, force, weight, ... variables of the basic mechanical system. The decision or design vector $x$ involves the nominal values of certain sizing, geometrical or material variables. Moreover, the random vector $a = a(\omega)$ contains the random load components of the structure and describes, in addition, the random variations of the materials and the manufacturing process.

Especially, the reliability analysis of a mechanical structure is based on the fact that a structure can be represented as a combination of certain series and parallel mechanical subsystems. Hence, for a given, fixed design or input vector $x$, the systems failure domain $V \subset \mathbb{R}^\nu$ can be represented [109, 140] by

$$V = \bigcup_{i=1}^{N} V_i \text{ with } V_i := \bigcap_{k=1}^{K_i} V_{ik}, \quad (3.4a)$$

where the $(i,k)$–th failure mode domain $V_{ik}$ is given by

$$V_{ik} := \{a \in \mathbb{R}^\nu : g_{ik}(a,x) < 0\}, \tag{3.4b}$$

cf. (2.38a,b). Here, $g_{ik} = g_{ik}(a,x)$ is the so–called limit state functionof the $(i,k)$–th failure mode. Typical examples are

$$g_{ik}(a,x) := -\left|y_{ik}(a,x)\right| + y_{ik}^{\max}(a), k = 1,\ldots,K_i, \tag{3.4c}$$

where $y_{ik} = y_{ik}(a,x)$ denotes the stress component of the $k$-th parallel element, $k = 1,\ldots,K_i$, within the $i$-th serial substructure, $i = 1,\ldots,N$. Furthermore, $y_{ik}^{\max} = y_{ik}^{\max}(a), k = 1,\ldots,K_i,\ i = 1,\ldots,N$, are the maximum stresses of the structural elements.

Obviously, the failure probability $p_{fi} = p_{fi}(x)$ of the $i$–th failure domain $V_i$,

$$p_{fi} := \mathcal{P}(V_i) = \mathcal{P}\left(g_{ik}\left(a(\omega),x\right) < 0, 1 \le k \le K_i\right), \tag{3.4d}$$

is a probability function on the type (3.1a). Furthermore, according to Section 2.5.1, the system failure probability

$$p_f = p_f(x) := \mathcal{P}(V_1 \cup V_2 \cup \ldots \cup V_N)$$

as well as several upper and lower Bonferroni-type bounds consist of probability functions of the type (3.1a,b).

According to Section 2.1, reliability based optimization (RBO) problems are deterministic substitute problems based on some probability functions:

I) **Reliability maximization.** Find an input vector $x \in D_0$ (with the feasible domain $D_0 \subset \mathbb{R}^r$) fulfilling certain deterministic constraints on $x$ and

$$P(x) \longrightarrow \text{maximum ( or } \psi\Big(P_1(x),\ldots,P_M(x)\Big) \to \max \text{ )} \tag{3.5a}$$

with a certain function $\psi$;

II) **Reliability constraints.** Given certain minimum probabilities (reliabilities) $\alpha_i, 1 \le i \le m, \alpha$, resp., find $x \in D_0$ such that

$$P_i(x) \ge \alpha_i, i = 1,\ldots,m, \text{ and/or } P(x) \ge \alpha, \tag{3.5b}$$

and a certain objective function is minimized (maximized).

Furthermore, besides the (RBO) problems (3.5a,b), in many considerations of systems reliability [20, 140], the following problem, occurs:

III) **Sensitivity of reliabilities.** Determine the rate of change of the probability functions $P(x), P_i(x), 1 \le i \le m$, with respect to the input variables $x_k, k = k_1,\ldots,k_n$.

Hence, in all three cases one needs informations on the first – and also higher – order derivatives of the probability functions $P(x), P_i(x), 1 \le i \le m$, with respect to the decision variables or inputs $x_k, k = 1, \ldots, r$.

According to definition (3.1a,b), the probability functions $P = P(x), P_i = P_i(x)$ can be represented by parameter-dependent integrals in $\mathbb{R}^\nu$. However, formulas for the differentiation of parameter–dependent multiple integrals are well known since long time in fluid dynamics. Here, differentiation formulas for parameter–dependent multiple integrals, similar to Leibniz's formula [21] for definite integrals in $\mathbb{R}^1$, were developed, see [136]. These formulas are needed for the computation of the rate of change $\frac{dN}{dt}$, with respect to time $t$, of a certain property $N(t)$ like mass, momentum, energy, etc., of a physical system within a time varying system boundary, describing the so–called **control volume**. The control volume concept is used [136] to derive continuity, momentum, and energy equations, as well as to solve many other problems in fluid dynamics.

Quite similar differentiation formulas were derived later on for applications in different fields of stochastics, see e.g. [20, 145].

More constructive differentiation methods for probability functions and parameter–dependent integrals are based [79, 83, 85–87] on the following new techniques which are described in this chapter:

I) Transformation Method
II) Stochastic Completion and Transformation Method
III) Orthogonal Function Series Expansion
IV) Combination of (I) - (III).

Next to the Transformation Method is presented by means of differention of probability functions arising in chance constrained programming and structural reliability analysis, see Example 3.1.

## 3.2 Transformation Method: Differentiation by Using an Integral Transformation

In many important cases integral formulas for the first and also higher order derivatives of probability functions and parameter-dependent integrals can be obtained by using a certain *integral transformation*.

The basic principle is demonstrated now in case of the important probability functions

$$P(x) := \mathcal{P}\Big(A(\omega)x \le b(\omega)\Big) \tag{3.6a}$$

$$P_i(x) := \mathcal{P}\Big(A_i(\omega)x \le b_i(\omega)\Big), i = 1, \ldots, m, \tag{3.6b}$$

occuring in the chance constrained programming approach to linear programs with random data [51,52,69]

Here, $(A,b) = \big(A(\omega), b(\omega)\big)$ denotes a random $m \times (r+1)$-matrix with rows $(A_i, b_i), i = 1, \ldots, m$, having a known probability distribution. Obviously, setting in (3.2)

$$y(a,x) := Ax - b \text{ with } a := (A,b), \tag{3.7a}$$

$$y_{li} := -\infty \text{ with " } < \text{ " in the left inequality,} \tag{3.7b}$$

$$y_{ui} := 0 \text{ with " } \leq \text{ " in the right inequality,} \tag{3.7c}$$

$P(x), P_i(x)$ defined by (3.6a), (3.6b), resp., can be represented by (3.1a), (3.1b). Suppose that the random matrix $\big(A(\omega), (\omega)\big)$ has a probability density

$$f = f(A,b) = f(a_1, \ldots, a_r, b),$$

where $a_k$ denotes the $k$–th column of $A$. We find

$$P(x) = \int\limits_{\sum\limits_{k=1}^{r} a_k x_k \leq b} f_i(a_{i1} m \ldots, a_{ir}, b_i) \prod_{k=1}^{r} da_{ik} db_i. \tag{3.8a}$$

Furthermore, $P_i(x), i = 1, \ldots, m$, can be represented by

$$P_i(x) = \int\limits_{\sum\limits_{k=1}^{r} a_{ik} x_k \leq b_i} f_i(a_{i1}, \ldots, a_{ir}, b_i) \prod_{k=1}^{r} da_{ik} db_i, \tag{3.8b}$$

where

$$f_i = f_i(A_i, b_i) = f_i(a_{i1}, \ldots, a_{ir}, b_i)$$

designates the (marginal) density of the random (row) vector $\big(A_i(\omega), b_i(\omega)\big)$. Assume that the integrals in (3.8a) and (3.8b) exist and are finite for each $x$ under consideration.

We want to find now $\frac{\partial P}{\partial x_k}(x)$ at points

$$x = x(t) := (x_1^0, \ldots, x_{k-1}^0, t, x_k^0, \ldots, x_r)', \; t \in I, \tag{3.9}$$

where $x_j^0, j \neq k$, are given fixed values for the components $x_j, j \neq k$, and $I$ is a certain interval in $\mathbb{R}$ with $0 \notin I$. Consider in $\mathbb{R}^{m(r+1)}$ the 1-1–transformation $a = (a_1, \ldots, a_r, b) = T_x^{(k)}(q), q = (q_1, \ldots, q_r, q_{r+1})$, defined column-wise by

$$a_k := \frac{1}{x_k} q_k, a_j := q_j, 1 \le j \le r, j \ne k, b := q_{r+1}. \tag{3.10a}$$

Applying transformation (3.10a) to the multiple integral in (3.8a), cf. [113], $P(x)$ may be represented by

$$P(x) = \int\limits_{\sum\limits_{j \ne k} q_j x_j + q_k \le q_{r+1}} f(q_1, \ldots, q_{k-1}, \frac{q_k}{x_k}, q_{k+1}, \ldots, q_r, q_{r+1}) \frac{1}{|x_k|^m} \prod_{j=1}^{r+1} dq_j. \tag{3.10b}$$

Furthermore, for the probability function $P_i(x)$ we consider the integral transformation $(A_i, b_i) = T_x^{(i,k)}(q_{i1}, \ldots q_{ir}, q_{ir+1})$ in $\mathbb{R}^{r+1}$ defined by

$$a_{ik} := \frac{1}{x_k} q_{ik}, a_{ij} := q_{ij}, 1 \le j \le r, j \ne k, b_i := q_{ir+1}, \tag{3.11a}$$

$x_k \ne 0$. We find

$$P_i(x) = \int\limits_{\sum\limits_{j \ne k} q_{ij} x_j + q_{ik} \le q_{ir+1}} f_i\left(q_{i1}, \ldots, q_{ik-1}, \frac{q_{ik}}{x_k}, q_{ik+1}, \ldots, q_{ir}, q_{ir+1}\right) \frac{1}{|x_k|} \prod_{j=1}^{r+1} dq_{ij} \tag{3.11b}$$

$i = 1, \ldots, m$. Obviously, by using the integral transformations $T_x^{(k)}, T_x^{(i,k)}$, resp., we end up with a region of integration $\tilde{B}_k, \tilde{B}_{ik}$, being independent of the argument $x_k$. Thus under weak assumptions, the derivatives of $P$ and $P_i$ with respect to $x_k$ can be obtained now simply by interchanging differentiation and integration. For the formulation of corresponding regularity assumptions we need, in the case of $P(x)$, the following notations:

$$\tilde{B}_k := \left\{ (q_1, \ldots, q_r, q_{r+1}) \in (\mathbb{R}^m)^{r+1} : \sum_{j \ne k} a_j x_j^\circ + q_k \le q_{r+1} \right\} \tag{3.12a}$$

$$\begin{aligned} h_k &= h_k(q_1, \ldots, q_r, q_{r+1}; x) := \\ &-\frac{1}{x_k} \left\{ \nabla a_k f\left(q_1, \ldots, q_{k-1}, \frac{q_k}{x_k}, q_{k+1}, \ldots, q_r, q_{r+1}\right)^T \frac{q_k}{x_k} \right. \\ &\left. + m f\left(q_1, \ldots, q_{k-1}, \frac{q_k}{x_k}, q_k, \ldots, q_r, q_{r+1}\right) \right\} \frac{1}{|x_k|^m}, \end{aligned} \tag{3.12b}$$

where the existence of the gradient $\nabla a_k f$ of the density $f$ with respect to $a_k$ is presupposed for all arguments under consideration.

Having these prerequisites, we can formulate the following differentiation formulas, c.f. Lemma 2.3:

**Theorem 3.1.** *For a given, fixed integer $k, 1 \le k \le r$, and given, fixed components $x_1^\circ, \dots, x_{k-1}^\circ, x_{k+1}^\circ, \dots, x_r^0$, define the vectors $x = x(t), t \in \mathbb{R}$, by (3.9). Suppose that for a given interval $I \subset \mathbb{R}$ with $0 \notin I$ the multiple integral in (3.8a) or (3.10b) exists and is finite for each $x(t)$ with $t \in I$. Furthermore, suppose that the function $h_k\Big(q_1, \dots, q_r, q_{r+1}; x(t)\Big)$, defined by (3.12b), exists for all $t \in I$ and has an integrable majorant, i.e. a nonnegative measurable function $H_k = H_k(q_1, \dots, q_{r+1})$, definite at least on $\tilde{B}_k$, with*

$$\left|h_k\Big(q_1, \dots, q_r, q_{r+1}; x(t)\Big)\right| \le H_k(q_1, \dots, q_r, q_{r+1}) \text{ for all } t \in I \tag{3.12c}$$

and

$$\int\limits_{\tilde{B}_k} H_k(q_1, \dots, q_r, q_{r+1}) \prod_{j=1}^{r+1} dq_j < +\infty. \tag{3.12d}$$

Then, $\dfrac{\partial P}{\partial x_k}(x)$ exists for each $x \in \{x(t) : t \in I\}$ and is given, cf (3.7a), by

$$\begin{aligned} \tfrac{\partial P}{\partial x_k}(x) &= \int\limits_{\tilde{B}_k} h_k(q_1, \dots, q_{r+1}; x) \prod_{j=1}^{r+1} dq_j \\ &= -\tfrac{1}{x_k} \int\limits_{\sum_{j=1}^{r} a_j x_j \le b} \operatorname{div}_{a_k}\Big(a_k f(A, b)\Big)\, dA db \end{aligned} \tag{3.13a}$$

where $\operatorname{div}_{a_k}$ denotes the divergence with respect to the vector $a_k$.

*Proof.* Under the above assumptions we may interchange [4, 113] for each $x \in \{x(t) : t \in I\}$ differentiation with respect to $x_k$ and integration in (3.10b), hence,

$$\begin{aligned} \frac{\partial P}{\partial x_k}(x) &= \frac{\partial}{\partial x_k} \int\limits_{\tilde{B}_k} f\left(q_1, \dots, q_{k-1}, \frac{q_k}{x_k}, q_{k+1}, \dots, q_r, q_{r+1}\right) \frac{1}{|x_k|^m} \prod_{j=1}^{r+1} dq_j \\ &= \int\limits_{\tilde{B}_k} h_k(q_1, \dots, q_r, q_{r+1}; x) \prod_{j=1}^{r+1} dq_j. \end{aligned}$$

The second part of formula (3.13a) is obtained by applying the back–transformation $q = T_x^{(k)-1}\binom{a}{b}$ defined by

$$q_k := x_k a_k, q_j := a_j, 1 \le j \le r, j \ne k, q_{r+1} := b.$$

Obviously, the derivative of $P_i$ can be obtained in the same way: Indeed, according to (3.11b), we first define

$$\tilde{B}_{ik} := \left\{ (q_{i1}, \ldots, q_{ir}, q_{ir+1})' \in \mathbb{R}^{r+1} : \sum_{j \neq k} q_{ij} x_j^0 + q_{ik} \leq q_{ir+1} \right\} \tag{3.12a'}$$

$$\begin{aligned} h_{ik} &= h_{ik}(q_{i1}, \ldots, q_{ir}, q_{ir+1}; x) \\ &:= -\frac{1}{x_k} \left\{ \frac{\partial f_i}{\partial a_{ik}} \left( q_{i1}, \ldots, q_{ik-1}, \frac{q_{ik}}{x_k}, \; q_{ik+1}, \ldots, q_{ir}, q_{ir+1} \right) \frac{q_{ik}}{x_k} \right. \\ &\quad \left. + f_i \left( q_{i1}, \ldots, q_{ik-1}, \frac{q_{ik}}{x_k}, q_{ik+1}, \ldots, q_{ir}, q_{ir+1} \right) \right\} \frac{1}{|x_k|}, \end{aligned} \tag{3.12b'}$$

where $x = x(t), t \in I$, is again defined by (3.9). Furthermore, we suppose that $h_{ik}$ has an integrable majorant $H_{ik}$, hence,

$$|h_{ik}(q_{i1}, \ldots, q_{ir}, q_{ir+1}; x(t)| \leq H_{ik}(q_{i1}, \ldots, q_{ir}, q_{ir+1}) \text{ for all } t \in I \tag{3.12c'}$$

and

$$\int_{\tilde{B}_{ik}} H_{ik}(q_{i1}, \ldots, q_{ir}, q_{ir+1}) \prod_{j=1}^{r+1} dq_j < +\infty. \tag{3.12d'}$$

Replacing now $\tilde{B}_k, h_k, H_k$ in Theorem 3.1 by $\tilde{B}_{ik}, h_{ik}, H_{ik}$, resp., the following consequence from Theorem 3.1 (for $m := 1$) is obtained:

**Corollary 3.1.** *Suppose that conditions (3.12a') – (3.12d') hold. Then* $\frac{\partial P_i}{\partial x_k}(x)$ *exists for each* $x \in \{x(t) : t \in I\}$, *and is given by*

$$\begin{aligned} \frac{\partial P_i}{\partial x_k}(x) &= \int_{\tilde{B}_{ik}} h_{ik}(q_{i1}, \ldots, q_{ir+1}; x) \prod_{j=1}^{r+1} dq_{ij} \\ &= -\frac{1}{x_k} \int_{\sum_{j=1}^{r} a_{ij} x_j \leq b_i} \frac{\partial}{\partial a_{ik}} (a_{ik} f_i(A_i, b_i)) \; dA_i db_i. \end{aligned} \tag{3.13b}$$

*Note*

i) The condition "$0 \notin I$" in Theorem 3.1 and Corollary 3.1 can be removed, of course, by considering the limit $x_k \to 0$. Hence, if $P, P_i$, resp., is continuously differentiable at $x^0 = (x_1^0, \ldots, x_{k-1}^0, 0, x_{k+1}^0, \ldots, x_r^0)'$, then

$$\frac{\partial P}{\partial x_k}(x^0) = \lim_{t \to 0} \frac{\partial P}{\partial x_k}\big(x(t)\big), \; \frac{\partial P_i}{\partial x_k}(x^0) = \lim_{t \to 0} \frac{\partial P_i}{\partial x_k}\big(x(t)\big), \tag{3.13c}$$

where $\frac{\partial P}{\partial x_k}\big(x(t)\big), \frac{\partial P_i}{\partial x_k}\big(x(t)\big)$, resp., is given by Theorem 3.1 and Corollary 3.1.

ii) If $x_k \neq 0$ for $k \in K$, where $K$ is an arbitrary, fixed subset of $\{1, \dots, r\}$, then, instead of (3.10a), (3.11a), resp., we can use the transformation $\binom{a}{b} = T_x^K(q), (A_i, b_i) = T_x^{i,K}(q_{i1}, \dots, q_{ir+1})$ defined by

$$a_k := \frac{1}{x_k} q_k, k \in K, a_j := q_j, j \notin K, b := q_{r+1}, \tag{3.10a'}$$

$$a_{ik} := \frac{1}{x_k} q_{ik}, k \in K, a_{ij} := q_{ij}, j \notin K, b_i := q_{ir+1} \tag{3.11a'}$$

Repeating the arguments leading to formulas (3.13a,b), under corresponding assumptions as in Theorem 3.1, Corollary 3.1, resp., we also find the *higher order partial derivatives* $D_{\underline{l}}P, D_{\underline{l}}P_i$ of $P, P_i$ with respect to the variables $x_{l_1}, x_{l_2}, \dots, x_{l_s}$:

$$D_{\underline{l}}P(x) = (-1)^s \frac{1}{\prod_{t=1}^{s} x_{l_t}} \int_{B(x)} \left( \sum_{i_1=1}^{m} \cdots \sum_{i_s=1}^{m} \frac{\partial^s}{\prod_{t=1}^{s} \partial a_{i_t l_t}} \left( \prod_{t=1}^{s} a_{i_t l_t} f(A, b) \right) \right) \times dA db, \tag{3.14a}$$

$$D_{\underline{l}}P_i(x) = (-1)^s \frac{1}{\prod_{t=1}^{s} x_{l_t}} \int_{B_i(x)} \frac{\partial^s}{\prod_{t=1}^{s} \partial a_{il_t}} \left( \prod_{t=1}^{s} a_{il_t} f_i(A_i, b_i) \right) dA_i db_i, \tag{3.14b}$$

where $\underline{l} := (l_1, l_2, \dots, l_s)$ and $B(x), B_i(x)$ denote the same region of integration as in (3.8a) and (3.13a), (3.8b) and (3.13b), respectively.

### 3.2.1 Representation of the Derivatives by Surface Integrals

Using the Gaussian divergence theorem [144], we obtain integral representations of $\frac{\partial P}{\partial x_k}, \frac{\partial P_i}{\partial x_k}$ having integrands *without involving derivatives of the densities* $f, f_i$ which is more advantageous in some cases, see Section 3.5.

We consider first the simpler case (3.13b). For given $x$, fulfilling the assumptions of Corollary 3.1, $\frac{\partial P_i}{\partial x_k}$ can be represented by

$$\frac{\partial P_i}{\partial x_k}(x) = -\frac{1}{x_k} \int_{B_i(x)} \operatorname{div} v^{(i,k)}(A_i, b_i) \, dA_i \, db_i, \tag{3.15a}$$

where $B_i(x) := \left\{ (A_i, b_i) \in \mathbb{R}^{r+1} : (A_i, b_i)\hat{x} \le 0 \right\}$ with $\hat{x} := \binom{x}{-1}$, and the $r$-vector field $v^{(i,k)} = v^{(i,k)}(A_i, b_i)$ is defined by

$$v_j^{(i,k)}(A_i, b_i) := \begin{cases} a_{ik} f_i(A_i, b_i) & , j = k \\ 0 & , j \neq k \end{cases} \tag{3.15b}$$

Obviously, $\mathcal{O}_i(x) = \left\{(A_i, b_i) \in \mathbb{R}^{r+1} : (A_i, b_i)\hat{x} = 0\right\}$ is the hypersurface (here a hyperplane) of the domain of integration $B_i(x)$, and at a point $(A_i, b_i) \in \mathcal{O}_i(x)$ the unit outward normal vector to $\mathcal{O}_i(x)$ reads

$$n(A_i, b_i) := \frac{\hat{x}}{\|\hat{x}\|}. \tag{3.15c}$$

Thus, the divergence theorem yields

$$\begin{aligned} \frac{\partial P_i}{\partial x_k}(x) &= -\frac{1}{x_k} \int\limits_{\mathcal{O}_i(x)} v^{(i,k)}(A_i, b_i)^T n(A_i, b_i)\, d0 \\ &= -\frac{1}{\|\hat{x}\|} \int\limits_{\mathcal{O}_i(x)} a_{ik} f_i(A_i, b_i)\, d0. \end{aligned} \tag{3.15d}$$

Since the hyperplane $\mathcal{O}_i(x)$ can be represented by the equation

$$b_i = \phi_i(A_i), A_i \in \mathbb{R}^r, \text{ with } \phi_i(A_i) := A_i x, \tag{3.15e}$$

the surface element $d0$ at each $(A_i, b_i) \in \mathcal{O}_i(x)$ reads

$$d0 = \left(1 + \sum_{j=1}^{r} \left(\frac{\partial \phi_i}{\partial a_{ij}}\right)^2\right)^{1/2} dA_i = \|\hat{x}\|\, dA_i. \tag{3.15f}$$

Hence, from (3.15d) and (3.15f) we get also the following differentiation formula:

**Corollary 3.2.** *If the assumptions of Corollary 3.1 hold, then*

$$\frac{\partial P_i}{\partial x_k}(x) = -\int a_{ik} f_i(A_i, A_i x)\, dA_i. \tag{3.16}$$

*Note.* The integrand of (3.16) *does not involve derivatives of the density* $f_i$.

We want to derive now a corresponding representation for the more complicated formula (3.13a). Note first that

$$\frac{\partial P}{\partial x_k}(x) = -\frac{1}{x_k} \int\limits_{B(x)} \text{div}\, v^{(k)}(a)\, da, \tag{3.17a}$$

where $a := (A_1, b_1, \ldots, A_m, b_m)^T$, c.f. (3.7a), and $v^{(k)}(a) = \Big(v_{11}^{(k)}(a), \ldots, v_{1r}^{(k)}(a)$ $v_{1r+1}^{(k)}(a), \ldots, v_{m1}^{(k)}(a), \ldots, v_{mr}^{(k)}(a), v_{mr+1}^{(k)}(a)\Big)^T$ is the $m(r+1)$-vector field defined by

$$v_{ij}^{(k)}(a) := \begin{cases} a_{ik} f(A, b) & , j = k, i = 1, \ldots, m \\ 0 & , \text{ else.} \end{cases} \tag{3.17b}$$

The hypersurface $\mathcal{O}(x)$ of the domain of integration

$$B(x) = \left\{ a \in \mathbb{R}^{(r+1)} : (A_i, b_i)\hat{x} \le 0, i = 1, \ldots, m \right\} \tag{3.17c}$$

where $\hat{x} = \binom{x}{-1}$, is given by the union $\mathcal{O}(x) = \bigcup_{i=1}^{m} \mathcal{O}^i(x)$ of the partial hypersurfaces

$$\mathcal{O}^i(x) := \{a \in B(x) : (A_i, b_i)\hat{x} = 0\}, i = 1, \ldots, m, \tag{3.17d}$$

which have also the representation

$$b_i = \phi_i(a^i), a^i \in B^i(x), \text{ with } \phi_i(a^i) := A_i x, \tag{3.17e}$$

where $a^i := (A_1, b_1, \ldots, A_{i-1}, b_{i-1}, A_i, A_{i+1}, b_{i+1}, \ldots, A_m, b_m)^T$ and $B^i(x) := \{a^i \in \mathbb{R}^{m(r+1)-1} : (A_j, b_j)\hat{x} \le 0, j \ne i\}$. According to (3.17d), the unit outward normal vector $n(a) = n^i(a)$ to $\mathcal{O}(x)$ at $a \in \mathcal{O}^i(x)$ is given by

$$n^i(a) = \frac{1}{\|\hat{x}\|}(0, \ldots 0, \ldots, x_1, x_2, \ldots, x_r, -1, \ldots, 0, \ldots, 0)^T, \tag{3.17f}$$

where $n^i(a)$ has zero entries up to the positions $(i,j), j = 1, \ldots, r, r+1$, with the entries $x_1, \ldots, x_r, -1$, respectively. Having (3.17b-f), according to the divergence theorem we find

$$\frac{\partial P}{\partial x_k}(x) = -\frac{1}{x_k} \int_{\mathcal{O}(x)} v^{(k)}(a)^T n(a)\, d0 = -\frac{1}{x_k} \sum_{i=1}^{m} \int_{\mathcal{O}^i(x)} v^{(k)}(a)^T n^i(a)\, d0 \tag{3.17g}$$

$$= -\frac{1}{\|\hat{x}\|} \sum_{i=1}^{m} \int_{\mathcal{O}^i(x)} a_{ik} f(A, b)\, d0.$$

Because of the representation (3.17d,e) of $\mathcal{O}^i(x)$, the surface element $d0$ on $\mathcal{O}^i(x)$ at each $a = (a^i, b_i) \in \mathcal{O}^i(x)$, i.e. $b_i = \phi_i(a^i), a^i \in B^i(x)$, is given by

$$d0 = \left(1 + \sum_{i,j=1}^{r} \left(\frac{\partial \phi_i}{\partial a_{ij}}\right)^2 + \sum_{j \ne i} \left(\frac{\partial \phi_i}{\partial b_i}\right)^2\right)^{1/2} da_i = \|\hat{x}\|\, da^i. \tag{3.17h}$$

A representation of $\frac{\partial P}{\partial x_k}(x)$ *without involving derivatives of the density* $f(A, b)$ is obtained now from (3.17g) and (3.17h):

**Corollary 3.3.** *If the assumptions in Theorem 3.1 hold, then*

$$\frac{\partial P}{\partial x_k}(x) = -\sum_{i=1}^{m} \int_{B^i(x)} a_{ik} f(A, 1, b_1, \ldots, A_{i-1}, b_{i-1}, A_i, A_i x, A_{i+1}, \tag{3.18}$$

$$b_{i+1}, \ldots, A_m, b_m)\, da^i.$$

## 3.3 The Differentiation of Structural Reliabilities

In structural reliability and design, see Section 2.1.1 and 2.6, the probability of survival (safety) of a structure (structural system) can be represented, assuming elastic behavior, by

$$P(x) = \mathcal{P}\left(\eta\Big(u\big(a(\omega), x\big)\Big) \geq 0\right). \tag{3.19a}$$

Here, $u = u(a, x)$ denotes the $m$-vector of the basic displacement variables $u_i = u_i(a, x), 1 \leq i \leq m$, depending on the $r$-vector $x$ of design variables $x_k$ and the random parameter $\nu$-vector

$$a = a(\omega) := \begin{pmatrix} p(\omega) \\ F(\omega) \end{pmatrix}. \tag{3.19b}$$

Here $F = F(\omega)$ denotes the random load $m$-vector of the structure and $p = p(\omega)$ is a $(\nu - m)$-vector of further stochastic structural parameters such as material parameters (e.g. elastic moduli), manufacturing errors. Moreover, $\eta : \mathbb{R}^m \to \mathbb{R}^{m_1}$ is a given vector function selected such that the inequality

$$\eta\big(u(a, x)\big) \geq 0 \tag{3.19c}$$

describes the relevant behavioral constraints, e.g. certain displacement and stress constraints.

From structural mechanics [55,93] we know that the displacement vector $u(a, x)$ is given by

$$u(a, x) := K(p, x)^{-1} F, \tag{3.19d}$$

where $K = K(p, x)$ denotes the stiffness $m \times m$ matrix of the structure. Assume in the following – without restrictions – that $p(\omega), F(\omega)$ are stochastic independent random vectors having probability densities $f_1 = f_1(p), f_2 = f_2(F)$. We find

$$\begin{aligned} P(x) &= \mathcal{P}\left(\eta\Big(u\big(a(\omega), x\big)\Big) \geq 0\right) \\ &= \int\limits_{\eta\left(K(p,x)^{-1}F\right) \geq 0} f_1(p) f_2(F)\, dp dF. \end{aligned} \tag{3.20}$$

Hence, we consider – for given vector $x$ – the transformation $T_x$ given by

$$\begin{pmatrix} p \\ F \end{pmatrix} = T_x(q) := \begin{pmatrix} q_1 \\ K(q_1, x) q_2 \end{pmatrix}, q = \begin{pmatrix} q_1 \\ q_2 \end{pmatrix}. \tag{3.21a}$$

Since $K(p, x)$ is positive (semi) definite, the absolute value of the functional determinant of $T_x$ reads

$$\left|\det\left(\frac{\partial T_x}{\partial q}(q)\right)\right| = \det\Big(K(q_1, x)\Big). \tag{3.21b}$$

Applying (3.21a), (3.21b) to the integral in (3.20), we get

$$P(x) = \int\limits_{\tilde{B}} f_1(q_1) f_2\Big(K(q_1, x)q_2\Big) \det\Big(K(q_1, x)\Big)\, dq_1 dq_2, \tag{3.22a}$$

where

$$\tilde{B}. = \left\{\begin{pmatrix} q_1 \\ q_2 \end{pmatrix} \in \mathbb{R}^{\nu_1} \times \mathbb{R}^m : \eta(q_2) \geq 0\right\} \tag{3.22b}$$

cf. (3.10b), (3.11b). Since the domain of integration $\tilde{B}$ in (3.22a) does not depend on the design vector $x$, we may proceed now as in the development of the differentiation formulas (3.13a), (3.13b): Under some additional weak assumptions we may differentiate (3.22a) by interchanging differentiation and integration. Since

$$\frac{\partial}{\partial x_k} \det\Big(K(q_1, x)\Big) = \det\Big(K(q_1, x)\Big) tr\Big(K(q_1, x)^{-1} \frac{\partial K}{\partial x_k}(q_1, x)\Big), \tag{3.22c}$$

where "tr" designates the trace of a matrix, we set

$$\begin{aligned} h_k(q_1, q_2; x) := f_1(q_1) \Big\{ \nabla f_2\Big(K(q_1, x)q_2\Big)^T \frac{\partial K}{\partial x_k}(q_1, x) q_2 \\ + f_2\Big(K(q_1, x)q_2\Big) tr\Big(K(q_1, x)^{-1} \frac{\partial K}{\partial x_k}(q_1, x)\Big) \Big\} \det\Big(K(q_1, x)\Big). \end{aligned} \tag{3.22d}$$

Clearly, the existence of the gradient $\nabla f_2(F)$ of $f_2(F)$ is presupposed here for all arguments under consideration.

**Theorem 3.2.** *For a given fixed integer $k, 1 \leq k \leq r$, and given components $x_l^0, l \neq k$, define $x = x(t)$ again by (3.9). Suppose that for a given interval $I \subset \mathbb{R}$ the multiple integral (3.20) or (3.22a) exists and is finite for each $x = x(t)$ with $t \in I$. Furthermore, suppose that function $h_k\Big(q_1, q_2; x(t)\Big)$, defined by (3.22d), exists for all $t \in I$ and has an integrable majorant, i.e. a nonnegative measurable function $H_k = H_k(q_1, q_2)$, defined at least on $\tilde{B}$, such that*

$$|h_k\Big(q_1, q_2; x(t)\Big)| \leq H_k(q_1, q_2) \text{ for all } t \in I \tag{3.22e}$$

*and*

$$\int\limits_{\tilde{B}} H_k(q_1, q_2) dq_1 dq_2 < +\infty. \tag{3.22f}$$

*Then $\frac{\partial P}{\partial x_k}(x)$ exists for each $x \in \Big\{x(t) : t \in I\Big\}$ and is given by*

$$\frac{\partial P}{\partial x_k}(x) = \int\limits_{\tilde{B}} h_k(q_1, q_2; x)\, dq_1 dq_2 \tag{3.23a}$$

$$= \int\limits_{\eta\left(K(p,x)^{-1}F\right)\geq 0} f_1(p)\Big\{\nabla f_2(F)^T \frac{\partial K}{\partial x_k}(p,x)K(p,x)^{-1}F$$

$$+ f_2(F)tr\left(K(p,x)^{-1}\frac{\partial K}{\partial x_k}(p,x)\right)\Big\}\, dpdF$$

$$= - \int\limits_{\eta\left(u(a,x)\right)\geq 0} \mathrm{div}_F \left(f(a)\left(\frac{\partial u}{\partial F}(a,x)\right)^{-1}\frac{\partial u}{\partial x_k}(a,x)\right) da,$$

where $a = \binom{p}{F}$, $f(a) = f_1(p)f_2(F)$, and $u = u(a,x)$ is the vector of displacement variables defined by (3.19d).

Proof. Under the above assumptions we may interchange [4,113] for each $x \in \left\{x(t) : t \in I\right\}$ differentiation with respect to $x_k$ and integration in (3.22a), hence,

$$\frac{\partial P}{\partial x_k}(x) = \frac{\partial}{\partial x_k}\int\limits_{\tilde{B}} f_1(q_1)f_2\left(K(q_1,x)q_2\right)\det\left(K(q_1,x)\right)\, dq_1 dq_2 \tag{3.23b}$$

$$= \int\limits_{\tilde{B}} h_k(q_1, q_2; x)\, dq_1 dq_2.$$

The second and third equation in (3.23a) is obtained then from (3.22c,d), and by applying the back–transformation

$$q = \begin{pmatrix} q_1 \\ q_2 \end{pmatrix} = T_x^{-1}\begin{pmatrix} p \\ F \end{pmatrix} = \begin{pmatrix} p \\ K(p,x)^{-1}F \end{pmatrix}.$$

*Note.* By iteration of the above procedure also the higher order partial derivatives $D_{\underline{l}}P(x), \underline{l} = (l_1, \ldots, l_s)$ of $P(x)$ with respect to $x_{l_1}, \ldots, x_{l_s}$ can be obtained, cf. Section 3.2.

## 3.4 Extensions

### 3.4.1 More General Response (State) Functions

In generalization of the vector–valued functions $y(a,x) = Ax - b$ and $y(a,x) = \eta\left(u(a,x)\right) = \eta\left(K(p,x)^{-1}F\right)$ treated in Sections 3.2 and 3.3, we consider the class of $m$–vector functions $y = y(a,x)$ given by

$$y(a,x) := \eta\Big(Q^{(0)}(x;a_0^1)a_0^2, Q^{(1)}(x,a_1^1)a_1^2, \ldots, Q^{(\rho)}(x,a_\rho^1)a_\rho^2\Big). \tag{3.24a}$$

Here,

i) $\eta = \eta(q_0, q_1, \ldots, q_\rho)$ is a given $m$–vector function on $\mathbb{R}^{\nu_{02}} \times \mathbb{R}^{\nu_{12}} \times \ldots \times \mathbb{R}^{\nu_{\rho 2}}$, where $\nu_{l2}, l = 0,1,\ldots,\rho$, are given integers such that $\nu_{l2} \geq 1, l = 0,1,\ldots,\rho$.
ii) $Q^{(l)} = Q^{(l)}(x, a_l^1)$ is a regular $\nu_{l2} \times \nu_{l2}$ matrix for each $l = 0,1,\ldots,\rho$ and each $(x, a_l^1)$ in a convex subset of $\mathbb{R}^r \times \mathbb{R}^{\nu_{l1}}$, where $\nu_{l1} \geq 0$ is a given integer and $\nu_{l1} = 0$ means that $Q^{(l)}$ is independent of $a$,
iii) $(a_l^1, a_l^2), l = 0,1,\ldots,\rho$, is a partition of $a$ into disjoint $\nu_{l1}-, \nu_{l2}-$ subvectors of $a$.

Important special cases of (3.24a) are given as follows:

$$y(a,x) = y_0(x) + \sum_{l=1}^{\rho} \varphi_l(x) H_l a_l = y_0(x) + \sum_{l=1}^{\rho} H_l\Big(\varphi_l(x) a_l\Big), \tag{3.24b}$$

In (3.24b) $y_0 = y_0(x)$ is a given $m$–vector function, $\varphi_l = \varphi_l(x), l = 1,\ldots,\rho$, are given real valued functions and $H_l, l = 1,\ldots,\rho$, are given $m \times \nu_l$ matrices. Furthermore, $a_1, \ldots, a_\rho$ is a partition of $a$ into $\nu_l$–subvectors. An interesting special version of (3.24b) reads

$$y(a,x) = y_0(x) + \sum_{l=1}^{\rho} \varphi_l(x) a_l \tag{3.24c}$$

where $\nu_1 = \cdots \nu_\rho = m$.

It is easy to see that a large class of functions $y = y(\omega, x)$ depending on a random element $\omega \in (\Omega, \mathcal{A}, P)$ can be modelled – at least approximatively (with arbitrary accuracy) – by functions of the type (3.24b), (3.24c). Indeed, if the components $y_i(\omega, x), i = 1,\ldots,m$, of $y(\omega, x)$ are approximated by Taylor polynomials of a certain order at a reference point $x^0$, then

$$y(\omega, x) \approx \sum_{\underline{l} \in \Lambda} a_{\underline{l}} \varphi_{\underline{l}}(x), \tag{3.25a}$$

where $\Lambda$ is a certain set of (vector) indices,

$$a_{\underline{l}} := \begin{pmatrix} D_{\underline{l}} y_1(\omega, x^0) \\ D_{\underline{l}} y_2(\omega, x^0) \\ \vdots \\ D_{\underline{l}} y_m(\omega, x^0) \end{pmatrix}, \varphi_{\underline{l}}(x) := \prod_{t=1}^{s} (x_{l_t} - x_{l_t}^0), \tag{3.25b}$$

and $D_{\underline{l}}$ denotes the derivative with respect to the variables $x_{l_1}, x_{l_2}, \ldots, x_{l_s}$ with $(l_1, l_2, \ldots, l_s) =: \underline{l}$.

In case (3.24a), the derivatives of the probability function

$$P(x) = P\Big(y_l \le y(a(\omega), x) \le y_u\Big) \tag{3.26}$$

can be obtained again easily by using the integral transformation $T_x$ defined by

$$\begin{pmatrix} a_l^1 \\ a_l^2 \end{pmatrix} := \begin{pmatrix} q_l^1 \\ Q^{(l)}(x, q_l^1)^{-1} q_l^2 \end{pmatrix}, l = 0, 1, \ldots, \varrho. \tag{3.27}$$

If the random vector $a = a(\omega)$ has a probability density $f = f(a)$, then (3.27) yields

$$P(x) = \int_{\tilde{B}} f\left(a_0^1, Q^{(0)}(x, q_0^1)^{-1} q_0^2, \ldots\right. \tag{3.28a}$$

$$\left. \ldots, q_\rho^1, Q^{(\rho)}(x, q_\rho^1)^{-1} q_\rho^2\right) \prod_{l=0}^{\rho} \frac{dq_l^1 dq_l^2}{|\det Q^{(l)}(x, q_l^1)|}$$

$$= \int \left( \int_{\tilde{B}^2} f\left(q_0^1, Q^{(0)}(x, q_0^1)^{-1} q_0^2, \ldots\right.\right.$$

$$\left.\left. \ldots, q_\rho^1, Q^{(\rho)}(x, q_\rho^1) q_\rho^2\right) \prod_{l=0}^{\rho} \frac{dq_l^2}{|\det Q^{(l)}(x, q_l^1)|} \right) \prod_{l=0}^{\rho} dq_l^1 .$$

In the above we set

$$\tilde{B} := \left\{ \begin{pmatrix} q^1 \\ q^2 \end{pmatrix} \in \mathbb{R}^\nu : q^1 \in \mathbb{R}^{\nu_1}, y_l \le \eta\left(q_0^2, q_1^2, \ldots, q_\rho^2\right) \le y_u \right\}, \tag{3.28b}$$

$$\tilde{B}^2 := \left\{ q^2 \in \mathbb{R}^{\nu_2} : y_l \le \eta(q_0^2, q_1^2, \ldots, q_\rho^2) \le y_u \right\} \tag{3.28c}$$

with $q^1 := (q_0^{1^T}, q_1^{1^T}, \ldots, q_\rho^{1^T})^T, \nu_1 := \sum_{l=0}^{\rho} \nu_{l1}, q^2 := (q_0^{2^T}, q_1^{2^T}, \ldots, q_\rho^{2^T})^T, \nu_2 := \sum_{l=0}^{\rho} \nu_{l2}, \nu = \nu_1 + \nu_2$.

For the differentiation of the right hand side of (3.28a) we need the following formulas, cf. (3.22c),

$$\frac{\partial Q^{(l)-1}}{\partial x_k} = -Q^{(l)-1} \frac{\partial Q^{(l)}}{\partial x_k} Q^{(l)-1}, \tag{3.29}$$

$$\frac{\partial}{\partial x_k} \det Q^{(l)} = \det Q^{(l)} tr \left( Q^{(l)-1} \frac{\partial Q^{(l)}}{\partial x_k} \right).$$

Since $Q^{(l)}(x, q_l^1), l = 0, 1, \ldots, \rho$, are regular (by assumption) for all arguments under consideration, we find

$$sgn\left(\det Q^{(l)}(x, q_l^1)\right) = \epsilon_l, l = 0, 1, \ldots, \rho, \text{ for all } (x, q),$$

where $\epsilon_l, l = 0, 1, \ldots, \rho$, are fixed values from $\{-1, 1\}$. Hence, under assumptions on $y = y(a, x)$ and $f = f(a)$ corresponding to the assumptions in Theorem 3.1, Corollary 3.1 and Theorem 3.2, we find by interchanging differentiation and integration in (3.28a) and using (3.29)

$$\frac{\partial P}{\partial x_k}(x) = \int\limits_{\tilde{B}} h_k\left(q_0^1, q_0^2, \ldots, q_\rho^1, q_\rho^2; x\right) \prod_{l=0}^{\rho} dq_l^1\, dq_l^2, \tag{3.30a}$$

where

$$\begin{aligned} h_k(q_0^1, q_0^2, \ldots, q_\rho^1, q_\rho^2; x) := -\Bigg\{ & \sum_{l=0}^{\rho} \nabla_{a_l^2}\left(q_0^1, Q^{(0)}(x, a_0^1)^{-1} q_0^2, \ldots,\right. \\ & \left. q_\rho^1, Q^{(\rho)}(x, q_\rho^1)^{-1}\right) Q^{(l)}(x, q_l^1)^{-1} Q^{(l)}(x, a_l^1)^{-1} q_l^2 \\ & + f\left(q_0^1, Q^{(0)}(x, q_0^1)^{-1} q_0^2, \ldots, q_\rho^1, Q^{(\rho)}(x, q_\rho^1)^{-1} q_\rho^2\right) \\ & \times \sum_{l=0}^{\rho} tr\left(q^{(l)}(x, q_l^1)^{-1} \frac{\partial Q^{(l)}}{\partial x_k}(x, q_l^1)\right)\Bigg\} \prod_{l=0}^{\rho} \frac{1}{|\det Q^{(l)}(x, q_l^1)|} \end{aligned} \tag{3.30b}$$

Then, by inverse transformation, cf. (3.27), we find

$$\begin{aligned} \frac{\partial P}{\partial x_k}(x) = & -\int\limits_{B(x)} \Bigg\{ \sum_{l=0}^{\rho} \Bigg( \nabla_{a_l^2} f(a_0^1, a_0^2, \ldots, a_\rho^1, a_\rho^2)' Q^{(l)}(x, a_l^1)^{-1} \frac{\partial Q^{(l)}}{\partial x_k}(x, a_l^1) a_l^2 \\ & + f(a_0^1, a_0^2, \ldots, a_\rho^1, a_\rho^2) tr\left(Q^{(l)}(x, a_l^1)^{-1} \frac{\partial Q^{(l)}}{\partial x_k}(x, a_l^1)\right)\Bigg)\Bigg\}\, da \qquad (3.31a) \\ = & -\int\limits_{B(x)} \left\{ \sum_{l=0}^{\rho} \operatorname{div}_{a_l^2}\left( f(a) Q^{(l)}(x, a_l^1)^{-1} \frac{\partial Q^{(l)}}{\partial x_k}(x, a_l^1) a_l^2 \right)\right\} da, \end{aligned}$$

where

$$B(x) = \{a \in \mathbb{R}^\nu : y_l \le y(a, x) \le y_u\}. \tag{3.31b}$$

Using thc Gaussian divergence theorem, (3.31a) may be also represented, for given fixed vector $x$, by

$$\frac{\partial P}{\partial x_k}(x) = -\int\limits_{\mathcal{O}(x)} v^{(k)}(a; x)^T n(a; x)\, d0, \tag{3.32a}$$

where the hypersurface $\mathcal{O}(x)$ is the boundary of $B(x), n(a; x)$ is the unit outward normal vector to $\mathcal{O}(x)$ at a point $a \in \mathcal{O}(x)$; moreover, the ($\nu-$)vector field $v^{(k)} = v^{(k)}(a; x)$ is given by

$$v^{(k)}(a;x) := f(a)\begin{pmatrix} 0 \\ Q^{(0)}(x;a_0^1)^{-1}\frac{\partial Q^{(0)}}{\partial x_k}(x,a_0^1)a_0^2 \\ 0 \\ Q^{(1)}(x,a_1^1)^{-1}\frac{\partial Q^{(1)}}{\partial x_k}(x,a_1^1)a_1^2 \\ \vdots \\ 0 \\ Q^{(\rho)}(x,a_\rho^1)^{-1}\frac{\partial Q^{(\rho)}}{\partial x_k}(x,a_\rho^1)a_\rho^2 \end{pmatrix}. \tag{3.32b}$$

We observe that (3.32b) does not involve any derivatives of the density $f = f(a)$. Because of (3.31b) the boundary hypersurface $\mathcal{O}(x)$ of $B(x)$ can be represented by

$$\mathcal{O}(x) = \bigcup_{t=1}^{\tau} \mathcal{O}_t(x), \tag{3.33a}$$

where the hypersurfaces $\mathcal{O}_t(x), t = 1, \ldots, \tau$, with an integer $\tau = \tau(x)$, are given by

$$\mathcal{O}_t(x) = \{a \in \mathbb{R}^\nu : y_{i_t}(a,x) = y_{\lambda_t i_t}\} \tag{3.33b}$$

with integers $i_t = i_t(x) \in \{1,\ldots,m\}$ and indices $\lambda_t = \lambda_t(x) \in \{l,u\}, t = 1,2,\ldots,\tau$. Hence

$$n(a;x) = \epsilon_t \frac{\nabla_a y_{i_t}(a;x)}{\|\nabla_a y_{i_t}(a;x)\|} \text{ for } a \in \mathcal{O}_t(x), t = 1,2,\ldots,\tau, \tag{3.33c}$$

where $\epsilon_t = \epsilon_t(x)$ is an integer with $\epsilon_t \in \{-1,1\}$. We have

$$\nabla_a y_i(a,x) := \begin{pmatrix} \nabla_{a_0^1} y_i(a,x) \\ \nabla_{a_0^2} y_i(a,x) \\ \nabla_{a_1^1} y_i(a,x) \\ \nabla_{a_1^2} y_i(a,x) \\ \vdots \\ \nabla_{a_\rho^1} y_i(a,x) \\ \nabla_{a_\rho^2} y_i(a,x) \end{pmatrix} =$$

$$= \begin{pmatrix} \nabla_{a_0^1} y_i(a,x) \\ Q^{(0)}(x,a_1^1)^T \nabla_{q_0}\eta_i\Big(Q^{(0)}(x,a_0^1)a_0^2,\ldots,Q^{(\rho)}(x,a_\rho^1)a_\rho^2\Big) \\ \nabla_{a_1^1} y_i(a,x) \\ Q^{(1)}(x,a_1^1)^T \nabla_{q_1}\eta_i\Big(Q^{(0)}(x,a_0^1)a_0^2,\ldots,Q^{(\rho)}(x,a_\rho^1)a_\rho^2\Big) \\ \vdots \\ \nabla_{a_\rho^1} y_i(a,x) \\ Q^{(\rho)}(x,a_\rho^1)^T \nabla_{q_\rho}\eta_i\Big(Q^{(0)}(x,a_0^1)a_0^2,\ldots,Q^{(\rho)}(x,a_\rho^1)a_\rho^2\Big) \end{pmatrix}. \tag{3.33d}$$

Hence, equations (3.33a) - (3.33d) yield

$$\frac{\partial P}{\partial x_k}(x) = -\sum_{t=1}^{\tau} \int\limits_{\mathcal{O}_t(x)} f(a)\frac{\partial y_{i_t}}{\partial x_k}(a,x)\frac{\epsilon_t d0}{\|\nabla_a y_{i_t}(a,x)\|} \tag{3.34a}$$

$$= -\sum_{t=1}^{\tau} \int\limits_{\mathcal{O}_t(x)} \left(f(a)\frac{\partial y_{i_t}}{\partial x_k}(a,x)\frac{\nabla_a y_{i_t}(a,x)}{\|\nabla_a y_{i_t}(a,x)\|^2}\right)^T d\underline{0},$$

where

$$d\underline{0} := \epsilon_t \frac{\nabla_a y_{i_t}(a,x)}{\|\nabla_a y_{i_t}(a,x)\|} d0 \tag{3.34b}$$

is the *directed surface element* at a point $a \in \mathcal{O}_t(x)$.

In the important special case $i_t = i_0$ for all $t = 1, \ldots, \tau$ and a fixed integer $i_0, 1 \leq i_0 \leq m$, which occurs e.g. for $m = 1$, we get again with the divergence theorem

$$\frac{\partial P}{\partial x_k}(x) = -\int\limits_{\mathcal{O}(x)} \left(f(a)\frac{\partial y_{i_0}}{\partial x_k}(a,x)\frac{\nabla_a y_{i_0}(a,x)}{\|\nabla_a y_{i_0}(a,x)\|^2}\right) d\underline{0} \tag{3.35}$$

$$= -\int\limits_{B(x)} \operatorname{div}_a \left(f(a)\frac{\partial y_{i_0}}{\partial x_k}(a,x)\frac{\nabla_a y_{i_0}(a,x)}{\|\nabla_a y_{i_0}(a,x)\|^2}\right) da$$

*Remark 3.1.* Related representations by means of an integral over the sum of certain lower dimensional surface integrals can be obtained if (3.31a) is written first as follows:

$$\frac{\partial P}{\partial x_k}(x) = -\int \left( \int\limits_{B(x|a^1)} c^{(k)}(a^2; a^1, x)da^2 \right) da^1, \tag{3.31a'}$$

where $a^j := (a_0^{j^T}, a_1^{j^T}, \ldots, a_\rho^{j^T})^T, j = 1, 2, B(x|a^1) := \{a^2 : y_l \leq \eta\left(Q^{(0)}(x, a_0^1)\right.$ $\left.a_0^2, \ldots, Q^{(\rho)}(x, a_\rho^1)a_\rho^2\right) \leq y_u\}$, and $c^{(k)}$ is the integrand in (3.31a).

## 3.5 Computation of Probabilities and its Derivatives by Asymptotic Expansions of Integral of Laplace Type

### 3.5.1 Computation of Structural Reliabilities and its Sensitivities

The computation of reliabilities of mechanical structures is a well established method, see e.g. [18, 19, 28, 48]. In the following we examine the potential of this technique to yield also the corresponding sensitivities, i.e. the derivatives of the probabilities of survival with respect to certain design variables or deterministic system parameters $x_k$.

According to (3.22a) and (3.23a) the probability function $D_{(0)}P(x) := P(x)$ and its partial derivative $D_{(k)}P(x) := \frac{\partial P}{\partial x_k}(x)$ can jointly be represented by the formula

$$D_{(l)}P(x) = \int_{\tilde{B}} c_l(q_1, q_2; x) f_1(q_1) dq_1 dq_2, l = 0, 1, \dots, r. \tag{3.36a}$$

In the above formula $c_{\underline{l}}(q_1, q_2; x)$ denotes the multiplicator of $f_1(q_1)$ in the integrand of (3.22a), the multiplicator of $f_1(q_1)$ in (3.22d), respectively. Note that for the higher order partial derivatives $D_{\underline{l}}P(x), \underline{l} := (l_1, l_2, \dots, l_s)$, we get, under corresponding assumptions,

$$D_{\underline{l}}P(x) = \int_{\tilde{B}} c_{\underline{l}}(q_1, q_2; x) f_1(q_1)\, dq_1 dq_2 \tag{3.36b}$$

with a more complicated function $c_{\underline{l}} = c_{\underline{l}}(q_1, q_2; x)$, cf. (3.14a,b).

In order to return to the original variables $a = \binom{p}{F}$ without involving the argument $x$, we consider in (3.36a) the integral transformation

$$\begin{pmatrix} q_1 \\ q_2 \end{pmatrix} = S\begin{pmatrix} p \\ F \end{pmatrix} := \begin{pmatrix} p \\ K(p_0, x_0)^{-1}F \end{pmatrix}, \tag{3.37}$$

where $(p_0, x_0) \in \mathbb{R}^{\nu-m} \times \mathbb{R}^m$ is any *fixed pair* of vectors such that $K(p_0, x_0)$ is positive definite. Application of (3.37) to (3.36a) yields

$$\begin{aligned} D_{(l)}P(x) &= \int_{S^{-1}(\tilde{B})} c_l\left(p, K(p_0, x_0)^{-1}F; x\right) \frac{1}{\det\left(K(p_0, x_0)\right)} f_1(p)\, dpdF \\ &= \int_{S^{-1}(\tilde{B})} C_l(p, F; x) f_1(p) f_2(F)\, dpdF, \end{aligned} \tag{3.38a}$$

where

$$C_l(p, F; x) := c_l\left(p, K(p_0, x_0)^{-1}F; x\right) \frac{1}{\det\left(K(p_0, x_0)\right)} \cdot \frac{1}{f_2(F)}, \tag{3.38b}$$

$$S^{-1}(\tilde{B}) := \left\{ \begin{pmatrix} p \\ F \end{pmatrix} \in \mathbb{R}^{\nu-m} \times \mathbb{R}^m : \eta\left(K(p_0, x_0)^{-1}F\right) \geq 0 \right\}. \tag{3.38c}$$

Clearly, a corresponding formula holds for $D_l P(x)$, see (3.36b).

We represent then the random vector $a(\omega) = \begin{pmatrix} p(\omega) \\ F(\omega) \end{pmatrix}$ by

$$\begin{pmatrix} p(\omega) \\ F(\omega) \end{pmatrix} = \Gamma\left(z(\omega)\right) = \begin{pmatrix} \Gamma_p\left(z_p(\omega)\right) \\ \Gamma_F\left(z_F(\omega)\right) \end{pmatrix}, \tag{3.39}$$

where

i) $\Gamma(z) = \begin{pmatrix} \Gamma_p(z_p) \\ \Gamma_F(z_F) \end{pmatrix}$ is a sufficiently smooth 1–1-transformation in $\mathbb{R}^{\nu-m} \times \mathbb{R}^m$, and

ii) $z(\omega) = \begin{pmatrix} z_p(\omega) \\ z_F(\omega) \end{pmatrix}$ is a $N(0, I)$–normal distributed random $\nu$–vector.

This yields

$$\begin{aligned} D_{(l)}P(x) &= E1_{S^{-1}(\tilde{B})}\Big(a(\omega)\Big)C_l\Big(a(\omega); x\Big) \\ & E1_{S^{-1}(\tilde{B})}\Big(\Gamma\big(z(\omega)\big)\Big)\, C_l\Big(\Gamma\big(z(\omega)\big); x\Big) \qquad (3.40a) \\ &= \int\limits_{\Gamma^{-1}\left(S^{-1}(\tilde{B})\right)} C_l\Big(\Gamma(z); x\Big)(2\pi)^{-\nu/2} \exp\left(-\frac{1}{2}\|z\|^2\right) dz, \end{aligned}$$

where $1_M$ denotes the characteristic function of a set $M$ and

$$\Gamma^{-1}\Big(S^{-1}(\tilde{B})\Big) = \left\{ \begin{pmatrix} z_p \\ z_F \end{pmatrix} \in \mathbb{R}^{\nu-m} \times \mathbb{R}^m : \eta\Big(K(p_0, x_0)^{-1}\Gamma_F(z_F)\Big) \geq 0 \right\}. \tag{3.40b}$$

Of course, a corresponding representation holds also for $D_{\underline{l}}P(x)$, cf. (3.36b).

Let $z^* = \begin{pmatrix} z_P^* \\ z_F^* \end{pmatrix}$ denote the projection of the origin $0 = \begin{pmatrix} 0_p \\ 0_F \end{pmatrix}$ in $\mathbb{R}^{\nu-m} \times \mathbb{R}^m$ onto the closed set $\Gamma^{-1}\Big(S^{-1}(\tilde{B})\Big)$. According to (3.40b) we find $z_p^* = 0$, and $z_F^*$ is the projection of $0_F$ onto

$$Z_F := \left\{ z_F \in \mathbb{R}^m : \eta\Big(K(p_0, x_0)^{-1}\Gamma_F(z_F)\Big) \geq 0 \right\}. \tag{3.40c}$$

Moreover, let

$$\beta^* := \|z^*\| = \|z_F^*\|. \tag{3.40d}$$

In the following we suppose that

$$\beta^* \text{ is "large" (i.e. } P\Big(\|z(\omega)\| \leq \beta^*\Big) \text{ is close to 1).} \tag{3.41}$$

*Remark 3.2.* If $0_F \notin Z_F$, i.e. $\beta^* > 0$, condition (3.41) can be fulfilled in many cases by an appropriate selection of the vectors $p_0, x_0$ in the stiffness matrix $K(p_0, x_0)$. If $0 \in \text{int } Z_F$ and therefore $\beta^* = 0$, then condition (3.41) can be reached by replacing first $\tilde{B}$ by its complement $\tilde{B}^c$ and considering therefore $P_1(x) := 1 - P(x)$ instead of $P(x)$. It is easy to see that the case of multiple projections of 0 onto $\Gamma^{-1}\Big(S^{-1}(\tilde{B})\Big)$ can be reduced to the case of a unique projection, see Section 3.5.2.

Under the assumptions (3.41) we apply to (3.40a) the integral transformation in $\mathbb{R}^\nu$ defined by

$$z := \beta^* w. \tag{3.42}$$

We obtain

$$D_{(l)}P(x) = (2\pi)^{-\nu/2}\beta^{*^{\nu}} I(\beta^{*2}), \tag{3.43a}$$

where the integral $I = I(\lambda)$ of *Laplace type* [14] is given by

$$I(\lambda) = \int\limits_{w \in \frac{1}{\beta^*}\Gamma^{-1}\left(S^{-1}(\tilde{B})\right)} C_l\Big(\Gamma(\beta^* w); x\Big) \exp\left(\lambda\Big(-\frac{1}{2}\|w\|^2\Big)\right) dw. \tag{3.43b}$$

Note that a corresponding representation can be derived also for $D_{\underline{l}}P(x)$.

Since $\beta^*$ is large by assumption (3.41), the integral (3.43b) can be evaluated now *analytically* by using the theory of asymptotic expansions [14]:

**Theorem 3.3.** *Under the above assumptions the following expansions holds (being asymptotically exact for $\beta^* \to +\infty$):*

$$D_{(l)}P(x) \approx \frac{1}{(2\pi)^{1/2}\beta^*} C_l\Big(\Gamma(z^*); x\Big) \frac{1}{|J(w^*)|^{1/2}} \exp\left(-\frac{1}{2}\beta^{*2}\right), \tag{3.44a}$$

where $J(w^*)$ is given by

$$J(w^*) = w^{*T} Q(w^*) w^*. \tag{3.44b}$$

Here $Q(w^*)$ is the matrix of cofactors [8] of the matrix

$$\tilde{Q}(w^*) := -\Big(I + \mu^* \nabla^2 \psi(w^*)\Big), \tag{3.44c}$$

where $\psi(w) = \psi(w; w^*)$ is a function such that the hypersurface (boundary) of $\frac{1}{\beta^*}\Gamma^{-1}\left(S^{-1}(\tilde{B})\right)$ in a neighbourhood $U(w^*)$ of $w^* = \frac{z^*}{\beta^*}$ is given by

$$\partial\left(\frac{1}{\beta^*}\Gamma^{-1}\Big(S^{-1}(\tilde{B})\Big)\right) \cap U(w^*) = \{w \in U(w^*) : \psi(w; w^*) = 0\}. \tag{3.44d}$$

Moreover, $\mu^*$ is determined by the equation

$$1 = \|w^*\| = \mu^* \|\nabla\psi(w^*)\|. \tag{3.44e}$$

*Remark 3.3.* i) The matrix $(cof\ a_{st})$ of cofactors of a quadratic matrix $A$ is defined by $cof\ a_{st} := (-1)^{s+t} \det A^{(s,t)}$, where $A^{(s,t)}$ is the matrix resulting from $A$ by deleting the $s$–th row and $t$–th column of $A$.

ii) If $\frac{1}{\beta^*}\Gamma^{-1}\left(S^{-1}(\tilde{B})\right)$ has a plane boundary in a certain neighbourhood of $w^*$, then $Q(w^*) = (-1)^{\nu-1}I$.

iii) A corresponding asymptotic expansion holds – of course – also for $D_{\underline{l}}P(x)$.

For the *numerical computation of* $D_{(l)}P(x)$ we have now the (asymptotically exact) expansion (3.44a) with formulas (3.38b) and (3.44b-e). On the other hand, because of the *mean value representation* (3.40a) of $D_{(l)}P(x)$ and a corresponding representation of $D_{\underline{l}}P(x)$, these derivatives can be estimated, of course, also by ordinary sampling techniques.

### 3.5.2 Numerical Computation of Derivatives of the Probability Functions Arising in Chance Constrained Programming

The technique of asymptotic expansion of integrals applied in Section 3.5.1 to the computation of reliabilities and its sensitivities of mechanical structures is considered now for the computation of the probability functions and its derivatives treated in Section 3.2.

For simplification, in the following we assume that $x_j \neq 0$ for all $j = 1, 2, \ldots, r$. Hence, using transformation (3.10a'), corresponding to representations (3.10b) of $D_{(0)}P := P$ and (3.13a) of $D_{(k)}P := \frac{\partial P}{\partial x_k}, k \geq 1$, for $D_{(l)}P, l = 0, 1, \ldots, r$, we have the joint representation

$$D_{(l)}P(x) = \int\limits_{\sum_{j=1}^{r} q_j \leq q_{r+1}} h_l(q_1, \ldots, q_r, q_{s+1}; x) \prod_{j=1}^{r+1} dq_j, \tag{3.45}$$

where $h_0$ is the integrand corresponding to (3.10b) for $l = 0$; and for $l = k \geq 1$, the function $h_k$ is defined by a modification of (3.12b) based on (3.10a').

Following now the integral transformation (3.37), we consider the transformation

$$q_j := \xi_j^0 a_j, j = 1, \ldots, r, q_{r+1} := b \tag{3.46}$$

with a fixed vector $\xi^0 := (\xi_1^0, \ldots, \xi_r^0)^T$ such that $\xi_j^0 \neq 0$ for all $j = 1, \ldots, r$. Applying (3.46) to the integral (3.45), for $l = 0, 1, \ldots, r$ we find

$$\begin{aligned} D_{(l)}P(x) &= \int\limits_{A\xi^0 \leq b} C_l(A, b; x) f(A, b)\, dAdb & \text{(3.47a)} \\ &= E 1_{[A\xi^0 \leq b]}\Big(A(\omega), b(\omega)\Big) C_l\Big(A(\omega), b(\omega); x\Big) \end{aligned}$$

where

$$C_l(A, b; x) := h_l(\xi_1^0 a_1, \ldots, \xi_r^0 a_r, b; x) \frac{1}{f(A, b)} \prod_{j=1}^{r} |\xi_j^0|^m \tag{3.47b}$$

and $1_{[A\xi^0 \leq b]}$ denotes the characteristic function of the set $[A\xi^0 \leq b] := \Big\{(A, b) \in \mathbb{R}^{m(r+1)} : A\xi^0 \leq b\Big\}$. Note that in (3.47a,b) we have now again a *fixed* region of integration, cf. (3.22a,b). Furthermore, we observe that (3.47a) contains a *mean value representation* of $D_{(l)}P(x)$.

Consider now the representation

$$\Big(A(\omega), b(\omega)\Big) = \Gamma\Big(Z(\omega), z(\omega)\Big), \tag{3.48}$$

cf. (3.39), where $\Gamma$ is a continuous $1-1$–transformation in the space of $m \times (r+1)$ matrices, and $\Big(Z(\omega), z(\omega)\Big)$ is a random $m \times (r+1)$ matrix having stochastically independent normal distributed elements with mean zero and variance 1. Corresponding to (3.40a) we find

$$D_{(l)}P(x) = \int\limits_{\Gamma^{-1}([A\xi^0 \le b])} C_l\Big(\Gamma(Z,z);x\Big)(2\pi)^{-m(r+1)/2}\exp\left(-\frac{1}{2}\|(Z,z)\|^2\right)dZdz, \tag{3.49a}$$

where $\|(Z,z)\|$ denotes the Euclidean norm of $(Z,z)$ and

$$\Gamma^{-1}\Big([A\xi^0 \le b]\Big) = \left\{(Z,z) \in \mathbb{R}^{m(r+1)} : \Gamma(Z,z)\begin{pmatrix}\xi^0\\-1\end{pmatrix} \le 0\right\} \tag{3.49b}$$

If the $m \times (r+1)$ random matrix $\Big(A(\omega), b(\omega)\Big)$ has a joint normal distribution with mean $(\overline{A}, \overline{b})$ and an $m(r+1) \times m(r+1)$ covariance matrix $R$, the transformation $(A,b) = \Gamma(Z,z)$ reads

$$\Gamma(Z,z) := \begin{pmatrix} (\overline{A}_1, \overline{b}_1) + & (Z_1, z_1, \ldots, Z_m, z_m)V^{(1)} \\ \vdots & \vdots \quad \vdots \quad \vdots \\ (\overline{A}_m, \overline{b}_m) + & (Z_1, z_1, \ldots, Z_m, z_m)V^{(m)} \end{pmatrix}. \tag{3.50a}$$

Here, $(\overline{A}_i, (\overline{b}_i), (Z_i, z_i), i = 1, \ldots, m$, are the rows of $(\overline{A}, \overline{b}), (Z, z)$, resp., and $V^{(i)}, i = 1, \ldots, m$, are $m(r+1) \times (r+1)$ submatrices of the $m(r+1) \times m(r+1)$ matrix $V$ such that $R = V^T V$. Hence, in case (3.50a) for (3.49b) we find

$$\Gamma^{-1}\left([A\xi^0 \le b]\right) = \Big\{(Z,z) \in \mathbb{R}^{m(r+1)} : \overline{A}_i\xi^0 - \overline{b}_i \tag{3.50b}$$
$$+ (Z_1, z_1, \ldots, Z_m, z_m)V^{(i)}\begin{pmatrix}\xi^0\\-1\end{pmatrix} \le 0, 1 \le i \le m\Big\}.$$

Let denote $(Z^*, z^*)$ the projection of the origin $(O, o)$ of $\mathbb{R}^{mr} \times \mathbb{R}^m$ onto the closed subset $\Gamma^{-1}\Big([A\xi^0 \le b]\Big)$ of $\mathbb{R}^{mr} \times \mathbb{R}^m$; multiple projections of $(O, o)$ onto this set are denoted by $(Z^{*j}, z^{*j}), j = 1, \ldots, L$. Corresponding to (3.40b-d), here we define

$$\beta^* := \|(Z^*, z^*)\| \quad \left(= \|(Z^{*j}, z^{*j})\|, 1 \le j \le L\right), \tag{3.50c}$$

where $\|(Z,z)\|$ denotes the Euclidean norm of the matrix $(Z,z)$. If $\Gamma$ is defined by (3.50a), $(Z^*, z^*)$ can be computed *analytically*. E.g., if $m = 1$, then

$$(Z^*, z^*)^T = -(\overline{A}\xi^0 - \overline{b})\frac{V\hat{\xi}^0}{\|V\hat{\xi}^0\|^2} \text{ with } \hat{\xi}^0 := \begin{pmatrix}\xi^0\\-1\end{pmatrix},$$
$$\beta^* = \|(Z^*, z^*)\| = \frac{|\overline{A}\xi^0 - \overline{b}|}{\|V\hat{\xi}^0\|}. \tag{3.51}$$

We consider now the following two cases:

*Case A:*

$$(O,o) \notin \Gamma^{-1}\Big([A\xi^0 \le b]\Big) \Leftrightarrow \Gamma(O,o)\begin{pmatrix}\xi^0\\ -1\end{pmatrix} \not\le 0; \tag{3.52a}$$

if $\Gamma$ is defined by (3.50a), (3.52a) holds if and only if

$$\overline{A}\xi^0 \not\le \overline{b}. \tag{3.52b}$$

*Case B:*

$$(O,o) \in \text{ int}\left(\Gamma^{-1}\Big([A\xi^0 \le b]\Big)\right), \tag{3.52c}$$

i.e. $(O,o)$ is contained in the interior of

$$G := \Gamma^{-1}\Big([A\xi^0 \le b]\Big).$$

Obviously, (3.52b) holds if $\Gamma(O,o)\begin{pmatrix}\xi^0\\ -1\end{pmatrix} < 0$. If $\Gamma$ is defined by (3.50a), then this is guaranteed by

$$\overline{A}\xi^0 < \overline{b}. \tag{3.52d}$$

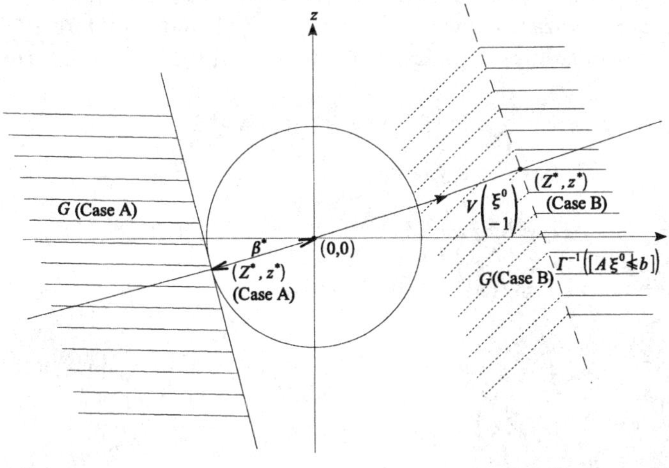


**Fig. 3.1.** $m = r = 1, \Gamma$ given by (3.50a), $G := \Gamma^{-1}\Big([A\xi^0 \le b]\Big)$

Next to we show that Case B can be reduced to Case A: For the complement $G^c$ of $G$ we get

$$G^c = \Gamma^{-1}\Big([A\xi^0 \not\le b]\Big) = \bigcup_{i=1}^{m} \Gamma^{-1}\Big([A_i\xi^0 > b_i]\Big).$$

If (3.52c) holds, then

$$(O,o) \notin cl(G^c) = cl\left(\Gamma^{-1}\left([A\xi^0 \nleq b]\right)\right), \tag{3.52e}$$

where "*cl*" denotes the closure of a set. Since condition (3.52e) corresponds to (3.52a), we consider the probability function

$$P^c(x) := P\Big(A(\omega)x \nleq b(\omega)\Big). \tag{3.53a}$$

Since $P(x) + P^c(x) = 1$ for all $x \in \mathbb{R}^r$, we have

$$D_l P(x) = -D_l P^c(x) \text{ for arbitrary } l. \tag{3.53b}$$

Furthermore, it is easy to see that

$$D_{(l)} P^c(x) = \int\limits_{A\xi^0 \nleq b} C_l(A, b; x) f(A, b)\, dA db, \tag{3.53c}$$

where $C_l(A, b; x)$ is given by (3.47b). Hence, because of (3.53a-c) and (3.52e), in Case B the asymptotic expansion of $D_{(l)}P(x) = -D_{(l)}P^c(x)$ can be obtained by the same methods as in Case A which is treated now.

Because of (3.52a) we have $(Z^*, z^*) \neq (O, o)$ and therefore $\beta^* > 0$, cf. (3.50c). Corresponding to (3.42), here we aply the $1-1$–transformation in $\mathbb{R}^{m(r+1)}$

$$(Z, z) := \beta^*(W, w) \tag{3.54}$$

to the integral (3.49a). This yields

$$D_{(l)}P(x) = (2\pi)^{-m(r+1)/2} \beta^{*m(r+1)} I(\beta^{*2}), \tag{3.55a}$$

where the integral $I = I(\lambda), \lambda \geq 0$, given by

$$I(\lambda) := \int\limits_{\frac{1}{\beta^*}\Gamma^{-1}\left([A\xi^0 \leq b]\right)} C_l\left(\Gamma\Big(\beta^*(W, w)\Big); x\right) \exp\left(-\frac{1}{2}\lambda \|(W, w)\|^2\right), \tag{3.55b}$$

is an integral of *Laplace type*, cf. (3.43a,b).

Under the condition

$$\beta^* \text{ is "large"}, \tag{3.56a}$$

cf. (3.41), we can approximate $I(\beta^{*2})$ by

$$I(\beta^{*2}) \approx \tilde{I}(\beta^{*2}), \tag{3.56b}$$

where $\tilde{I} = \tilde{I}(\lambda)$ is an asymptotic expansion [14], $I(\lambda) \sim \tilde{I}(\lambda)$ as $\lambda \to \infty$, of the integral $I(\lambda)$. According to [14], the asymptotic behavior of $I(\lambda)$ for

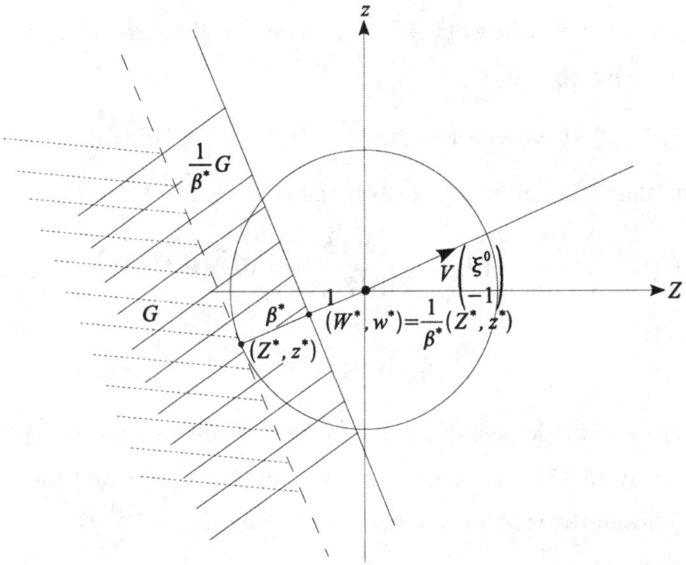


**Fig. 3.2.** $m = r = 1, \Gamma$ given by (3.51), $G = \Gamma^{-1}\left([A\xi^0 \le b]\right)$

$\lambda \to \infty$ depends on the shape of the hypersurface of $\frac{1}{\beta^*}G$ in a neighbourhood $U_\epsilon(W^*, w^*), U_\epsilon(W^{*j}, w^{*j}), j = 1, \dots, L$, of the projection(s)

$$(W^*, w^*) = \frac{1}{\beta^*}(Z^*, z^*), (W^{*j}, w^{*j}) = \frac{1}{\beta^*}(Z^{*j}, z^{*j}), 1 \le j \le \iota,$$

resp., of $(O, o)$ onto $\frac{1}{\beta^*}G$. The case of multiple projections $(W^{*j}, w^{*j}), j = 1, \dots, L$, can be reduced to the case of unique projection by partitioning $\frac{1}{\beta^*}G$ into subdomains $D^j, j = 1, \dots, L$, such that $\bigcup_{j=1}^L D^j = \frac{1}{\beta^*}G$ and $(W^{*j}, w^{*j})$ is the unique projection of $(O, o)$ onto $D^j$. Consequently,

$$I(\lambda) = \sum_{j=1}^{L} I_j(\lambda), \tag{3.57a}$$

where

$$I_j(\lambda) := \int_{D^j} C_l\left(\Gamma\left(\beta^*(W, w)\right); w\right) \exp\left(-\frac{1}{2}\lambda\|(W, w)\|^2\right) dW dw. \tag{3.57b}$$

Thus, in the following we suppose that $(W^*, w^*)$ is the unique projection of $(O, o)$ onto $\frac{1}{\beta^*}G$. Let now $\psi(W, w) = \psi(W, w; W^*, w^*)$ denote a function such that the hypersurface of $\frac{1}{\beta^*}G$ in a certain neighbourhood $U_\epsilon(W^*, w^*)$ of $(W^*, w^*)$ is given by

$$\{(W, w) \in U_\epsilon(W^*, w^*) : \psi(W, w; W^*, w^*) = 0\}. \tag{3.58}$$

Assuming that $\psi$ is sufficiently smooth, define the $m(r+1) \times m(r+1)$ matrix

$$H(W^*, w^*) := -\left(I + \frac{1}{\|\nabla\psi(W^*, w^*)\|}\nabla^2\psi(W^*, w^*)\right). \tag{3.59a}$$

Note that

$$\tilde{H}(W^*, w^*) = -I \tag{3.59b}$$

if $\frac{1}{\beta^*}G$ has a plane hypersurface in a certain neighbourhood of $(W^*, w^*)$. If $\Gamma$ is given by (3.50a), equation (3.59b) holds for $m = 1$ and for $m > 1$ if $(W^*, w^*)$ lies in the interior of a boundary hyperplane of $\frac{1}{\beta^*}G$.

Let then

$$H(W^*, w^*) = (\mathrm{cof}\ \tilde{h}_{st})_{s,t=1,\dots,m(r+1)} \tag{3.59c}$$

denote the matrix of cofactors of $\tilde{H}(W^*, w^*) = (\tilde{h}_{st})$. If (3.59b) holds, then

$$H(W^*, w^*) = (-1)^{m(r+1)-1}I. \tag{3.59d}$$

Using (3.55a), (3.55b) and (3.56a), (3.56b), asymptotic expansion of $I(\lambda)$ as $\lambda \to \infty$, see [14], yields the following approximation of $D_{(l)}P(x)$:

**Theorem 3.4.** *If the above assumptions hold, we have the expansion (being asymptotically exact for $\beta^* \to \infty$)*

$$D_{(l)}P(x) \sim \frac{1}{(2\pi)^{1/2}\beta^*}\frac{C_l\left(\Gamma(Z^*, z^*); x\right)}{|J(W^*, w^*)|^{1/2}}\exp\left(-\frac{1}{2}\beta^{*2}\right), \tag{3.60a}$$

where

$$J(W^*, w^*) := (W_1^*, w_1^*, \dots, W_m^*, w_m^*)H(W^*, w^*)(W_1^*, w_1^*, \dots, W^*, w_m^*)^T. \tag{3.60b}$$

*Example 3.2.* If $m = 1$ and $\Gamma$ is given by (3.50a), then, cf. (3.50b), (3.50c), (3.51), (3.52a,b), $|J(W^*, w^*)| = 1$ and therefore

$$D_{(l)}P(x) \sim \frac{1}{\sqrt{2\pi}}\frac{1}{\beta^*}\exp\left(-\frac{1}{2}\beta^{*2}\right)C_l\left((\overline{A}, \overline{b}) - \beta^*\frac{\hat{\xi}^{0^T}R}{(\hat{\xi}^{0T}R\hat{\xi}^0)^{1/2}}; x\right) \tag{3.61a}$$

for "large"

$$\beta^* = \frac{\overline{A}\xi^0 - \overline{b}}{(\hat{\xi}^{0\prime}R\xi^0)^{1/2}}. \tag{3.61b}$$

## 3.6 Integral Representations of the Probability Function $P(x)$ and its Derivatives

In the following we suppose that for all $x$ under consideration the random $m$–vector

$$y_x(\omega) := y\Big(a(\omega), x\Big) = \Big(y_1\Big(a(\omega), x\Big), \ldots, y_m\Big(a(\omega), x\Big)\Big)^T \tag{3.62a}$$

has a probability density

$$f = f(y; x). \tag{3.62b}$$

The density $f(y; x)$ exists under weak assumptions:

**Lemma 3.1.** *Let $m \le \nu$. For given $x$, suppose that the Jacobian $\nabla_a y(a, x)$ of the mapping $a \longrightarrow y(a, x)$ exists and has rank $\nabla_a y(a, x) = m$ for all $a \in \mathbb{R}^\nu$ up to a set of $P_{a(\cdot)}$–measure zero, where $P_{a(\cdot)}$ denotes the probability distribution of $a = a(\omega)$. If the random parameter vector $a = a(\omega)$ has a probability density $f_0 = f_0(a)$, then the density $f = f(y; x)$ exists.*

*Proof.* First a vector $\eta = \eta(a, x) := \Big(y_{m+1}(a, x), \ldots, y_\nu(a, x)\Big)^T$ of suitable smooth functions $y_{m+1}(a, x), \ldots, y_\nu(a, x)$ is added to the given functions $y_1(a, x), \ldots, y_m(a, x)$ such that rank $\nabla_a \tilde{y}(a, x) = \nu$ for all $a \in \mathbb{R}^\nu$ up to a $P_{a(\cdot)}$–measure zero, where $\tilde{y}(a, x) := \Big(y_1(a, x), \ldots, y_m(a, x),\ y_{m+1}(a, x), \ldots,$ $y_\nu(a, x)\Big)^T$, see [113], (S. 226). Consider then for a given, fixed vector $x$ the 1–1–transformation

$$T_x = T_x(a) := \tilde{y}(a, x). \tag{3.63a}$$

The density $\tilde{f} = \tilde{f}(\tilde{y}; x)$ of $\tilde{y} = \tilde{y}\big(a(\omega), x\big)$ follows by means of the transformation rule for densities

$$\tilde{f}(\tilde{y}; x) = f_0\Big(T_x^{-1}(\tilde{y})\Big) \frac{1}{|\det \nabla_a T_x\,(T_x^{-1}(\tilde{y}))\,|}. \tag{3.63b}$$

Finally, as a marginal density of $\tilde{f}(\tilde{y}; x)$, $f(y, x)$ is given by

$$f(y; x) = \int_{-\infty}^{+\infty} \cdots \int_{-\infty}^{+\infty} \tilde{f}\left(\begin{pmatrix} y \\ \eta \end{pmatrix}; x\right)\, d\eta. \tag{3.63c}$$

Besides the density representation by means of (3.63a-c), the probability density of $y_x(\omega) = y\Big(a(\omega), x\Big)$ can also be described by using known integral representations of derivatives for parameter-dependent integrals, cf. [136]. Indeed, if $a = a(\omega)$ has a density $f_0 = f_0(a)$, then the distribution function

$$F(\eta; x) := P\Big(y\Big(a(\omega), x\Big) \le \eta\Big), \eta \in \mathbb{R}^m, \tag{3.63d}$$

of $y_x(\omega)$ is a special probability function being represented by the integral

$$F(\eta;x) = \int\limits_{y(a,x)\leq\eta} f_0(a)\,da. \tag{3.63e}$$

Hence, the following result is obtained by applying one of the above mentioned differentiation formulas first to (3.63e) and then to the resulting integral representation of the density $f(y,x)$ :

**Lemma 3.2.** *Let $m = 1$ and suppose that $a(\omega)$ has a density $f_0 = f_0(a)$. Furthermore, assume that the derivatives and integrals under consideration exist and are finite. Then*

$$\begin{aligned} f(\eta;x) &= \int\limits_{y(a,x)=\eta} f_0(a)\frac{dS}{\|\nabla_a y(a,x)\|} \\ &= \int\limits_{y(a,x)\leq\eta} div\left(f_0(a)\frac{\nabla_a y(a,x)}{\|\nabla_a y(a,x)\|^2}\right) da, \end{aligned} \tag{3.64}$$

*where $dS$ denotes the surface element of $S = S_{x,\eta} := \Big\{a \in \mathbb{R}^\nu : y(a,x) = \eta\Big\}$. Moreover,*

$$\frac{\partial f}{\partial x_k}(\eta;x) = -\int\limits_{y(a,x)\leq\eta} div\left\{ div\left(f_0(a)\frac{\partial y}{\partial x_k}(a,x)\frac{\nabla_a y(a,x)}{\|\nabla_a y(a,x)\|^2}\right) \times\frac{\nabla_a y(a,x)}{\|\nabla_a y(a,x)\|^2}\right\} da. \tag{3.65}$$

*Remark 3.4.* Higher order derivatives of $f(y;x)$ with respect to $x$ can be obtained in an iterative way, cf. (3.64) and (3.65).

The case $m > 1$ can be treated by using more general differentiation formulas. On the other hand, very simple representations of approximative density representations can be found by means of *stochastic completion technique*, see also [79,85,87]: Here, the distribution function $F(y;x)$ is approximated first by

$$\tilde{F}(\eta;x) := P\Big(y\big(a(\omega),x\big) + \delta(\omega) \leq \eta\Big), \eta \in \mathbb{R}^m, \tag{3.66a}$$

where the stochastic completion term $\delta = \delta(\omega)$ is a random $m$–vector being independent of $y_x(\omega)$ and having a smooth probability density $\varphi = \varphi(\delta)$. Since

$$\tilde{F}(\eta;n) = EF\big(\eta - \delta(\omega);x\big) = E\Phi\Big(\eta - y\big(a(\omega),x\big)\Big), \tag{3.66b}$$

where $\Phi = \Phi(\delta)$ denotes the distribution function of $\delta(\omega)$, approximations of $f(y;x)$ and its derivatives can be obtained as follows:

**Lemma 3.3.** *Suppose that the expectations under consideration exist and differentiation and expectations may be interchanged. Approximating $f = f(\eta; x)$ by*

$$\tilde{f}(\eta; x) := \frac{\partial^m}{\partial \eta_1 \cdots \partial \eta_m} \tilde{F}(y; x),$$

we have

$$\tilde{f}(\eta; x) = E\varphi\Big(\eta - y\big(a(\omega), x\big)\Big) \tag{3.67}$$

$$\frac{\partial \tilde{f}}{\partial x_k}(\eta; x) = -E\sum_{i=1}^{m} \frac{\partial \varphi}{\partial \delta_i}\Big(\eta - y\big(a(\omega), x\big)\Big) \frac{\partial y_i}{\partial x_k}\big(a(\omega), x\big). \tag{3.68}$$

Furthermore, $\tilde{f}(\eta; x) \to f(\eta, x)$ and $\frac{\partial \tilde{f}}{\partial x_k}(\eta; x) \to \frac{\partial f}{\partial x_k}(\eta; x)$ as $\delta(\omega) \to 0$ a.s., provided that $f(\cdot; x), \frac{\partial f}{\partial y_k}(\cdot;)$ are continuous.

*Remark 3.5.* Approximations to higher order partial derivatives of $f = f(y; x)$ with respect to $x$ can be obtained in the same way.

Under the above assumptions the probability function $P = P(x)$ defined by (3.1a,b) can be represented now by the multiple integral

$$P(x) = \int_{y_l}^{y_u} f(y; x)\, dy \tag{3.69a}$$

having the very simple *fixed* domain of integration

$$B = \{y \in \mathbb{R}^m : y_l \leq y \leq y_u\}. \tag{3.69b}$$

In order to compute the partial derivative $\frac{\partial P}{\partial x_k}(x)$ with respect to $x_k$ at points

$$x = x(t) = \left(x_1^0, \ldots, x_{k-1}^0, t, x_k^0, \ldots, x_r\right)^T, t \in I, \tag{3.70}$$

where $x_j^0, j \neq k$, are given fixed values for the components $x_j, j \neq k$, and $I \subset \mathbb{R}$ is certain finite interval, we need, cf. Section 3.2, Section 2.2, the following first assumption on the density $f = f(y; x)$ :

$\frac{\partial f}{\partial x_k}\big(y; x(t)\big)$ exists for all $t \in I$ and has an integrable majorant $h = h(y)$ on $B$.

Hence,

$$\left|\frac{\partial f}{\partial x_k}\big(y; x(t)\big)\right| \leq h(y) \text{ for all } t \in I \text{ almost everywhere on } B \tag{3.71}$$

with a nonnegative, integrable function $f(y)$.

*Remark 3.6.* If $B$ is a bounded domain, condition (3.71) holds e.g. if the function $(y,t) \to \frac{\partial f}{\partial x_k}\big(y, x(t)\big)$ is piecewise continuous on $B \times \overline{I}$, where $I$ denotes the closure of $I$. Indeed, put $h(y) := \sup\left\{\left|\frac{\partial f}{\partial x_k}\big(y,x(t)\big)\right| : y \in B, t \in \overline{I}\right\}$.

If (3.71) holds, we may interchange, cf. [4,113] und Lemma 2.3, differentiation and integration in (3.69a). Hence, for all $x \in \{x(t) : t \in I\}$ the derivative $\frac{\partial P}{\partial x_k}(x)$ exists, and is given by

$$\frac{\partial P}{\partial x_k}(x) = \int_{y_l}^{y_u} \frac{\partial f}{\partial x_k}(y;x)\, dy. \tag{3.72a}$$

Iterating the above procedure, under corresponding assumptions also the higher order derivatives

$$D_{\underline{l}}P(x) := \frac{\partial^s P(x)}{\partial x_{l_1}\partial x_{l_2}\dots\partial x_s} \text{ with } \underline{l} := (l_1, l_2, \dots, l_s) \tag{3.72b}$$

of $P$ may be obtained:

$$D_{\underline{l}}P(x) = \int_{y_l}^{y_u} D_{\underline{l}}f(y;x)\, dy, \tag{3.72c}$$

cf. (3.14a,b).

Estimation and approximation of probability densitives and more general functions by *orthogonal function series expansions* is a well established technique, see e.g. [120,126–128,143]. Hence, in the following we consider estimates/approximations of $P = P(x)$ and its derivatives based on *orthogonal function series expansions* of $f(\cdot;x), \frac{\partial f}{\partial x_k}(\cdot;x), D_{\underline{l}}f(\cdot;x)$ with respect to the orthonormal systems of

- Hermite functions ($y_x(\omega)$ has range $\mathbb{R}^m$, unknown range, resp.)
- trigonometric functions ($y_x(\omega)$ has a bounded range)
- Legendre functions ($y_x(\omega)$ has a bounded range)
- Laguerre functions ( $y_x(\omega)$ is bounded from the left, right, resp.).

## 3.7 Orthogonal Function Series Expansions I: Expansions in Hermite Functions, Case $m = 1$

In the following, let $x$ denote an arbitrary, but fixed $r$-vector. For simplification we consider here first the case of a *scalar* response function $y(a,x) = y_1(a,x)$.

If the random variable $y_x(\omega) = y\Big(a(\omega), x\Big)$ takes values from $-\infty$ to $+\infty$ or if it is difficult to ascertain its range, we consider expansions in Hermite functions.

Define [120, 128] the Hermite polynomials $H_j$ by

$$H_j(y) := (-1)^j e^{y^2} \frac{d^j}{dy^j} e^{-y^2}, j = 0, 1, 2, \dots, \tag{3.73a}$$

e.g.

$$\begin{aligned} H_0(y) &= 1 \\ H_1(y) &= 2y \\ H_2(y) &= 4y^2 - 2 \\ H_3(y) &= 8y^3 - 12y \\ H_4(y) &= 16y^4 - 48y^2 + 12 \\ H_5(y) &= 32y^5 - 160y^3 + 120y \\ &\dots, \end{aligned} \tag{3.73b}$$

and the normalized Hermite functions $\varphi_j$ by

$$\varphi_j(y) := \left(\pi^{1/2} 2^j j!\right)^{-\frac{1}{2}} e^{-\frac{y^2}{2}} H_j(y), j = 0, 1, 2, \dots . \tag{3.73c}$$

It is known [120, 128, 143] that

$$\varphi_j \in L^1(\mathbb{R}) \cap L^2(\mathbb{R}), j = 0, 1, 2, \dots , \tag{3.73d}$$

and the sequence

$$(\varphi_j) \text{ forms a complete, orthonormal system in } L^2(\mathbb{R}). \tag{3.73e}$$

Furthermore, there exists a uniform constant $d_2$, see [128], such as

$$|\varphi_j(y)| < d_2 \text{ for all } y \in \mathbb{R} \text{ and each } j = 0, 1, 2, \dots . \tag{3.73f}$$

Assume that

$$f(\cdot; x) \in L_2(\mathbb{R}) \tag{3.74a}$$

and

$$\frac{\partial f}{\partial x_k}(\cdot; x) \in L_2(\mathbb{R}) \tag{3.74b}$$

$$D_{\underline{l}} f(\cdot; x) \in L_2(\mathbb{R}). \tag{3.74c}$$

Then these functions can be expanded in the orthogonal series

$$f(y;x) = \sum_{j=0}^{\infty} c_j(x)\varphi_j(y) \tag{3.75a}$$

$$\frac{\partial f}{\partial x_k}(y;x) = \sum_{j=0}^{\infty} c_{kj}(x)\varphi_j(y) \tag{3.75b}$$

$$D_{\underline{l}}f(y;x) = \sum_{j=0}^{\infty} c_{\underline{l}j}(x)\varphi_j(y), \tag{3.75c}$$

where the (Fourier) coefficients $c_j = c_j(x), c_{kj} = c_{kj}(x), c_{\underline{l}j} = c_{\underline{l}j}(x)$, resp., are defined [120, 128, 143] by

$$c_j(x) = \int f(y;x)\varphi_j(y)\,dy \tag{3.76a}$$

$$c_{kj}(x) = \int \frac{\partial f}{\partial x_k}(y;x)\varphi_j(y)\,dy \tag{3.76b}$$

and

$$c_{\underline{l}j}(x) = \int D_{\underline{l}}f(y;x)\varphi(y)\,dy. \tag{3.76c}$$

Where under assumptions (3.74a-c) the series (3.75a-c), resp., are convergent in the $L^2$–sense, for the pointwise convergence of the series towards $f(\cdot;x), \frac{\partial f}{\partial x_k}(\cdot;x), D_{\underline{l}}f(\cdot;x)$ some additional assumptions [120, 128] are needed:

**Lemma 3.4.** *Let $g = g(y;x)$ denote one of the functions $f(\cdot;x), \frac{\partial f}{\partial x_k}, D_{\underline{l}}f(\cdot;x)$. Suppose that $g(\cdot;x) \in L^1(\mathbb{R}) \cap L^2(\mathbb{R})$. If $g(\cdot;x)$ is continuous and of bounded variation in an interval $[y_1, y_2]$, then the series in (3.75a), (3.75b), (3.75c), resp., converges uniformly to $g(\cdot;x)$ in any open subinterval of $[y_1, y_2]$.*

Since $f = f(y;x)$ is the probability density function of the random variable $y = y\Big(a(\omega), x\Big)$, see (3.62a,b), for $c_j(x)$ we have also the representation

$$c_j(x) = E\varphi_j\left(y\Big(a(\omega), x\Big)\right), j = 0, 1, 2, \ldots . \tag{3.77}$$

Moreover, if there exists an integrable function $h_k = h_k(y)$ on $\mathbb{R}$ such that, with $x = x(t)$ given by (3.70),

$$\left|\frac{\partial f}{\partial x_k}\Big(y; x(t)\Big)\varphi_j(y)\right| \leq h_k(y), y \in \mathbb{R}, \text{ for all } t \in I \text{ and } y \in \mathbb{R}, \tag{3.78a}$$

cf. (3.71) and (3.73f), then, by interchanging differentiation and expectation in (3.76a), from (3.76a), (3.76b) and (3.77) for $x \in \{x(t) : t \in I\}$ we obtain the important relation

$$\frac{\partial}{\partial x_k} c_j(x) = \int \frac{\partial f}{\partial x_k}(y;x)\varphi_j(y)\,dy = c_{kj}(x). \tag{3.78b}$$

Iterating this procedure, i.e. replacing (3.76a) by (3.76b) and $f(y;x)$ by $\frac{\partial f}{\partial x_k}(y;x)$ etc., under corresponding modifications of assumption (3.78a) we also find

$$D_{\underline{l}} E\varphi_j\Big(y\big(a(\omega),x\big)\Big) = D_{\underline{l}} c_j(x) = c_{\underline{l}j}(x). \tag{3.78c}$$

Because of (3.69a) and the series representation (3.75a) of $f(\cdot;x)$, for arbitrary finite bounds $y_l, y_u, -\infty < y_l < y_u < +\infty$, we get

$$\begin{aligned}
\left| P(x) - \sum_{j=0}^{n} c_j(x) \int_{y_l}^{y_u} \varphi_j(x)\,dy \right| &= \left| \int_{y_l}^{y_u} \Big( f(y;x) - \sum_{j=0}^{n} c_j(x)\varphi_j(y) \Big)\,dy \right| \\
&\le \int_{y_l}^{y_u} \left| f(y;x) - \sum_{j=0}^{n} c_j(x)\varphi_j(y) \right|\,dy \\
&\le (y_u - y_l)^{1/2} \left\| f(\cdot;x) - \sum_{j=0}^{n} c_j(x)\varphi_j(\cdot) \right\|_2 ,
\end{aligned} \tag{3.79a}$$

where $\|\cdot\|_2$ denotes the norm of $L^2(\mathbb{R})$. In the same way from (3.72a), (3.72c) and (3.75b), (3.75c), resp., we obtain

$$\begin{aligned}
&\left| \frac{\partial P}{\partial x_k}(x) - \sum_{j=0}^{n} c_{kj}(x) \int_{y_l}^{y_u} \varphi_j(y)dy \right| \\
&\le (y_u - y_l)^{1/2} \left\| \frac{\partial f}{\partial x_l}(\cdot;x) - \sum_{j=0}^{n} c_{\underline{l}j}(x)\varphi_i(\cdot) \right\|_2 ,
\end{aligned} \tag{3.79b}$$

$$\begin{aligned}
&\left| D_{\underline{l}} P(x) - \sum_{j=0}^{n} c_{\underline{l}j}(x) \int_{y_l}^{y_u} \varphi_j(y)dy \right| \\
&\le (y_u - y_l)^{1/2} \left\| D_{\underline{l}} f(\cdot;x) - \sum_{j=0}^{n} c_{\underline{l}j}(x)\varphi_j(\cdot) \right\|_2 .
\end{aligned} \tag{3.79c}$$

Since the series in (3.75a-c) are convergent in the $L^2$–norm, the above inequalities yield the following result:

**Theorem 3.5.** *Based on the assumptions concerning the existence of the density $f = f(\cdot;x)$ and its derivatives with respect to $x$ mentioned in the above, suppose that $P(x), \frac{\partial P}{\partial x_k}, (x), D_{\underline{l}} P(x)$, resp., can be represented for a given vector $x$ by (3.69a), (3.72a), (3.72c). Furthermore assume that the orthogonal series representation (3.75a), (3.75b), (3.75c), resp., holds. Then*

$P(x), \frac{\partial P}{\partial x_k}(x), D_{\underline{l}}P(x)$, *resp., can be represented for finite bounds* $-\infty < y_l < y_u < +\infty$ *by the following convergent series:*

$$P(x) = \sum_{j=0}^{\infty} c_j(x) \int_{y_e}^{y_u} \varphi_j(y)\, dy, \tag{3.80a}$$

$$\frac{\partial P}{\partial x_k}(x) = \sum_{j=0}^{\infty} c_{kj}(x) \int_{y_e}^{y_u} \varphi_j(y)\, dy, \tag{3.80b}$$

$$D_{\underline{l}}P(x) = \sum_{j=0}^{\infty} c_{\underline{l}j}(x) \int_{y_e}^{y_u} \varphi_j(y)\, dy. \tag{3.80c}$$

### 3.7.1 Integrals over the Basis Functions and the Coefficients of the Orthogonal Series

The integrals over the basis functions $\varphi_j$, see (3.73b,c), read

$$\int_{y_e}^{y_u} \varphi_j(y)\, dy = C_j \sum_{j=0}^{j} h_{ji} J_i, \tag{3.81a}$$

where $C_j, h_{ji}, 0 \le i \le j, j \ge 0$, are given, fixed coefficients. Moreover, for the definite integrals

$$J_i := \int_{y_l}^{y_u} e^{-\frac{y^2}{2}} y^i\, dy, i = 0, 1, \ldots, \tag{3.81b}$$

by partial integration we find the recursion:

$$J_i = y_l^{i-1} e^{-\frac{y_l^2}{2}} - y_u^{i-1} e^{-\frac{y_u^2}{2}} + (i-1)J_{i-2}, i \ge 2. \tag{3.81c}$$

Hence, the sequence $(J_i)$ can be obtained very easily from (3.81c) and

$$J_0 = \sqrt{2\pi}\left(\Phi(y_u) - \Phi(y_l)\right), \tag{3.81d}$$

$$J_1 = e^{-\frac{y_l^2}{2}} - e^{-\frac{y_u^2}{2}}, \tag{3.81e}$$

where $\Phi$ designates the distribution function of the $N(0,1)$-normal distribution. Hence, for $i = 2, 3$ we get e.g.

$$J_2 = y_l e^{-\frac{y_l^2}{2}} - y_u e^{-\frac{y_u^2}{2}} + \sqrt{2\pi}\left(\Phi(y_u) - \Phi(y_l)\right) \tag{3.81f}$$

$$J_3 = y_l^2 e^{\frac{y_l^2}{2}} - y_u^l e^{-\frac{y_u^2}{2}} + 2\left(e^{-\frac{y_l^2}{2}} - e^{-\frac{y_u^2}{2}}\right). \tag{3.81g}$$

Mean value representations for the Fourier coefficients $c_{kj}(x)$ and $c_{\underline{l}j}(x)$ can be obtained by interchanging differentiation and integration in the mean value representation (3.77) for $c_j(x)$:

If in addition to condition (3.78a) there is a function $\tilde{h}_k = \tilde{h}_k(a)$ on $\mathbb{R}^\nu$ such that $\int \tilde{h}_k(a) f_0(a)\, da < +\infty$ and

$$\left|\varphi_j'\left(y\left(a, x(t)\right)\right) \frac{\partial y}{\partial x_k}\left(a, x(t)\right)\right| \leq \tilde{h}_k(a) \text{ a.s. on } \mathbb{R}^\nu \text{ for all } t \in I, \tag{3.82}$$

where again $x = x(t), t \in I$, is given by (3.70), then (3.77) and (3.78b) yield for $x = x(t), t \in I$,

$$\begin{aligned} c_{kj}(x) &= \frac{\partial}{\partial x_k} E\varphi_j\left(y\left(a(\omega), x\right)\right) \\ &= E\varphi_j'\left(y\left(a(\omega), x\right)\right) \frac{\partial y}{\partial x_k}\left(a(\omega), x\right). \end{aligned} \tag{3.83}$$

Because of

$$\varphi_j'(y) = -y\varphi_j(y) + \sqrt{2j}\varphi_{j-1}(y) = \frac{\sqrt{j}}{\sqrt{2}}\varphi_{j-1}(y) - \frac{\sqrt{j+1}}{\sqrt{2}}\varphi_{j+1}(y), j \geq 0, \tag{3.84a}$$

cf. [120], where $\varphi_{-1}(y) \equiv 0$, and (3.73f) we get

$$|\varphi_j'(y)| \leq \frac{d_2}{\sqrt{2}}\left(\sqrt{j} + \sqrt{j+1}\right) \text{ for all } y \in \mathbb{R} \text{ and } j = 0, 1, 2, \ldots. \tag{3.84b}$$

Hence, also the derivatives $\varphi_j'$ of $\varphi_j$ are bounded on $\mathbb{R}$.

Furthermore, under corresponding assumptions from (3.83) and (3.78c) we get, see also (3.72c),

$$c_{\underline{l}j}(x) = ED_{\underline{l}}\varphi_j\left(y\left(a(\omega), x\right)\right), \underline{l} = (l_1, \ldots, l_s). \tag{3.85}$$

Obviously, $D_{\underline{l}}\varphi_j\left(y(a, x)\right)$ may be obtained by the chain rule, and (3.84a,b) yield also the boundedness of the higher derivatives $\varphi_j^{(t)}$ of the Hermite functions $\varphi_j$ for arbitrary values of $t$.

Based on the mean value representations (3.77), (3.83) and (3.85), the Fourier coefficients can be estimated by sampling methods. Important examples are the sample means

$$\hat{c}_j(x) := \frac{1}{n}\sum_{t=1}^{n} \varphi_j\left(y(a^t, x)\right), \tag{3.86a}$$

$$\hat{c}_j(x) := \frac{1}{n}\sum_{t=1}^{n} \varphi_j'\left(y(a^t, x)\right) \frac{\partial y}{\partial x_k}(a^t, x), \tag{3.86b}$$

$$\hat{c}_{\underline{l}j}(x) := \frac{1}{n}\sum_{t=1}^{n} D_{\underline{l}}\varphi_j\left(y(a^t, x)\right), \tag{3.86c}$$

resp., where $a^t, t = 1, \ldots, n$, denote independent realizations of $a = a(\omega)$.

Suppose that the moments

$$\mu_i(x) := E\left(y\left(a(\omega)j, x\right)\right)^i, i = 0, 1, 2, \ldots,$$

of $y_x(\omega) = y\left(a(\omega), x\right)$ and their derivatives with respect to $x$ up to a certain order are available. Then approximations to the Fourier coefficients $c_j(x), c_{kj}(x), c_{\underline{l}j}(x)$ can be obtained for given $x$

* by Taylor approximation $e^{-t} = T_N(t) + R_N(t)$ of $e^{-t}$ at $t = 0$ of a certain order $N$, or
* by Taylor approximation of the functions

$$\begin{aligned}
\psi_j(a) &:= \varphi_j\left(y(a, x)\right), \\
\psi_{kj}(a) &:= \varphi_j'\left(y(a, x)\right)\frac{\partial y}{\partial x_k}(a, x), \\
\varphi_{\underline{l}j}(a) &:= D_{\underline{l}}\varphi_j\left(y(a, x)\right),
\end{aligned}$$

resp., with respect to $a$ at the mean $\bar{a} := Ea(\omega)$; in practice mainly *second order* approximations ($N = 2$) are used in this case.

In the first case the Hermite functions $\varphi_j(y)$ occurring in $c_j(x), c_{kj}(x), c_{\underline{l}j}(x)$, cf. (3.73c), are approximated by the polynomial

$$\tilde{\varphi}_j(y) := C_j T_N\left(\frac{y^2}{2}\right) H_j(y) \tag{3.87a}$$

of order $2N + j$, where the error $\varphi_j - \tilde{\varphi}_j$ can be estimated by

$$|\varphi_j(y) - \tilde{\varphi}_j(y)| \le \frac{1}{N!} |\varphi_j(y)| \int_0^{\frac{1}{2}y^2} s^N e^s \, ds. \tag{3.87b}$$

Now $c_j(x)$ is approximated, see (3.77), by

$$\begin{aligned}
\tilde{c}_j(x) &:= EC_jT_N\left(\frac{1}{2}y^2\left(a(\omega), x\right)\right) H_j\left(y\left(a(\omega), x\right)\right) \qquad (3.87c) \\
&= C_j \sum_{k=0}^{2N+j} h_{ji}\mu_i(x),
\end{aligned}$$

where $h_{ji}, i = 0, 1, \ldots, 2N + j$, are certain coefficients. The approximation error can be estimated by

$$|c_j(x) - \tilde{c}_j(x)| \le \frac{d_2}{N!} \int_0^{+\infty} s^N e^s P\left(\left|y\left(a(\omega), x\right)\right| > \sqrt{2s}\right) ds, j = 0, 1, \ldots . \tag{3.87d}$$

For $c_{kj}(x)$ the corresponding approximation reads

$$\tilde{c}_{kj}(x) := \frac{1}{\sqrt{2}} E\left(\sqrt{j}\tilde{\varphi}_{j-1}\left(y\left(a(\omega), x\right)\right) - \sqrt{j+1}\tilde{\varphi}_{j+1}\left(y\left(a(\omega), x\right)\right)\right) \times \frac{\partial y}{\partial x_k}\left(a(\omega), x\right), \tag{3.87e}$$

where

$$|c_{kj}(x) - \tilde{c}_{kj}(x)| \le d_2 \frac{\sqrt{j} + \sqrt{j+1}}{\sqrt{2}N!} E\left|\frac{\partial y}{\partial x_k}\left(a(\omega), x\right)\right| \int_0^{\frac{1}{2}y^2\left(a(\omega), x\right)} s^N e^s \, ds. \tag{3.87f}$$

Moreover, similar approximations and error bounds can also be derived for $c_{\underline{l}j}(x)$.

In the second case, according to (2.31a-d), (2.32a-d), the Fourier coefficients are approximated (for $N = 2$) by

$$\tilde{c}_j(x) := \varphi_j\left(y\left(\overline{a}, x\right)\right) + \frac{1}{2} E\left(a(\omega) - \overline{a}\right)^T \left(\varphi_j''\left(y(\overline{a}, x)\right) \nabla_a y(\overline{a}, x) \nabla_a y(\overline{a}, x)^T + \varphi_j'\left(y(\overline{a}, x)\right) \nabla_a^2 y(\overline{a}, x)\right) \left(a(\omega) - \overline{a}\right). \tag{3.88a}$$

The error can be estimated by

$$|c_j(x) - \tilde{c}_j(x)| \le \frac{1}{2} \int_0^1 (1-t)^2 E \left\|\nabla_a^3 \varphi_j\left(y\left(\overline{a} + t\left(a(\omega) - \overline{a}\right)\right)\right)\right\| \times \|a(\omega) - \overline{a}\|^3 \, dt, \tag{3.88b}$$

where the derivative $\nabla_a^3 \varphi_j\left(y(a, x)\right)$ can be obtained by further differentiation of $\nabla_a^2 \varphi_j\left(y(a, x)\right)$ with respect to $a$, cf. (2.31b-d).

Obviously, related approximations and error bounds can be obtained also for $c_{kj}(x)$ and $c_{\underline{l}j}(x)$.

### 3.7.2 Estimation/Approximation of $P(x)$ and its Derivatives

Based on the expansions (3.75a-c), estimates of $f(y; x)$ and its derivatives $\frac{\partial f}{\partial x_k}(y; x), D_{\underline{l}} f(y; x)$, resp., can be defined, cf. [126–128], by

$$\hat{f}(y;x) := \sum_{j=0}^{q(n)} \hat{c}_j(x)\varphi_j(y), \tag{3.89a}$$

$$\frac{\hat{\partial f}}{\partial x_k}(y;x) := \sum_{j=0}^{q(n)} \hat{c}_{kj}(x)\varphi_j(y), \tag{3.89b}$$

$$\hat{D}_{\underline{l}}f(y;x) := \sum_{j=0}^{q(n)} \hat{c}_{\underline{l}j}(x)\varphi_j(y). \tag{3.89c}$$

In the above $q = q(n)$ is a certain integer, and $\hat{c}_j(x), \hat{(c)}_{kj}(x), \hat{c}_{\underline{l}j}(x)$ are unbiased estimators of the Fourier coefficients $c_j(x), c_{kj}(x), c_{\underline{l}j}(x)$, resp., as defined by (3.86a-c). Replacing in (3.69a) and (3.72a,c), the integrands $f(y;x), \frac{\partial f}{\partial x}(y;x), D_{\underline{l}}f(y;x)$, resp., by the above estimates for $P(x)$ and its derivatives, we get the estimators

$$\hat{P}(x) := \int_{y_l}^{y_u} \hat{f}(y;x)\,dy = \sum_{j=0}^{q(n)} \hat{c}_j(x) \int_{y_l}^{y_u} \varphi_j(y)\,dy, \tag{3.90a}$$

$$\frac{\hat{\partial P}}{\partial x_k}(x) := \int_{y_l}^{y_n} \frac{\hat{\partial f}}{\partial x_k}(y;x)\,dy = \sum_{j=0}^{q(n)} \hat{c}_{kj}(x) \int_{y_l}^{y_u} \varphi_j(y)\,dy; \tag{3.90b}$$

$$\hat{D}_{\underline{l}}P(x) := \int_{y_l}^{y_u} D_{\underline{l}}f(y;x)\,dy = \sum_{j=0}^{q(n)} \hat{c}_{\underline{l}j} \int_{y_l}^{y_u} \varphi_j(y)\,dy. \tag{3.90c}$$

Since $\hat{f}(\cdot;x), \frac{\hat{\partial f}}{\partial x_k}(\cdot;x), \hat{D}_{\underline{l}}f(\cdot;x) \in L^2(\mathbb{R})$, the mean square error of the estimates (3.90a-c) can be determined as follows:

$$E\Big(P(x) - \hat{P}(x)\Big)^2 = E\left|\int_{y_l}^{y_u} \Big(f(y;x) - \hat{f}(y;x)\Big)\,dy\right|^2 \tag{3.91a}$$

$$\le E\left(\int_{y_l}^{y_u} \Big|f(y;x) - \hat{f}(y;x)\Big|\,dy\right)^2 \le (y_u - y_l)E\int_{y_l}^{y_u} \Big(f(y;x) - \hat{f}(y;x)\Big)^2\,dy$$

$$\le (y_u - y_l)\ \mathrm{MISE}_{\hat{f}}.$$

The Mean Integrated Square Error of the density estimator $\hat{f}(y;x)$ is defined by

$$\mathrm{MISE}_{\hat{f}} := E\|f(\cdot;x) - \hat{f}(\cdot;x)\|_2^2 = E\int \Big(f(y;x) - \hat{f}(y;x)\Big)^2\,dy.$$

In the same way we get

$$E\left(\frac{\partial P}{\partial x_k}(x)-\frac{\hat{\partial P}}{\partial x_k}(x)\right)^2 \le (y_u-y_l)\text{MISE}_{\frac{\hat{\partial f}}{\partial x_k}} \tag{3.91b}$$

$$= (y_u-y_e)E\int\left(\frac{\partial f}{\partial x_k}(y;x)-\frac{\hat{\partial f}}{\partial x_k}(y;x)\right)^2 dy,$$

and

$$E\Big(D_{\underline{l}}P(x)-\hat{D}_{\underline{l}}P(x)\Big)^2 \le (y_u-y_e)\ \text{MISE}_{\hat{D}_{\underline{l}}f} \tag{3.91c}$$

$$= (y_u-y_e)E\int\Big(D_{\underline{l}}f(y;x)-\hat{D}_{\underline{l}}f(y;x)\Big)^2 dy.$$

Thus, we still have to consider the Mean Integrated Square Error of the estimators $\hat{f}, \frac{\hat{\partial f}}{\partial x_k}, \hat{D}_{\underline{l}}f$ defined by (3.89a-c). Because of

$$f(y;x)-\hat{f}(y;x)=\sum_{j=0}^{q(n)}\Big(c_j(x)-\hat{c}_j(x)\Big)\varphi_j(y)+\sum_{j=q(n)+1}^{\infty}c_j(x)\varphi_j(y)$$

and (3.73e) we get

$$\|f(\cdot;x)-\hat{f}(\cdot;x)\|_2^2=\sum_{j=0}^{q(n)}\Big(c_j(x)-\hat{c}_j(x)\Big)^2+\sum_{j=q(n)+1}^{\infty}c_j^2(x).$$

Hence, with (3.86a) we have

$$\text{MISE}_{\hat{f}}=\sum_{j=0}^{q(n)}\frac{1}{n}\ \text{var}\ \Big(\varphi_j\Big(y\Big(a(\cdot),x\Big)\Big)\Big)+\sum_{j=q(n)+1}^{\infty}c_j^2(x) \tag{3.92a}$$

$$=\frac{1}{n}\sum_{j=0}^{q(n)}E\ \Big(\varphi_j\Big(y\Big(a(\omega),x\Big)\Big)-c_j(x)\Big)^2+\sum_{j=q(n)+1}^{\infty}c_j^2(x).$$

In the same way with (3.86b), (3.86c), resp., we obtain

$$\text{MISE}_{\frac{\hat{\partial f}}{\partial x_k}}=\sum_{j=0}^{q(n)}\frac{1}{n}\ \text{var}\ \left(\varphi_j'\Big(y\Big(a(\cdot),x\Big)\Big)\frac{\partial y}{\partial x_k}\Big(a(\cdot),x\Big)\right)+\sum_{j=q(n)+1}^{\infty}c_{kj}^2(x) \tag{3.92b}$$

$$=\frac{1}{n}\sum_{j=0}^{q(n)}E\ \left(\varphi_j'\Big(y\Big(a(\omega),x\Big)\Big)\frac{\partial y}{\partial x_k}\Big(a(\omega),x\Big)-c_{kj}(x)\right)^2$$

$$+\sum_{j=q(n)+1}^{\infty}c_{kj}^2(x),$$

$$\mathrm{MISE}_{\hat{D}_{\underline{l}}f} = \sum_{j=0}^{q(n)} \frac{1}{n} \operatorname{var}\left(D_{\underline{l}}\varphi_j\left(y\left(a(\omega), x\right)\right)\right) + \sum_{j=q(n)+1}^{\infty} c_{\underline{l}j}^2(x) \tag{3.92c}$$
$$= \frac{1}{n}\sum_{j=0}^{q(n)} E\left(D_{\underline{l}}\varphi_j\left(y\left(a(\omega), x\right)\right) - c_{\underline{l}j}(x)\right)^2 + \sum_{j=q(n)+1}^{\infty} c_{\underline{l}j}^2(x).$$

Using (3.73f) and (3.84a), (3.84b), for $\mathrm{MISE}_{\hat{f}}$ and $\mathrm{MISE}_{\widehat{\frac{\partial f}{\partial x_k}}}$ we find then the bounds

$$\mathrm{MISE}_{\hat{f}} \le d_2^2 \frac{q(n)+1}{n} + \sum_{j=q(n)+1}^{\infty} c_j^2(x), \tag{3.93a}$$

$$\mathrm{MISE}_{\widehat{\frac{\partial f}{\partial x_k}}} \le 2d_2^2 \frac{\left(q(n)+1\right)\left(q(n)+2\right)}{n} E\left(\frac{\partial y}{\partial x_k}\left(a(\omega), x\right)\right)^2 \tag{3.93b}$$
$$+ \sum_{j=q(n)+1}^{\infty} c_{kj}^2(x),$$

and similar bounds can be derived for $\mathrm{MISE}_{\hat{D}_{\underline{l}}f}$. Consequently, we have the following consistency result:

**Theorem 3.6.** *For a given, fixed vector $x$ suppose that $f(\cdot; x) \in L^2(\mathbb{R})$, $\frac{\partial f}{\partial x_k}(\cdot; x) \in L^2(\mathbb{R})$, resp., and the second moments under consideration are finite.*

*a) If $q = q(n)$ is chosen such that $q(n) \to \infty$ and $\frac{q(n)}{n} \to 0$ as $n \to \infty$, then $\mathrm{MISE}_{\hat{f}} \to 0$ and $E\left(P(x) - \hat{P}(x)\right)^2 \to 0$ as $n \to \infty$.*

*b) If $q = q(n)$ is selected such that $q(n) \to \infty$ and $\frac{q(n)^2}{n} \to 0$ as $n \to \infty$, then $\mathrm{MISE}_{\widehat{\frac{\partial f}{\partial x_k}}} \to 0$ and $E\left(\frac{\partial P}{\partial x_k}(x) - \widehat{\frac{\partial P}{\partial x_k}}(x)\right)^2 \to 0$ as $n \to \infty$.*

*Proof.* The assertions follow from (3.91a,b), (3.92a,b), (3.93a,b), resp., and the fact that $\sum_{j=0}^{\infty} c_j^2(x) = \|f(\cdot; x)\|_2^2$, $\sum_{j=0}^{\infty} c_{kj}^2(x) = \left\|\frac{\partial f}{\partial x_k}(\cdot; x)\right\|_2^2$, resp., is a convergent series.

*Remark 3.7.* Similar consistency properties can be obtained also for the estimator $\hat{D}_{\underline{l}}f$, cf. (3.90c).

For the consideration of the convergence rate of the estimators (3.89a-c) and (3.90a-c), depending essentially on the second term in the equations (3.92a-c), we need the following result [128]:

**Lemma 3.5.** *Let $p > 1$ be a given integer. For $g = g(y;x)$ denoting $f(y;x)$, $\frac{\partial f}{\partial x_k}(y;x), D_{\underline{l}} f(y;x)$, resp., suppose that the functions*

$$y \longrightarrow y^i \frac{\partial^{p-1}}{\partial y^{p-1}} g(y;x), i = 0, 1, \ldots, p, \tag{3.94a}$$

*are integrable. Then the function*

$$\tilde{g}^{(p)}(y;x) := e^{\frac{y^2}{2}} \frac{\partial^p}{\partial y^p} \left( e^{-\frac{y^2}{2}} g(y;x) \right) \tag{3.94b}$$

*exists and is integrable, and the Fourier coefficients $b_j(x) = c_j(x), b_j(x) = c_{kj}(x), b_j(x) = c_{\underline{l}j}(x)$, resp., are bounded by*

$$|b_j(x)| \le d_3(p,x) \prod_{l=1}^{p} \Big( 2(j+l) \Big)^{-\frac{1}{2}}, j \ge 1, \tag{3.95a}$$

*where*

$$d_3(p,x) := d_2 \|\tilde{g}^{(p)}(\cdot;x)\|_1. \tag{3.95b}$$

*Proof.* (cf. [128]). Using the relation $H'_{j+1}(y) = 2(j+1)H_j(y)$, by iteration starting from the definition (3.76a-c) we get

$$b_j(x) = (-1)^p \prod_{l=1}^{p} \Big( 2(j+l) \Big)^{-\frac{1}{2}} \int \varphi_{j+p}(y) \tilde{g}^{(p)}(y;x)\, dy$$

which yields the assertion, cf. (3.73f).

If $g(y;x) = f(y;x), g(y;x) = \frac{\partial f}{\partial x_k}(y;x)$, resp., fulfills - for an integer $p > 1$ - the assumptions in the above lemma, then

$$\begin{aligned} \sum_{j=q(n)+1}^{\infty} b_j^2(x) &\le d_3^2(p,x) 2^{-p} \sum_{j=q(n)+1}^{\infty} j^{-p} \\ &\le d_3^2(p,x) 2^{-p} \int_{q(n)}^{+\infty} t^{-p} dt = d_3^2(p,x) 2^{-p} \frac{q(n)^{-(p-1)}}{p-1}. \end{aligned}$$

Consequently, according to (3.93a), (3.93b), resp., we obtain

$$\mathrm{MISE}_{\hat{f}} \le d_2^2 \frac{q(n)+1}{n} + d_3^2(p,x) \frac{2^{-p}}{p-1} q(n)^{-(p-1)} \tag{3.96a}$$

and

$$
\begin{aligned}
\text{MISE}_{\frac{\partial f}{\partial x_k}} \le\; & 2d_2^2 \frac{\Big(q(n)+1\Big)\Big(q(n)+2\Big)}{n} E\left(\frac{\partial y}{\partial x_k}\Big(a(\omega),x\Big)\right)^2 \\
& + d_3^2(p,x)\frac{2^{-p}}{p-1}q(n)^{-(p-1)}.
\end{aligned}
\tag{3.96b}
$$

This yields the following *convergence rates:*

**Theorem 3.7.** *If* $f(\cdot;x)$, $\frac{\partial f}{\partial x_k}(\cdot;x)$, *resp., satisfies the above assumptions with an integer* $p \ge 1$, *then*

$$
\begin{aligned}
E\Big(P(x)-\hat{P}(x)\Big)^2 &\le (y_u - y_l)\left(d_2^2\frac{q(n)+1}{n} + \sum_{j=q(n)+1}^{\infty} c_j^2(x)\right) \\
&\le (y_u - y_l)\left(d_2^2\frac{q(n)+1}{n}\right. \\
&\quad \left. + d_3^2(p,x)\frac{2^{-p}}{p-1}q(n)^{-(p-1)}\right),
\end{aligned}
\tag{3.97a}
$$

*and*

$$
\begin{aligned}
E\left(\frac{\partial P}{\partial x_k}(x) - \frac{\hat{\partial} P}{\partial x_k}(x)\right)^2 &\le (y_u - y_l)\left(2d_2^2\frac{\Big(q(n)+1\Big)\Big(q(n)+2\Big)}{n}\right. \\
&\quad \left. \times E\left(\frac{\partial y}{\partial x_k}\Big(a(\omega),x\Big)\right)^2 + \sum_{j=q(n)+1}^{\infty} c_{kj}^2(x)\right) \\
&\le (y_u - y_l)\left(2d_2^2\frac{\Big(q(n)+1\Big)\Big(q(n)+2\Big)}{n}\right. \\
\times E\left(\frac{\partial y}{\partial x_k}\Big(a(\omega),x\Big)\right)^2 &\left. + d_3^2(p,x)\frac{2^{-p}}{p-1}q(n)^{-(p-1)}\right)
\end{aligned}
\tag{3.97b}
$$

Concerning the selection of $q = q(n)$ we have the following consequence:

**Corollary 3.4.** *Let the assumptions of Theorem 3.7 be fulfilled.*

*a) If* $q(n) = 0(n^{1/p})$, *then* $E\Big(P(x)-\hat{P}(x)\Big)^2 = 0\Big(n^{-(p-1)/p}\Big)$.

*b) If* $q(n) = 0(n^{1/2p})$, *then* $E\left(\frac{\partial P}{\partial x_k}(x) - \frac{\hat{\partial} P}{\partial x_k}(x)\right)^2 = 0(n^{-(p-1)/p})$.

*Proof.* a) The first part is an immediate consequence of (3.97a).

b) If $q(n) = 0(n^{1/2p})$, then $\frac{\Big(q(n)+1\Big)\Big(q(n)+2\Big)}{n} = 0(n^{1/p})$, and the rest follows as in a).

### 3.7.3 The Integrated Square Error (ISE) of Deterministic Approximations

Corresponding to the estimators (3.89a), (3.89b) and (3.90a), (3.90b) of $f(y;x)$, $\frac{\partial f}{\partial x_k}(y;x)$ and $P(x)$, $\frac{\partial P}{\partial x_k}(x)$, resp., *deterministic approximations* of these functions can be defined by

$$\tilde{f}(y;x) := \sum_{j=0}^{q(n)} \tilde{c}_j(x)\varphi_j(y), \tag{3.98a}$$

$$\frac{\tilde{\partial f}}{\partial x_k}(y;x) := \sum_{j=0}^{q(n)} \tilde{c}_{kj}(x)\varphi_j(y) \tag{3.98b}$$

and

$$\tilde{P}(x) := \int_{y_l}^{y_u} \tilde{f}(y;x)\,dy = \sum_{j=0}^{q(n)} \tilde{c}_j(x) \int_{y_l}^{y_u} \varphi_j(y)\,dy, \tag{3.99a}$$

$$\frac{\tilde{\partial P}}{\partial x_k}(x) := \int_{y_l}^{y_u} \frac{\tilde{\partial f}}{\partial x_k}(y;x)\,dy = \sum_{j=0}^{q(n)} \tilde{c}_{kj}(x) \int_{y_l}^{y_u} \varphi_j(y)\,dy, \tag{3.99b}$$

where $\tilde{c}_j(x), \tilde{c}_{kj}(x)$ are the approximative Fourier coefficients defined by (3.87c), (3.88a) and (3.87e), respectively. Thus, for the integrated square error $\text{ISE}_{\tilde{f}}$, $\text{ISE}_{\frac{\tilde{\partial f}}{\partial x_k}}$ of $\tilde{f}(y;x)$, $\frac{\tilde{\partial f}}{\partial x_k}(y;x)$, resp., we get

$$\text{ISE}_{\tilde{f}} := \|f(\cdot;x) - \tilde{f}(\cdot;x)\|_2^2 = \sum_{j=0}^{q(n)} \Big(c_j(x) - \tilde{c}_j(x)\Big)^2 + \sum_{j=q(n)+1}^{\infty} c_j^2(x), \tag{3.100a}$$

and

$$\text{ISE}_{\frac{\tilde{\partial f}}{\partial x_k}} := \|\frac{\partial f}{\partial x_k}(\cdot;x) - \frac{\tilde{\partial f}}{\partial x_k}(\cdot;x)\|_2^2 = \sum_{j=0}^{q(n)} \Big(c_{kj}(x) - \tilde{c}_{kj}(x)\Big)^2 + \sum_{j=q(n)+1}^{\infty} c_{kj}^2(x). \tag{3.100b}$$

We have, cf. (3.91a,b),

$$\left(P(x) - \tilde{P}(x)\right)^2 \leq (y_u - y_l)\ \mathrm{ISE}_{\tilde{f}}, \tag{3.101a}$$

$$\left(\frac{\partial P}{\partial x_k}(x) - \frac{\tilde{\partial} P}{\partial x_k}(x)\right)^2 \leq (y_u - y_l)\ \mathrm{ISE}_{\frac{\tilde{\partial} f}{\partial x_k}}. \tag{3.101b}$$

Thus, bounds for $\mathrm{ISE}_{\tilde{f}}$, $\mathrm{ISE}_{\frac{\tilde{\partial} f}{\partial x_k}}$ and therefore for the error $|P(x) - \tilde{P}(x)|$, $\left|\frac{\partial \Gamma}{\partial x_k}(x) - \frac{\tilde{\partial} \Gamma}{\partial x_k}(x)\right|$ follow, cf. Theorem 3.7, from the error bounds (3.87d) and (3.88b), (3.87f), resp., and using again Lemma 3.5.

## 3.8 Orthogonal Function Series Expansions II: Expansions in Hermite Functions, Case $m > 1$

The results known from the univariate case ($m = 1$) can be transfered to the multivariate case ($m > 1$) by replacing the univariate Hermite functions $\varphi_j$ by the product functions

$$\varphi_\lambda := \prod_{i=1}^{m} \varphi_{\lambda_i}(y_i), y \in \mathbb{R}^m. \tag{3.102a}$$

Here, $\lambda = (\lambda_1, \ldots, \lambda_m)^T \in \mathbb{Z}_+^m$ denotes a multiple index, and $\varphi_{\lambda_i}$ is one of the univariate Hermite functions given by (3.73c). Hence,

$$\varphi_\lambda \in L^1(\mathbb{R}^m) \cap L^2(\mathbb{R}^m) \text{ for all } \lambda \in \mathbb{Z}_+^m, \tag{3.102b}$$

cf. (3.73d), and it is known [126, 127] that

$$(\varphi_\lambda) \text{ forms a complete, orthonormal set in } L^2(\mathbb{R}^m), \tag{3.102c}$$

see (3.73e).

Assume again, cf. (3.74a-c), that

$$f(\cdot; x) \in L^2(\mathbb{R}^m), \quad \frac{\partial f}{\partial x_k}(\cdot; x) \in L^2(\mathbb{R}^m) \tag{3.103a}$$

$$D_{\underline{l}} f(\cdot; x) \in L^2(\mathbb{R}^m). \tag{3.103b}$$

Then the following expansions in series of orthogonal functions in $L^2(\mathbb{R}^m)$ hold:

$$f(y; x) = \sum_{\|\lambda\|^2=0}^{\infty} c_\lambda(x) \varphi_\lambda(y) \tag{3.104a}$$

$$\frac{\partial f}{\partial x_k}(y; x) = \sum_{\|\lambda\|^2=0}^{\infty} c_{k\lambda}(x) \varphi_\lambda(y) \tag{3.104b}$$

$$D_{\underline{l}} f(y; x) = \sum_{\|\lambda\|^2=0}^{\infty} c_{\underline{l}\lambda}(x) \varphi_\lambda(y), \tag{3.104c}$$

cf. (3.75a-c); the corresponding Fourier coefficients are defined, see (3.76a-c), by

$$c_\lambda(x) := \int f(y;x)\varphi_\lambda(y)\,dy \tag{3.105a}$$

$$c_{k\lambda}(x) := \int \frac{\partial f}{\partial x_k}(y;x)\varphi_\lambda(y)\,dy \tag{3.105b}$$

$$c_{\underline{l}\lambda}(x) := \int D_{\underline{l}}f(y;x)\varphi_\lambda(y)\,dy. \tag{3.105c}$$

Obviously, cf. (3.77),

$$c_\lambda(x) = E\varphi_\lambda\left(y\left(a(\omega),x\right)\right), \tag{3.106a}$$

and under assumptions corresponding to (3.78a), (3.82), resp., we get, cf. (3.78b) and (3.83), (3.78c) and (3.85),

$$\begin{aligned} c_{k\lambda} &= \frac{\partial}{\partial x_k}c_\lambda(x) = E\frac{\partial}{\partial x_k}\varphi_\lambda\left(y\left(a(\omega),x\right)\right) && (3.106\text{b})\\ &= \sum_{i=1}^{m} E\prod_{j\neq i}\varphi_{\lambda_j}\left(y_j\left(a(\omega),x\right)\right)\varphi'_{\lambda_i}\left(y_i\left(a(\omega),x\right)\right)\frac{\partial y_i}{\partial x_k}\left(a(\omega),x\right), \end{aligned}$$

$$c_{\underline{l}\lambda}(x) = D_{\underline{l}}c_\lambda(x) = ED_{\underline{l}}\varphi_\lambda\left(y\left(a(\omega),x\right)\right). \tag{3.106c}$$

The inequalities (3.79a-c) can be generalized easily to the present case: Replacing the index "$j$" by the multiple index $\lambda$, we find

$$\begin{aligned} &\left|P(x) - \sum_{\|\lambda\|^2=0}^{n} c_\lambda(x)\int_{y_l}^{y_u}\varphi_\lambda(y)\,dy\right| \\ &\leq \mu_0\left\|f(\cdot;x) - \sum_{\|\lambda\|^2=0}^{n} c_\lambda(x)\varphi_\lambda(\cdot)\right\|_2, && (3.107\text{a})\end{aligned}$$

$$\begin{aligned} &\left|\frac{\partial P}{\partial x_k}(x) - \sum_{\|\lambda\|^2=0}^{n} c_{k\lambda}(x)\int_{y_l}^{y_u}\varphi_\lambda(y)\,dy\right| \\ &\leq \mu_0\left\|\frac{\partial f}{\partial x_k}(\cdot;x) - \sum_{\|\lambda\|^2=0}^{n} c_{k\lambda}(x)\varphi_\lambda(\cdot)\right\|_2, && (3.107\text{b})\end{aligned}$$

$$\begin{aligned} &\left|D_{\underline{l}}P(x) - \sum_{\|\lambda\|^2=0}^{n} c_{\underline{l}\lambda}(x)\int_{y_l}^{y_u}\varphi_\lambda(y)\,dy\right| \\ &\leq \mu_0\left\|D_{\underline{l}}f(\cdot;x) - \sum_{\|\lambda\|^2=0}^{n} c_{\underline{l}\lambda}(x)\varphi_\lambda(\cdot)\right\|_2, && (3.107\text{c})\end{aligned}$$

where $\mu_0 := \left(\prod_{i=1}^{m}(y_{ui} - y_{li})\right)^{1/2}$ with $y_l, y_u \in \mathbb{R}^m$. Hence, we have now the following generalization of Theorem 3.5:

**Theorem 3.8.** *Under the above assumptions, the probability function $P(x)$ and its derivatives can be represented by the following convergent series:*

$$P(x) = \sum_{\|\lambda\|^2=0}^{\infty} c_\lambda(x) \int_{y_l}^{x_u} \varphi_\lambda(y)\, dy, \tag{3.108a}$$

$$\frac{\partial P}{\partial x_k}(x) = \sum_{\|\lambda\|^2=0}^{\infty} c_{k\lambda}(x) \int_{y_l}^{y_u} \varphi_\lambda(y)\, dy, \tag{3.108b}$$

$$D_{\underline{l}}P(x) = \sum_{\|\lambda\|^2=0}^{\infty} c_{\underline{l}\lambda}(x) \int_{y_l}^{y_U} \varphi_\lambda(y)\, dy. \tag{3.108c}$$

Because of

$$\int_{y_l}^{y_u} \varphi_\lambda(y)\, dy = \int_{y_{l_1}}^{y_{u_1}} \cdots \int_{y_{l_m}}^{y_{u_m}} \prod_{i=1}^{m} \varphi_{\lambda_i}(y_i)\, dy_i = \prod_{i=1}^{m} \int_{y_{l_i}}^{y_{u_i}} \varphi_{\lambda_i}(y_i)\, dy_i, \tag{3.109}$$

the integrals in (3.108a-c) can be obtained also by means of the simple relations (3.81a-e) derived for the case $m = 1$.

Since the Fourier coefficients $c_\lambda(x), c_{k\lambda}(x), c_{\underline{l}\lambda}(x)$, resp., are given by the expectations (3.106a) - (3.106c), consistent estimators $\hat{c}_\lambda(x), \hat{c}_{k\lambda}(x), \hat{c}_{\underline{l}\lambda}(x)$ of $c_k(x), c_{k\lambda}(x), c_{\underline{l}\lambda}(x)P(x)$ and therefore also consistent estimators $\hat{P}(x)$, $\frac{\hat{\partial} P}{\partial x_k}(x)$, $\hat{D}_{\underline{l}}P(x)$ of $P(x)$ and its derivatives can be obtained similar to the case $m = 1$.

## 3.9 Orthogonal Function Series Expansions III: Expansions in Trigonometric, Legendre and Laguerre Series

Since the generalization to the case $m \geq 1$ can be carried out relatively easy, see Sections 3.7 and 3.8, in the following we concentrate to the case $m = 1$.

### 3.9.1 Expansions in Trigonometric and Legendre Series

Expansions in trigonometric and Legendre series are suitable if we know in advance that

$$y_0 < y\big(a(\omega), x\big) < y_1 \text{ a.s. for all } x \in \Delta, \tag{3.110a}$$

where $y_0 < y_1$ are given, fixed bounds, and $\Delta$ is an open neighborhood of a point $x_0$ at which the probability function or one of its derivatives should be calculated.

Up to the rate of convergence for estimators based on Hermite expansions, given by Theorem 3.7, the formulas and results from Section 3.7 can be transferred directly to the present situation by replacing the Hermite functions (3.73c) by the *trigonometric functions*

$$\varphi_{c0}(y) := \frac{1}{\sqrt{y_1 - y_0}}, \varphi_{cj}(y) := \left(\frac{2}{y_1 - y_0}\right)^{1/2} \cos\left(j\pi\frac{y - y_0}{y_1 - y_0}\right), \tag{3.110b}$$

$$\varphi_{sj} := \left(\frac{2}{y_1 - y_0}\right)^{1/2} \sin\left(j\pi\frac{y - y_0}{y_1 - y_0}\right), j = 1, 2, \ldots,$$

or by the *Legendre polynomials*

$$\varphi_j(y) := \left(\frac{2j+1}{y_1 - y_0}\right)^{1/2} \frac{1}{j!(y_1 - y_0)} \frac{d^j}{dy^j}\left((y - y_0)(y - y_1)\right), \tag{3.110c}$$

where $y_0 \leq y \leq y_1$. Since the range of $y_x(\omega)$ is bounded here, the assumptions (3.78a) and (3.82) can be weakened considerably!

### 3.9.2 Expansions in Laguerre Series

Expansions in Laguerre series are of interest if it is known a priori that $y_x(\omega)$ is bounded from one side. In the case

$$y_x(\omega) = y\big(a(\omega), x\big) \geq 0 \text{ a.s.} \tag{3.111a}$$

the corresponding sequence of *Laguerre functions* reads

$$\varphi_j(y) = C_j e^{\frac{y}{2}} y^{-\frac{\alpha}{2}} \frac{d^j}{dy^j}(y^{j+\alpha}e^{-y}), y \geq 0, j = 0, 1, \ldots, \tag{3.111b}$$

where $\alpha > -1$ is a given, fixed parameter, and $C_j, j = 0, 1, \ldots,$ are normalizing constants. Corresponding to Section 3.9.1, the formulas and results from Section 3.7, up to Theorem 3.7, can be transferred directly to this case by replacing the Hermite functions (3.73c) by the Laguerre functions (3.111b).

# Part III

# Deterministic Descent Directions

# 4

# Deterministic Descent Directions and Efficient Points

## 4.1 Convex Approximation

According to Section 1.2 and 2.1 we consider here deterministic substitute problems of the type

$$\min F(x) \text{ s.t. } x \in D \tag{4.1a}$$

with the convex feasible domain $D$ and the expected total cost function

$$F(x) := EG_0\Big(a(\omega), x\Big) + \Gamma(x), \tag{4.1b}$$

where

$$\Gamma(x) := E\gamma\left(y\Big(a(\omega), x\Big)\right). \tag{4.1c}$$

Here, $G_0 = G_0(a, x)$ is the primary cost function (e.g. setup, construction costs), and $y = y(a, x)$ is the $m_y$–vector of state functions of the underlying structure/system/process. The function $G_0 = G_0(a, x)$ and the vector function $y = y(a, x)$ depend on the $\nu$–vector $a$ of random model parameters and the $r$–vector $x$ of decision/design variables. Moreover, $\gamma = \gamma(y)$ is a loss function evaluating violations of the operating conditions $y(a, x) \in B$, see (2.1g). We assume in this chapter that the expectations in (4.1b,c) exist and are finite. Moreover, we suppose that the gradient of $F(x), \Gamma(x)$ exists, and can be obtained, see formulas (2.23b,c) by interchanging differentiation and expectation:

$$\nabla F(x) = E\nabla_x G_0\Big(a(\omega), x\Big) + \nabla\Gamma(x) \tag{4.2a}$$

with

$$\nabla\Gamma(x) = E\nabla_x y\Big(a(\omega), x\Big)^T \nabla\gamma\left(y\Big(a(\omega), x\Big)\right). \tag{4.2b}$$

The main goal of this chapter is to present several methods for the construction of i) deterministic *descent directions* $h = h(x)$ for $F(x), \Gamma(x)$ at certain points $x \in \mathbb{R}^r$, and ii) so–called *efficient points* $x^\circ$ of the stochastic optimization problem (4.1a-c). Note that a descent direction for the function $F$ at a point $x$ is a vector $h = h(x)$ such that

$$F(x + th) < F(x) \text{ for } 0 < t \leq t_0$$

with a constant $t_0 > 0$. The descent directions $h = h(x)$ to be constructed here depend not explicitly on the gradient $\nabla F, \nabla \Gamma$ of $F, \Gamma$, but on some structural properties of $F, \Gamma$. Moreover, efficient points $x^\circ \in D$ are feasible points of (4.1a-c) not admitting any feasible descent directions $h(x^\circ)$ at $x^\circ$. Hence, efficient points $x^\circ$ are candidates for optimal solutions $x^*$ of (4.1a-c). For the construction of a descent direction $h = h(x)$ at a point $x = x_0$ the objective function $F = F(x)$ is replaced first [69] by the mean value function

$$F_{x_0}(x) := E\left(G_0\Big(a(\omega), x_0\Big) + \nabla_x G_0\Big(a(\omega), x_0\Big)^T (x - x_0)\right) + \Gamma_{x_0}(x), \quad (4.3a)$$

with

$$\Gamma_{x_0}(x) := E\gamma\left(y\Big(a(\omega), x_0\Big) + \nabla_x y\Big(a(\omega), x_0\Big)(x - x_0)\right). \quad (4.3b)$$

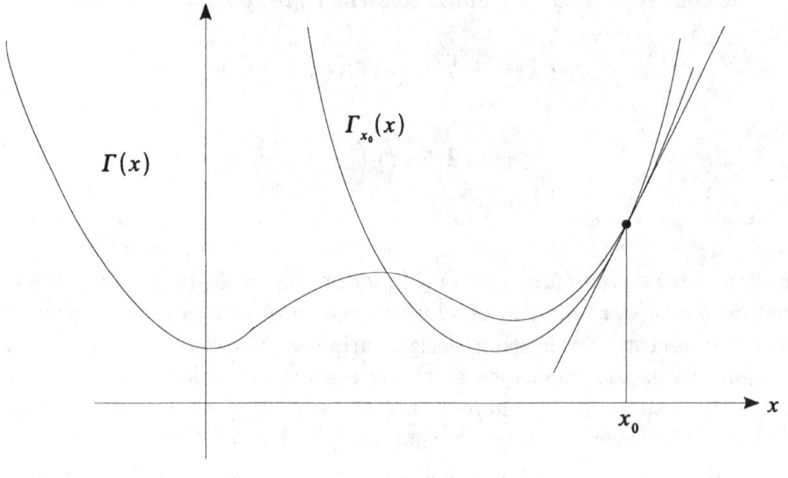


**Fig. 4.1.** Convex approximation $\Gamma_{x_0}$ of $\Gamma$ at $x_0$

Obviously, $F_{x_0}$ follows from $F$ by linearization of the primary cost function $G_0 = G_0\Big(a(\omega), x\Big)$ with respect to $x$ at $x_0$, and by "inner linearization" of the expected recourse cost function $\Gamma = \Gamma(x)$ at $x_0$.

Corresponding to (4.2b), the gradient $\nabla F_{x_0}(x)$ is given by

$$\nabla F_{x_0}(x) = \nabla_x G_0\Big(a(\omega, x_0\Big) + \nabla \Gamma_{x_0}(x) \quad (4.3c)$$

with

$$\nabla \Gamma_{x_0}(x) = E\nabla_x y\Big(a(\omega), x_0\Big)^T \nabla\gamma\left(y\Big(a(\omega), x_0\Big) + \nabla_x y\Big(a(\omega), x_0\Big)(x - x_0)\right). \tag{4.3d}$$

Under the general assumptions concerning $F$ and $\Gamma$, the following basic properties of $F_{x_0}$ hold:

**Theorem 4.1.** *I) If $\gamma$ is a convex cost function, then $\Gamma_{x_0}$ and $F_{x_0}$ are convex functions for each $x_0$.*
*II) At point $x_0$ the following equations hold:*

$$\Gamma_{x_0}(x_0) = \Gamma(x_0), F_{x_0}(x_0) = F(x_0) \tag{4.4a}$$
$$\nabla \Gamma_{x_0}(x_0) = \nabla \Gamma(x_0), \nabla F_{x_0}(x_0) = \nabla F(x_0). \tag{4.4b}$$

*III) Suppose that $\gamma$ fulfills the Lipschitz condition*

$$\Big|\gamma(y) - \gamma(z)\Big| \leq L\|y - z\|, y, z \in B \tag{4.4c}$$

*with a constant $L > 0$ and a convex set $B \subset \mathbb{R}^{m_y}$. If $y = y(a, x)$ and its linearization at $x_0$ fulfill*

$$y\Big(a(\omega), x\Big) \in B,\ y\Big(a(\omega), x_0\Big) + \nabla_x y\Big(a(\omega), x_0\Big)(x - x_0) \in B \ a.s. \tag{4.4d}$$

*for a point $x$, then*

$$|\Gamma(x) - \Gamma_{x_0}| \leq L\|x - x_0\| \cdot E \sup_{0 \leq \lambda \leq 1} \Big\|\nabla_x y\Big(a(\omega), \lambda x + (1 - \lambda)x_0\Big)$$
$$-\nabla_x y\Big(a(\omega), x_0\Big)\Big\|. \tag{4.4e}$$

*Proof.* a) The first term of $F_{x_0}$ is linear in $x$, and also the argument of $\gamma$ in $\Gamma_{x_0}$ is linear in $x$. Hence, $\Gamma_{x_0}, F_{x_0}$ are convex for each convex loss function $\gamma$ such that the expectations exist and are finite. b) The equations in (4.4a,b) follow from (4.1b,c), (4.3a,b) and from (4.2a,b), (4.3c,d) by putting $x = x_0$. The last part follows by applying the mean value theorem for vector valued functions, cf. [27].

*Remark 4.1.* a) If $\gamma = \gamma(y)$ is a convex function on $\mathbb{R}^{m_y}$, then (4.4c) holds on each bounded closed subset $B \subset \mathbb{R}^{m_y}$ with a local Lipschitz constant $L = L(B)$. b) An important class of convex loss functions $\gamma$ having a global Lipschitz constant $L = L_\gamma$ are the ***sublinear*** functions on $\mathbb{R}^{m_y}$. A function $\gamma$ is called sublinear if it is positive homogeneous and subadditive, hence,

$$\gamma(\lambda y) = \lambda\gamma(y), \lambda \geq 0, y \in \mathbb{R}^{m_y} \tag{4.5a}$$
$$\gamma(y + z) \leq \gamma(x) + \gamma(z), y, z \in \mathbb{R}^{m_y}. \tag{4.5b}$$

Obviously, a sublinear function is convex. Furthermore, defining the norm $\|\gamma\|$ of a sublinear function $\gamma$ on $\mathbb{R}^{m_y}$ by

$$\|\gamma\| := \sup\Big\{|\gamma(y)| : \|y\| \le 1\Big\}, \tag{4.5c}$$

we have

$$\Big|\gamma(y)\Big| \le \|\gamma\| \cdot \|y\|, y \in \mathbb{R}^{m_y} \tag{4.5d}$$

$$\Big|\gamma(y) - \gamma(z)\Big| \le \|\gamma\| \cdot \|y - z\|, y, z \in \mathbb{R}^{m_y}. \tag{4.5e}$$

Further properties of sublinear functions may be found in [69].

In the following we suppose now that in (4.1b,c) differentiation and expectation can be interchanged. Using Theorem 4.1, for the construction of descent directions of $F$ at a point $x_0$ we may proceed as follows:

**Theorem 4.2.** *a) Let $x_0 \in \mathbb{R}^r$ be a given point. If $h = h(x_0)$ is a descent direction for $F_{x_0}$ at $x_0$, then $h$ is also a descent direction for $F$ at $x_0$. b) Suppose that $z$ is a vector such that $F_{x_0}(z) < F_{x_0}(x_0)$. Then $h = z - x_0$ is a descent direction of $F_{x_0}$ at $x_0$. c) If $\Gamma_{x_0}(z) = \Gamma_{x_0}(x_0)$ for $z \ne x_0$, and $\Gamma_{x_0}$ is not constant on the line segment $\overline{x_0 z} := \Big\{\lambda z + (1-\lambda)x_0 : 0 \le \lambda \le 1\Big\}$, then $h = z - x_0$ is a descent direction of $\Gamma_{x_0}$ at $x_0$.*

*Proof.* a) Using (4.4b), we have $\nabla F(x_0)^T h = \nabla F_{x_0}(x_0)^T h < 0$. Hence, $h$ is also a descent direction of $F$ at $x_0$. b) Since $F_{x_0}$ is a differentiable, convex function on $\mathbb{R}^{m_y}$, we obtain

$$\nabla F_{x_0}(x_0)^T h = \nabla F_{x_0}(x_0)^T (z - x_0) \le F_{x_0}(z) - F_{x_0}(x_0) < 0.$$

Thus, $h$ is a descent direction of $F_{x_0}$ at $x_0$. c) In this case there is a vector $w = \lambda z + (1-\lambda)x_0 = x_0 + \lambda(z - x_0), 0 < \lambda < 1$, such that $\Gamma_{x_0}(w) < \Gamma_{x_0}(x_0)$. Thus,

$$\nabla \Gamma_{x_0}(x_0)^T \lambda h = \nabla \Gamma_{x_0}(x_0)^T (w - x_0) \le \Gamma_{x_0}(w) - \Gamma_{x_0}(x_0) < 0$$

and therefore $\nabla \Gamma_{x_0}(x_0)^T h < 0$.

Deterministic descent directions $h$ of $F$ at $x_0$ can be obtained then in two ways:

*I) Separate linear inequality:*

**Corollary 4.1.** *For a given point $x_0 \in \mathbb{R}^r$, suppose that there is a vector $z \in \mathbb{R}^r$ such that*

$$E\nabla_x G_0\Big(a(\omega), x_0\Big)^T (z - x_0) < (\le) 0 \tag{4.6a}$$

$$\Gamma_{x_0}(z) < (=) \Gamma_{x_0}(x_0), \tag{4.6b}$$

*where in case "=" the function $\Gamma_{x_0}$ is not constant on $\overline{x_0 z}$, or (4.6a) holds with "<". Then $h = z - x_0$ is a descent direction for $F$ at $x_0$.*

*Proof.* The assertion follows from

$$F_{x_0}(z) - F_{x_0}(x_0) = E\nabla_x G_0\Big(a(\omega), x_0\Big)^T (z - x_0) + \Gamma_{x_0}(z) - \Gamma_{x_0}(x_0).$$

*II) Enlarged loss function:*
Defining the enlarged state function $\tilde{y} = \tilde{y}(a, x)$ and the enlarged loss function $\tilde{\gamma} = \tilde{y}(\tilde{\gamma})$ by

$$\tilde{y}(a, x) := \begin{pmatrix} G_0(a, x) \\ y(a, x) \end{pmatrix}, \tilde{\gamma}(\tilde{y}) = \tilde{\gamma}\begin{pmatrix} t \\ y \end{pmatrix} := t + \gamma(y), \tag{4.7a}$$

we have

$$F(x) = E\tilde{\gamma}\Big(\tilde{y}\big(a(\omega), x\big)\Big). \tag{4.7b}$$

Deterministic descent directions may be found then by applying Theorem 4.1 and 4.2 directly to the representation (4.7a,b) of $F$.

Obviously, the convex approximation $F_{x_0}$ of $F$ at $x_0$ can be represented by

$$F_{x_0}(x) = c_0 + \bar{c}^T x + E\gamma\Big(A(\omega)x - b(\omega)\Big), \tag{4.8a}$$

where

$$c_0 := EG_0\Big(a(\omega), x_0\Big) - E\nabla_x G_0\Big(a(\omega), x_0\Big)^T x_0 \tag{4.8b}$$

$$\bar{c} := E\nabla_x G_0\Big(a(\omega), x_0\Big). \tag{4.8c}$$

Furthermore, $\Big(A(\omega), b(\omega)\Big)$ is the random $m_y(r+1)$ matrix given by

$$A(\omega) := \nabla_x y\Big(a(\omega), x_0\Big) \tag{4.8d}$$

$$b(\omega) := -\Big(y\Big(a(\omega), x_0\Big) - \nabla_x y\Big(a(\omega), x_0\Big)x_0\Big). \tag{4.8e}$$

### 4.1.1 Approximative Convex Optimization Problem

According to the above results, the construction of deterministic descent directions of $F(x) = EG_0\Big(a(\omega), x\Big) + E\gamma\Big(y\Big(a(\omega), x\Big)\Big)$ at a point $x_0$, as well as the approximate solution of the (nonconvex) optimization problem (4.1a-c) can be reduced to a convex optimization problem

$$\min F(x) \text{ s.t. } x \in D \tag{4.9a}$$

with the convex objective

$$F(x) := \bar{c}^T x + Eu\Big(A(\omega)x - b(\omega)\Big). \tag{4.9b}$$

Here, $\bar{c}$ is a given, fixed $r$–vector of mean cost coefficients, and $\big(A(\omega), b(\omega)\big)$ is a random $m \times (r+1)$ matrix. Furthermore, $u = u(y)$ is a convex loss function such that the expectation in (4.9b) exists and is finite for all $x$ under consideration. We still suppose that $u$ has the following partial monotonicity property:

$$y_I \leq z_I, y_{II} = z_{II} \Rightarrow u(y) \leq u(z) \tag{4.10a}$$

$$y_I \leq z_I, y_I \neq z_I, y_{II} = z_{II} \Rightarrow u(y) < u(z), \tag{4.10b}$$

where $y, z \in \mathbb{R}^m, y_I := (y_i)_{i \in J}, y_{II} := (y_i)_{i \notin J}$ with a certain index set $J \subset \{1, \ldots, m\}$. Note that inequalities $a \leq b, a < b$ for vectors $a, b \in \mathbb{R}^m$ are defined componentwise.

The construction of descent directions for $F$ at a point $x_0$ depends on the type of the probability distribution of $\big(A(\omega), b(\omega)\big)$ and the functional properties of $u$. Thus, we need the following definitions:

**Definition 4.1.** *The loss function $u$ is called (partly) monotonous increasing or strictly (partly) monotonous increasing if $u$ fulfills (4.10a), (4.10b), respectively. Let $C^J, C^{JJ}$ denote the set of all convex functions $u$ on $\mathbb{R}^m$ fulfilling (4.10a), (4.10b), respectively. Moreover, let $\hat{C}^J, \hat{C}^{JJ}$ be the set of all strictly convex functions fulfilling (4.10a), (4.10b) respectively.*

In order to guarantee that the expectations under consideration exist and are finite, we take loss functions $u$ from the following subclasses:

**Definition 4.2.** *Let $C^J(P), C^{JJ}(P), \hat{C}^J(P), \hat{C}^{JJ}(P)$ designate the set of all $u \in C^J, C^{JJ}, \hat{C}^J, \hat{C}^{JJ}$ resp., such that the expectation, the integral with respect to the probability measure $P$,*

$$F_0(x) := Eu\Big(A(\omega)x - b(\omega)\Big) = \int u\Big(A(\omega)x - b(\omega)\Big)P(d\omega) \tag{4.11}$$

*exists and is finite for all $x \in \mathbb{R}^r$. Finally, let $C^J_{sep}, C^{JJ}_{sep}, \hat{C}^J_{sep}, \hat{C}^{JJ}_{sep}$ and $C^J_{sep}(P), C^{JJ}_{sep}(P), \hat{C}^J_{sep}(P), \hat{C}^{JJ}_{sep}(P)$ denote the subset of all separable elements (i.e. $u(z) := \sum_{i=1}^{m} u_i(z_i)$) of $C^J, \ldots, \hat{C}^{JJ}$ and $C^J(P), \ldots, \hat{C}^{JJ}(P)$, respectively.*

Note that $C^J \supset C^{JJ} \supset \hat{C}^{JJ}$ and $C^J \supset \hat{C}^J \supset \hat{C}^{JJ}$, where corresponding inclusions hold also for the subsets of separable or integrable $u$.

Solving now problem (4.9a,b), besides the well known difficulties in minimizing mean value functions, the loss function $u$ should be exactly known. However, in practice there is always some uncertainty about the appropriate selection of $u$, e.g. due to difficulties in assigning appropriate penalty costs $u(z)$ to the deviation $z = A(\omega)x - b(\omega)$ between the output $A(\omega)x$ of the system $x \to A(\omega)x$ and the target or upper bound $b(\omega)$. Consequently, in the following we want to modify [67,68,72,74,75,81,84] the approximate problem

(4.9a,b) in such a way that we need only some obvious functional properties (e.g. monotonicity, convexity) of $u$. Using certain moments of the random vector $\tilde{A}(\omega)x - \tilde{b}(\omega) := \begin{pmatrix} c(\omega)'x \\ A(\omega)x - b(\omega) \end{pmatrix}$, we find new substituting problems for (4.9a,b) and then also for (4.1a-c) having the following properties: i) Only certain moments of $\left(\tilde{A}(\omega), \tilde{b}(\omega)\right)$ are needed and ii) compromise solutions, called efficient solutions of (4.1a-c), are obtained which are valid for a large class of loss functions $u$.

## 4.2 Computation of Descent Directions in Case of Normal Distributions

In the following we suppose that the random $m \times (n+1)$ matrix $\left(A(\omega), b(\omega)\right)$ has a normal distribution with mean

$$(\bar{A}, \bar{b}) := E\left(A(\omega), b(\omega)\right) \tag{4.12a}$$

and covariance matrix

$$Q = \operatorname{cov}\left(A(\cdot), b(\cdot)\right) = (Q_{ij})_{i,j=1,\ldots,m}. \tag{4.12b}$$

The $(n+1) \times (n+1)$ matrix

$$Q_{ij} := \operatorname{cov}\left(\left(A_i(\cdot), b_i(\cdot)\right), \left(A_j(\cdot), b_j(\cdot)\right)\right) = Q_{ji}, \tag{4.12c}$$

denotes the covariance matrix of the $i$th and the $j$th row $(\left(A_i(\omega), b_i(\omega)\right)$, $\left(A_j(\omega), b_j(\omega)\right)$ of $\left(A(\omega), b(\omega)\right)$. Consequently, $\delta(\omega, x) := A(\omega)x - b(\omega)$ is normally distributed with mean $\bar{A}x - \bar{b}$ and covariance matrix

$$Q_x := \operatorname{cov}\left(A(\cdot)x - b(\cdot)\right) = \left(g_{ij}(x)\right)_{i,j=1,\ldots,m}. \tag{4.13a}$$

The functions $g_{ij}(x)$ are given by

$$g_{ij}(x) = \hat{x}^T Q_{ij} \hat{x} = \hat{x}^T Q_{ji} \hat{x} = \hat{x}^T \frac{Q_{ij} + Q_{ji}}{2} \hat{x} \text{ with } \hat{x} := \begin{pmatrix} x \\ -1 \end{pmatrix}; \tag{4.13b}$$

especially, the variance of $\delta_i(\omega, x) = A_i(\omega)x - b_i(\omega)$ is given by

$$g_{ii}(x) = \operatorname{var}\left(A(\cdot)x - b_i(\cdot)\right) = \hat{x}^T Q_{ii} \hat{x}. \tag{4.13c}$$

Derivative–free descent directions of the objective function $F$ can be obtained, as developed in [74, 75, 81], by using the relations given below. For sake of simplicity, we assume first that $u \in C^J_{sep}(P)$. In this case, with $\bar{A}_I := (\bar{A}_i)_{i \in J}, \bar{A}_{II} := (A_i)_{i \notin J}$, the relations read:

$$\bar{c}^T x \geq \bar{c}^T y \tag{4.14a}$$

$$\bar{A}_I x \geq \bar{A}_I y,\ \bar{A}_{II} x = \bar{A}_{II} y \qquad (4.14b),\ (4.14c)$$

$$g_{ii}(x) \geq g_{ii}(y),\ i = 1, \ldots, m. \tag{4.14d}$$

**Theorem 4.3.** *a) If any vectors $x, y \in \mathbb{R}^r$ fulfill relations (4.14a-d), then $F(x) \geq F(y)$ for each $u \in C^J_{sep}(P)$, b) Let $x, y \in \mathbb{R}^r$ be related according to (4.14a-d), then $F(x) > F(y)$ for each $u \in C^J_{sep}(P), C^{JJ}_{sep}(P), \hat{C}^J_{sep}(P)$, resp., if still the following additional condition holds:*

$$\bar{c}^T x > \bar{c}^T y \tag{4.14a'}$$

$$\bar{A}_i x > \bar{A}_i y \text{ for some } i \in J \tag{4.14b'}$$

$$g_{ii}(x) > g_{ii}(y) \text{ for some } i, 1 \leq i \leq m. \tag{4.14d'}$$

*Proof.* According to the assumptions, $u_i$ is a (strictly) monotoneous nondecreasing, (strictly) convex function for each $i \in J$. Furthermore, the objective $F$ reads:

$$F(x) = c^T x + \sum_{i=1}^{m} Eu_i\Big(\bar{A}_i x - \bar{b}_i + g_{ii}(x)\xi_i(\omega)\Big),$$

where $\xi_i = \xi_i(\omega)$ is a standard normal distributed random variable. This representation of $F$ yields then the assertion $F(x) \geq (>) F(y)$ in case of the relations (4.14a-d), with "=" in (4.14d), and (4.14a',b'). The proof for the general case (4.14d) and case (4.14d') is contained in the proof of the next theorem.

*Remark 4.2.* a) Obviously, relations (4.14a-d), (4.14a',b',d'), involving only first and second order moments of $\Big(A(\omega), b(\omega), c(\omega)\Big)$, are linear or quadratic in $y$, b) The constraint "$x \in D$" in (4.1a) is taken into account by adding to (4.14a-d) the condition

$$y \in D, \tag{4.14e}$$

where in the following $D$ may be any convex subset of $\mathbb{R}^r$.

For more general loss functions $u \in C^J(P), C^{JJ}(P), \hat{C}^J(P), \hat{C}^{JJ}(P)$, resp., relations (4.14a-e) must be replaced by the following stronger conditions:

$$\bar{c}^T x \geq \bar{c}^T y \tag{4.15a}$$

$$\bar{A}_I x \geq \bar{A}_I y, \bar{A}_{II} x = \bar{A}_{II} y \qquad (4.15b),\ (4.15c)$$

$$Q_x \succsim Q_y (\text{i.e. } Q_x - Q_y \text{ is positive semidefinite}) \tag{4.15d}$$

$$y \in D. \tag{4.15e}$$

Corresponding to Theorem 4.3, here we know the following result [74], [75]:

**Theorem 4.4.** *a) If $r$-vectors $x, y$ fulfill relations (4.15a-d), then $F(x) \geq F(y)$ for each $u \in C^J(P)$. b) Let $x, y \in \mathbb{R}^r$ be related according to (4.15a-d). Then $F(x) > F(y)$ for each $u \in C^J(P), C^{JJ}(P), \hat{C}^J(P)$, resp., if still the following additional condition holds:*

$$\bar{c}^T x > \bar{c}^T y \tag{4.15a'}$$
$$\bar{A}_i x > \bar{A}_i y \text{ for some } i \in J \tag{4.15b'}$$
$$Q_x \succsim Q_y, Q_x \neq Q_y. \tag{4.15d'}$$

*Proof.* According to (4.9b), (4.11) we have $F(x) = \bar{c}^T x + F_0(x)$ with $F_0(x) = Eu\Big(A(\omega)x - b(\omega)\Big)$, where $u \in C^J(P), C^{JJ}(P), \hat{C}^J(P)$, respectively. Thus, if (4.15a), (4.15a'), resp., holds, then $F(x) \geq (>) \bar{c}^T y + F_0(x)$, and in the rest of the proof we have to consider $F_0$ only. Defining

$$Y_x = Y_x(\omega) := A(\omega)x - b(\omega) - (\bar{A}x - \bar{b}) = \Big(A(\omega) - \bar{A}\Big)x - \Big(b(\omega) - \bar{b}\Big),$$

we find that

$$F_0(x) = Eu\Big(\bar{A}x - \bar{b} + Y_x(\omega)\Big) \geq (>) Eu\Big(\bar{A}y - \bar{b} + Y_x(\omega)\Big) =: \tilde{F}_0,$$

provided that (4.15b), (4.15b') holds and $u \in C^J(P), u \in C^{JJ}(P)$, respectively. Since $Y_x(\omega)$ is a $N(0, Q_x)$-normal distributed random $m$-vector, the characteristic functions $\hat{P}_x^{(0)}, \hat{P}_y^{(0)}$ of the distributions $P_x^{(0)}, P_y^{(0)}$ of $Y_x, Y_y$, resp., are given by

$$\hat{P}_x^{(0)}(z) := E\exp\Big(iY_x(\omega)^T z\Big) = \exp\left(-\tfrac{1}{2}z^T Q_x z\right),$$
$$\hat{P}_y^{(0)}(z) := E\exp\Big(iY_y(\omega)^T z\Big) = \exp\left(-\tfrac{1}{2}z^T Q_y z\right), z \in \mathbb{R}^m.$$

Supposing now that (4.15d), (4.15d'), resp., holds, we consider the characteristic function

$$\hat{K}(z) := \exp\left(-\frac{1}{2}z^T(Q_x - Q_y)z\right), \quad z \in \mathbb{R}^m,$$

of the $N(0, Q_x - Q_y)$-normal distribution $K$ on $\mathbb{R}^m$. We find

$$\hat{P}_y^{(0)}(z)\hat{K}(z) = \hat{P}_x^{(0)}(z), \ z \in \mathbb{R}^m,$$

hence, $P_x^{(0)}$ is the convolution $P_x^{(0)} = K * P_y^{(0)}$ of $K$ and $P_y^{(0)}$. Defining $\tilde{u}(z) := u(\bar{A}y - \bar{b} + z), z \in \mathbb{R}^m$, we obtain

$$\tilde{F}_0 = E\tilde{u}\Big(Y_x(\omega)\Big) = \int \tilde{u}(z)P_x^{(0)}(dz) = \int \tilde{u}(z)(K * P_y^{(0)})(dz)$$
$$= \int u(v + w)K(dw)P_y^{(0)}(dv) = \int J(v)P_y^{(0)}(dv),$$

where $J(v) := \int \tilde{u}(v+w)K(dw), v \in \mathbb{R}^m$. For any $u \in \mathcal{C}^J(P)$ we have that

$$\tilde{u}(v+w) - \tilde{u}(v) \geq g^T(v+w-v) = g^T w,\ w \in \mathbb{R}^m,$$

for all $v \in \mathbb{R}^m$ and a subgradient [69] $g \in \partial\tilde{u}(v)$. Consequently,

$$J(v) \geq \int \left(\tilde{u}(v) + g^T w\right) K(dw) = \tilde{u}(v) \text{ for all } v \in \mathbb{R}^m,$$

and therefore

$$\tilde{F}_0 = \int J(v) P_y^{(0)}(dv) \geq \int \tilde{u}(v) P_y^{(0)}(dv) = F_0(y).$$

If $u \in \hat{\mathcal{C}}^J(P)$, then

$$\tilde{u}(v+w) - \tilde{u}(v) > g^T w,\ w \neq 0,$$

for all $v \in \mathbb{R}^m$ and a subgradient $g \in \partial\tilde{u}(v)$. In case of (4.15d') we get

$$J(v) = \int \tilde{u}(v+w)K(dw) > \int \left(\tilde{u}(v) + g^T w\right) K(dw) = \tilde{u}(v) \text{ for all } v \in \mathbb{R}^m,$$

and therefore

$$\tilde{F}_0 = \int J(v) P_y^{(0)}(dv) > \int \tilde{u}(v) P_y^{(0)}(dv) = F_0(y),$$

which concludes the proof now.

There are several sufficient conditions for (4.15d) implying (4.14d). Using e.g. Gersgorin's circle theorem, we find that (4.15d), (4.15d'), is implied by

$$g_{ii}(x) - g_{ii}(y) \geq \sum_{\substack{j=1 \\ j \neq i}}^{m} \left| g_{ij}(x) - g_{ij}(y) \right|,\ i = 1, 2, \ldots, m, \tag{4.16a}$$

$$\text{(4.16a) holds with } g_{ii}(x) > g_{ii}(y) \text{ at least once,} \tag{4.16a'}$$

respectively. Clearly, (4.16a') implies (4.14d').

Though, for practical purposes, (4.16a,a'), with given $x$, can be solved for $y$ by certain search techniques, for analytical considerations we need the following representation of (4.16a): For each $i = 1, \ldots, m$, let $\sum_i$ designate the set of all $(m-1)$–vectors $\sigma_i = (\sigma_{i1}, \ldots, \sigma_{ii-1}, \sigma_{ii+1}, \ldots, \sigma_{im})^T$ such that $\sigma_{ij} \in \{-1, +1\}$ for all $j \neq i$. Obviously, (4.16a), (4.16a') is satisfied if and only if

$$g_{i\sigma_i}(x) \geq g_{i\sigma_i}(y) \text{ for all } \sigma_i \in \sum\nolimits_i, i = 1, \ldots, m, \tag{4.16b}$$

$$\text{(4.16b) holds with } g_{ii}(x) > g_{ii}(y) \text{ for some } 1 \leq i \leq m, \tag{4.16b'}$$

resp., where, cf. (4.13b,c),

$$g_{i\sigma_i}(x) := g_{ii}(x) - \sum_{j\neq i} \sigma_{ij} g_{ij}(x) = \hat{x}^T \Big( Q_{ii} - \sum_{j\neq i} \sigma_{ji} Q_{ij} \Big) \hat{x}$$

$$= \hat{x}^T \left( Q_{ii} - \sum_{j\neq i} \sigma_{ij} \frac{Q_{ij} + Q_{ji}}{2} \right) \hat{x}. \quad (4.16c)$$

With a given vector $x$, condition (4.16b), (4.16b'), resp., may contain at most $m2^{m-1}$ inequalities for $y$. However, in practice (4.16b), (4.16b') contain much fewer constraints for $y$, since mostly some of the covariance matrices $Q_{ij}(= Q_{ji}^T), j \neq i$, are zero.

A very interesting situation occurs certainly if for a given $i, 1 \leq i \leq m$,

$$Q_{ii}(\sigma_i) := Q_{ii} - \sum_{j\neq i} \sigma_{ij} \frac{Q_{ij} + Q_{ji}}{2} \text{ is positive (semi)definite for all } \sigma_i \in \sum\nolimits_i . \quad (4.16d)$$

If (4.16d) holds, then cf. (4.16c),

$$g_{i\sigma_i}(y) = \hat{y}^T Q_{ii}(\sigma_i) \hat{y}$$

is (strictly) convex for each $\sigma_i \in \sum_i$, see (4.16b).

### 4.2.1 Descent Directions of Convex Programs

If $n$–vectors $x, y$ are related such that $F(x) \geq F(y), y \neq x$, then $h = y - x$ is a descent direction for $F$ at $x$, provided that $F(x) \neq F(x)$ or $F$ is not constant on the line segment $xy = \{\lambda x + (1-\lambda)y : 0 \leq \lambda \leq 1\}$ joining $x$ and $y$. This suggests the following definition:

**Definition 4.3.** $C^J(P, D) := \{u \in C^J(P) : F(x) = \bar{c}^T x + Eu\big(A(\omega)x - b(\omega)\big)$ *is not constant on arbitrary line segments* $xy$ *of* $D\}$. $C^J_{sep}(P, D) := C^J_{sep}(P) \cap C^J(P, D)$.

Obviously, we have

$$F(x) = F_u(x) := Ev \begin{pmatrix} c(\omega)^T x \\ A(\omega)x - b(\omega) \end{pmatrix} \text{ with } v \begin{pmatrix} t \\ z \end{pmatrix} := t + u(z). \quad (4.17a)$$

By Lemma 2.2 in [75] we know then that $F$ is constant on a line segment $xy, y \neq x$, if and only if $F(x) = F(y)$ and ("a.s." means with probability one)

$$u\Big(\lambda\big(A(\omega)x - b(\omega)\big) + (1-\lambda)\big(A(\omega)y - b(\omega)\big)\Big)$$
$$= \lambda u\big(A(\omega)x - b(\omega)\big) + (1-\lambda) u\big(A(w)y - b(w)\big) \text{ a.s., } 0 \leq \lambda \leq 1. \quad (4.17b)$$

For convex functions $u$ being strictly convex with respect to a certain variable $z_\iota, 1 \leq \iota \leq m$, (4.17b) has the following consequences:

**Lemma 4.1.** *Let $u \in C^J(P)$ be a loss function such that for some index $1 \le \iota \le m$*

$$u\Big(\lambda z + (1-\lambda)w\Big) < \lambda u(z) + (1-\lambda)u(w) \text{ if } z_\iota \ne w_\iota \text{ and } 0 < \lambda < 1. \quad (4.17c)$$

*If (4.17b) holds, then $A_\iota(\omega)(y-x) = 0$ a.s. and $R_{\iota\iota}(y-x) = 0$, where $R_{\iota\iota}$ denotes the covariance matrix of $A_\iota(\omega)$.*

**Corollary 4.2.** *Let $u \in C^J(P)$ be a loss function fulfilling (4.17c). If $R_{\iota\iota}$ is regular for some $1 \le \iota \le m$, then $F = F_u$ is strictly convex on $\mathbb{R}^r$.*

Thus, we have the following inclusion:

**Corollary 4.3.** $C^J(P,D) \supset \{u \in C^J(P)$: *There exists at least one index $\iota, 1 \le \iota \le m$, such that $R_{\iota\iota}$ is regular and (4.17c) holds* $\}$.

Of course, a related inclusion holds also for $C^J_{sep}(P,D)$.
Now we have the following consequences from Theorem 4.3, 4.4:

**Corollary 4.4 (Construction of descent directions of $F$ at $x$ without using derivatives of $F$).**

*a) Let $u \in C^J_{sep}(P), u \in C^J(P)$, respectively. If $x, y \ne x$, are related according to (4.14a-d), (4.15a-d), resp., then $h := y - x$ is a descent direction for $F$ at $x$, provided that $F$ is not constant on $xy$ (e.g. $\bar{c}^T x > \bar{c}^T y$).*
*b) If $x, y \ne x$, fulfill relations (4.14a-d), (4.15a-d) and $u \in C^{JJ}_{sep}, C^{JJ}$, resp., then $h = y - x$ is a descent direction for $F$ at $x$, provided that $\bar{A}_i x > \bar{A}_i, y$ for some $i \in J$.*
*c) Let $x, y \ne x$, fulfill (4.14a-d), (4.15a-d), and consider $u \in \hat{C}^J_{sep}, \hat{C}^J$, respectively. Then $h = y - x$ is a descent direction for $F$ at $x$, provided that $g_{ii}(x) > g_{ii}(y)$ for some $i, 1 \le i \le m, Q_x \ne Q_y$, respectively.*

*Remark 4.3.* a) Feasible descent directions $h = y - x$ for $F$ at $x \in D$ are obtained if besides (4.14a-d), (4.15a-d) also (4.14e), (4.15e), resp., is taken into consideration. b) Besides descent directions for $F$, in the following we are also looking for necessary optimality conditions for (4.9a,b) without using derivatives of $F$. Obviously, if $x^*$ is an optimal solution of (4.9a,b), then $F$ admits no feasible descent directions at $x^*$.

Because of Theorem 4.3, 4.4, Corollary 4.4 and the above remark, we are looking now for solutions of (4.14a-d) and (4.15a-d).

Representing first the symmetric $(r+1) \times (r+1)$ matrix $\frac{1}{2}(Q_{ij} + Q_{ji})$, by

$$\frac{1}{2}(Q_{ij} + Q_{ji}) = \left(\begin{array}{c|c} R_{ij} & d_{ij} \\ \hline d_{ijT} & q_{ij} \end{array}\right), \quad (4.18a)$$

where $R_{ij}$ is a symmetric $r \times r$ matrix, $d_{ij} \in \mathbb{R}^r$ and $q_{ij} \in \mathbb{R}$, we get

$$g_{ij}(x) = x^T R_{ij} x - 2x^T d_{ij} + q_{ij}, \tag{4.18b}$$

see (4.13b). Since $g_{ii}$ is convex for each $i = 1, \dots, m$, we find

$$g_{ii}(y) \geq g_{ii}(x) + \nabla g_{ii}(x)^T (y - x), \tag{4.18c}$$

where in (4.18c) the strict inequality ">" holds for $y \neq x$, if $R_{ii}$ is positive definite. Thus, since (4.15a-d) implies (4.14a-d), we have the following lemma:

**Lemma 4.2 (Necessary conditions for (4.14a-d), (4.15a-d)).** *If $y \neq x$ fulfills (4.14a-d) or (4.15a-d) with given $x$, then $h = y - x$ satisfies the linear constraints*

$$\bar{c}^T h \leq 0 \tag{4.19a}$$

$$\bar{A}_I h \leq 0, \ \bar{A}_{II} h = 0 \qquad (4.19b), (4.19c)$$

$$(R_{ii}x - d_{ii})^T h \leq 0 (< 0, \ \textit{if } R_{ii} \textit{ is positive definite}), i = 1, \dots, m. \tag{4.19d}$$

*Remark 4.4.* If the cone $C_x$ of feasible directions for $D$ at $x \in D$ is represented by $C_x = \{h \in \mathbb{R}^r \backslash \{0\} : U_x^{(1)} h \leq 0, U_x^{(2)} h < 0\}$, where $U_x^{(1)}, U_x^{(2)}$, resp., are $\nu_x^{(1)} \times r, \nu_x^{(2)} \times r$ matrices with $\nu_x^{(1)}, \nu_x^{(2)} \geq 0$, then the remaining relations (4.14e), (4.15e) can be described by

$$U_x^{(1)} h \leq 0, \ U_x^{(2)} h < 0. \qquad (4.19e), \ (4.19f)$$

Note that $U_x^{(1)} = 0$ and $\nu_x^{(2)} = 0$ if $x \in \text{int } D$ (:= interior of $D$).

Using now theorems of the alternatives [66], for the existence of solutions $h \neq 0$ of (4.19a-f), we obtain this characterization:

**Lemma 4.3.** *a) Suppose that $R_{ii}$ is positive definite for $i \in K$, where $\emptyset \neq K \subset \{1, \dots, m\}$ or $\nu_x^{(2)} > 0$. Then, either (4.19a-f) has a solution $h (\neq 0)$, or there exist multipliers*

$$\gamma_i \geq 0, \ 0 \leq i \leq m, \ \lambda_I \in \mathbb{R}_+^{|J|}, \ \lambda_{II} \in \mathbb{R}^{m-|J|}, \ \pi^{(1)} \in \mathbb{R}_+^{\nu_x^{(1)}}, \pi^{(2)} \in \mathbb{R}_+^{\nu_x^{(2)}} \tag{4.20a}$$

*with $(\gamma_i)_{i \in K} \neq 0$ or $\pi^{(2)} \neq 0$ such that*

$$\gamma_0 \bar{c} + \sum_{i=1}^{m} \gamma_i (R_{ii}x - d_{ii}) + \bar{A}_I^T \lambda_I + \bar{A}_{II}^T \lambda_{II} + U_x^{(1)^T} \pi^{(1)} + U_x^{(2)^T} \pi^{(2)} = 0. \tag{4.20b}$$

*b) Suppose that $R_{ii}$ is singular for each $i = 1, 2, \dots, m$ and $\nu_x^{(2)} = 0$ (i.e. $U_x^{(2)}$ is cancelled). Then either (4.19a-f) has a solution $h (\neq 0)$ such that in (4.19a-f) the strict inequality "<" holds at least once in the $1 + |J| + m + \nu_x^{(1)}$ inequalities of (4.19a-f), or there exist multipliers*

$$\gamma_i > 0, \ i = 0, 1, \dots, m, \ \lambda_I > 0, \ \lambda_{II} \in \mathbb{R}^{M-|J|}, \pi^{(1)} > 0 \qquad (4.20a')$$

*such that (4.20b) holds.*

c) *If* $\nu_x^{(2)} = 0$, *then (4.19a-f) has a solution* $h \neq 0$ *such that "=" holds everywhere in (4.19a-f) if and only if rank* $\prod_x < r$ *for the* $(1 + 2m + \nu_x^{(1)}) \times r$ *matrix* $\prod_x$ *having the rows* $c_0^T, \overline{A}_i, i = 1, \ldots, m, (R_{ii}x - d_{ii})^T, i = 1, \ldots, m, U_{xj}^{(1)}, j = 1, \ldots, \nu_x^{(1)}$.

### 4.2.2 Solution of the Auxiliary Programs

For any given $x \in D$, solutions $y \neq x$ of (4.14a-e), (4.15a-e) can be obtained by random search methods, modified Newton methods, and by solving certain optimization problems related to (4.14a-e), (4.15a-e).

*Solution of (4.14a-e).* Here we have the program

$$\min f(y) \text{ s.t. .(4.14a-c), (4.14e),} \tag{4.21a}$$

where

$$f(y) := \max\Big\{g_{ii}(y) - g_{ii}(x) : 1 \leq i \leq m\Big\}. \tag{4.21b}$$

*Note.* Clearly, also any other selection of functions from $\{c^T y - c^T x, \bar{A}_i y - \bar{A}_i x, i \in J, g_{ii}(y) - g_{ii}(x), i = 1, \ldots, m\}$ can be used to define the objective function $f$ in (4.21a).

Obviously, (4.21a,b) is equivalent to

$$\min t \tag{4.21a'}$$

s.t.

$$\begin{aligned} \bar{c}^T y &\leq \bar{c}^T x && \text{(4.21b')} \\ \bar{A}_I y \leq \bar{A}_I x, \bar{A}_{II} y &= \bar{A}_{II} x && \text{(4.21c'), (4.21d')} \\ g_{ii}(y) - g_{ii}(x) &\leq t,\ i = 1, \ldots, m && \text{(4.21e')} \\ y &\in D. && \text{(4.21f')} \end{aligned}$$

It is easy to see that (4.21a,b), (4.21a'-f') are convex programs having for $x \in D$ always the feasible solution $y = x, (y, t) = (x, 0)$. Optimal solutions $y^*, f(y^*) \leq 0, (y^*, t^*), t^* \leq 0$, resp., exist under weak assumptions. Further basic properties of the above program are shown next:

**Lemma 4.4.** *Let* $x \in D$ *be any given feasible point.*

a) *If* $y^*, (y^*, t^*)$ *is optimal in (4.21a,b), (4.21a'-f'), resp., then* $y^*$ *solves (4.14a-d).*
b) *If* $y, (y, t)$ *is feasible in (4.21a,b), (4.21a'-f'), then* $y$ *solves (4.14a-d), provided that* $f(y) \leq 0, t \leq 0$*, respectively. If* $y$ *solves (4.14a-d), then* $y, (y, t), t_0 \leq t \leq 0$*, is feasible in (4.21a,b), (4.21a'-f'), resp., where* $t_0 := f(y) \leq 0$.

*c) (4.14a-d) has the unique solution $y = x$ if and only if (4.21a'-f') has the unique optimal solution $(y^*, t^*) = (x, 0)$.*

*Proof.* Assertions (a) and (b) are clear. Suppose now that (4.14a-d) yields $y = x$. Obviously, $(y^*, t^*) = (x, 0)$ is then optimal in (4.21a'-f'). Assuming that there is another optimal solution $(y^{**}, t^{**}) \neq (x, 0), t^* \leq 0$, of (4.21a'-f'), we know that $y^{**}$ solves (4.14a-d). Hence, we have $y^{**} = x$ and therefore $t^{**} < 0$ which is a contradiction to $y^{**} = x$. Conversely, assume that (4.21a'-f') has the unique optimal solution $(y^*, t^*) = (x, 0)$, and suppose that (4.14a-d) has a solution $y \neq x$. Since $(y, t), f(y) \leq t \leq 0$, is feasible in (4.21a'-f'), we find $t^* \leq t \leq 0$ and therefore $f(y) = 0$. Thus, $(y, 0) \neq (y^*, t^*)$ is also optimal in (4.21a'-f'), in contradiction to the uniqueness of $(y^*, t^*) = (x, 0)$.

According to Lemma 4.4, solutions $y$ of system (4.14a-d) can be determined by solving (4.21a'-f'). In the following we assume that

$$D = \Big\{x \in \mathbb{R}^r : g(x) \leq g_0\Big\}, \tag{4.22}$$

where $g(x) = \Big(g_i(x), \ldots, g_\nu(x)\Big)^T$ is a $\nu$–vector of convex functions on $\mathbb{R}^r, g_0 \in R^\nu$. Furthermore, suppose that (4.21a'-f') fulfills the Slater condition for each $x \in D$. This regularity condition holds e.g. (consider $(y, t) := (x, 1)$) if

$$g(y) := Gy \text{ with a } \nu \times r \text{ matrix } G. \tag{4.23}$$

The necessary and sufficient optimality conditions for (4.21a'-f') read:

$$\sum_{i=1}^{m} \gamma_i = 1 \tag{4.24a}$$

$$\sum_{i=1}^{m} \gamma_i(R_{ii}y - d_{ii}) = -\tfrac{1}{2}\Big(\gamma_0 c + \bar{A}_I^T \lambda_I + \bar{A}_{II}^T \lambda_{II} + \tfrac{\partial g}{\partial x}(y)^T \mu\Big) \tag{4.24b}$$

$$g_{ii}(y) - g_{ii}(x) - t \leq 0,\ i = 1, \ldots, m \tag{4.24c}$$

$$\gamma_i\Big(g_{ii}(y) - g_{ii}(x) - t\Big) = 0,\ \gamma_i \geq 0,\ i = 1, \ldots, m \tag{4.24d}$$

$$\bar{c}^T y - \bar{c}^T x \leq 0,\ \gamma_0(\bar{c}^T y - \bar{c}^T x) = 0,\ \gamma_0 \geq 0 \tag{4.24e}$$

$$\bar{A}_I y - \bar{A}_I x \leq 0,\ \lambda_I^T(\bar{A}_I y - \bar{A}_I x) = 0,\ \lambda_I \geq 0\ (\lambda_I \in \mathbb{R}^{|J|}) \tag{4.24f}$$

$$\bar{A}_{II} y - \bar{A}_{II} x = 0,\ \lambda_{II} \in \mathbb{R}^{m-|J|} \tag{4.24g}$$

$$g(y) - g_0 \leq 0,\ \mu^T\Big(g(y) - g_0\Big) = 0,\ \mu \geq 0\ (\mu \in \mathbb{R}^\nu). \tag{4.24h}$$

*Fundamental properties of optimal solutions of (4.21a'-f')*: For given $x \in D$, let $(y^*, t^*)$ be optimal in (4.21a'-f'). Hence, $t^* \leq 0, (y^*, t^*)$ is characterized by (4.24a-h), and $y^*$ solves (4.14a-d), see Lemma 4.4a. Moreover, one of the following cases will occur, see Corollary 4.4:

*Case i:* $t^* < 0$.

Here we have $g_{ii}(y^*) < g_{ii}(x), 1 \le i \le m$. Thus, $y^* \neq x$, and $h = y^* - x$ is a feasible descent direction for $F$ at $x$, provided that $u \in \hat{C}^J_{sep}(P)$ or $u \in C^J_{sep}(P)$ and $F = F_u$ is not constant on $xy^*$.

*Case ii.1*: $t^* = 0$ and $g_{ii}(y^*) < g_{ii}(x)$ for some $1 \le i \le m$ or $\bar{c}^T y^* < \bar{c}^T x$ or $\bar{A}_i y^* < \bar{A}_i x$ for some $i \in J$.

Here, $y^* \neq x$, and $h = y^* - x$ is a feasible descent direction for $F$ at $x$ if $u \in \tilde{C}^J_{sep}(P), u \in C^J_{sep}(P), u \in C^{JJ}_{sep}(p)$, respectively.

*Case ii.2*: $t^* = 0$ and $g_{ii}(y^*) = g_{ii}(x), 1 \le i \le m, \bar{c}^T y^* = \bar{c}^T x, \bar{A}_i y^* = \bar{A}_i x, i \in J$, with $y^* \neq x$.

Then $h = y^* - x$ is a feasible descent direction for $F$ at $x$ for all $u \in C^J_{sep}(P)$ such that $F = F_u$ is not constant on $xy^*$.

*Case ii.3*: $t^* = 0$ and $(y^*, t^*) = (x, 0)$ is the unique optimal solution of (4.21a'-f').

According to Lemma 4.4c, in Case ii.3 we know here that (4.14a-d) has only the trivial solution $y = x$. Hence, the construction of a descent direction $h = y - x$ by means of (4.14a-d) and Corollary 4.4 fails at $x$. Note that points $x \in D$ having this property are candidates for optimal solutions of (4.9a,b).

*Solution of (4.15a-e).* Let $x \in D$ be a given feasible point. Replacing for simplicity (4.15d) by (4.16a), for handling system (4.15a-e), we get the following optimization problem

$$\min \hat{f}(y) \text{ s.t. } (4.15\text{a-c}), (4.15\text{e}), \tag{4.25a}$$

where

$$\begin{aligned}\hat{f}(y) &:= \max\left\{ g_{ii}(y) - g_{ii}(x) + \sum_{j \neq i} \left| g_{ij}(x) - g_{ij}(y) \right| : \ 1 \le i \le m \right\} \quad (4.25\text{b})\\ &= \max\left\{ g_{i\sigma_i}(y) - g_{i\sigma_i}(x) : \sigma_i \in \textstyle\sum_i, \ 1 \le i \le m \right\},\end{aligned}$$

see the note to program (4.21a,b). In the above, $g_{i\sigma_i}$ is defined by (4.16c). An equivalent version of (4.25a,b) reads

$$\min t \tag{4.25a'}$$

s.t.

$$\begin{aligned}\bar{c}^T y &\le \bar{c}^T x && (4.25\text{b'})\\ \bar{A}_I y \le \bar{A}_I x, \ \bar{A}_{II} y &= \bar{A}_{II} x && (4.25\text{c'}), (4.25\text{d'})\\ g_{ii}(y) - g_{ii}(x) + \sum_{j \neq i} \left| g_{ij}(x) - g_{ij}(y) \right| &\le t, \ i = 1, \ldots, m && (4.25\text{e'})\\ y &\in D, && (4.25\text{f'})\end{aligned}$$

where condition (4.25e') has also the equivalent form

$$g_{i\sigma_i}(y) - g_{i\sigma_i}(x) \le t \ \text{ for all } \sigma_i \in \textstyle\sum_i, \ i = 1, \ldots, m. \tag{4.25e''}$$

It is easy to see that, for given $x \in D$, program (4.25a,b), (4.25a'-f'), resp., has always the trivial feasible solution $y = x, (y, t) = (x, 0)$. Moreover, optimal solutions $y^*, (y^*, t^*)$ of (4.25a,b), (4.25a'-f'), resp., exist under weak assumptions, where $t^* = \hat{f}(y^*) \leq 0$. If $g_{ii}(y^*) = g_{ii}(x)$ for an index $i$, then $g_{ij}(y^*) = g_{ij}(x)$ for all $j = 1, \ldots, m$ and $f(y^*) = t^* = 0$. If condition (4.16d) holds for every $i = 1, \ldots, m$, then (4.25a,b), (4.25a'-f') are convex. Further properties of (4.25a,b), (4.25a'-f') are given in the following:

**Lemma 4.5.** *Let $x \in D$ be a given point and consider system (4.15a-c,e), (4.16a).*

*a) If $y^*, (y^*, t^*)$ is optimal in (4.25a,b), (4.25a'-f'), resp., then $y^*$ solves (4.15a-c,e), (4.16a).*
*b) If $y, (y, t)$ is feasible in (4.25a,b), (4.25a'-f'), then $y$ solves (4.15a-c,e), (4.16a) provided that $\hat{f}(y) \leq 0, t \leq 0$, respectively. If $y$ solves (4.15a-c,e), (4.16a), then $y, (y, t), t_0 \leq t \leq 0$, is feasible in (4.25a,b), (4.25a'-f'), resp., where $t_0 := \hat{f}(y) \leq 0$.*
*c) (4.15a-c,e), (4.16a) has the unique solution $y = x$ if and only if program (4.25a'-f') has the unique optimal solution $(y^*, t^*) = (x, 0)$.*
*d) If $(y, t)$ is feasible in (4.25a'-f'), then $(y, t)$ is also feasible in (4.21a'-f').*

*Proof.* Assertions a), b), d) are clear, and c) can be shown as the corresponding assertion c) in Lemma 4.4.

As was shown above, (4.15a-c,e), (4.16a) can be solved by means of (4.25a'-f'). If (4.25a'-f') is convex, see (4.16d), then we may proceed as in (4.21a'-f'). Indeed, if $D$ is given by (4.22), and we assume that the Slater condition holds, then the necessary and sufficient conditions for an optimal solution $(y^*, t^*)$ of (4.25a'-f') read:

$$\sum_{i=1}^{m} \sum_{\sigma_i \in \sum_i} \gamma_{i\sigma_i} = 1 \tag{4.26a}$$

$$\sum_{i=1}^{m} \sum_{\sigma_i \in \sum_i} \gamma_{i\sigma_i} \left\{ \left( R_{ii} - \sum_{j \neq i} \sigma_{ij} R_{ij} \right) y - \left( d_{ii} - \sum_{j \neq i} \sigma_{ij} d_{ij} \right) \right\} =$$
$$-\frac{1}{2} \left( \gamma_0 c + \bar{A}_I^T \lambda_I + \bar{A}_{II}^T \lambda_{II} + \frac{\partial g}{\partial x}(y)^T \mu \right) \tag{4.26b}$$

$$g_{i\sigma_i}(y) - g_{i\sigma_i}(x) - t \leq 0, \; \sigma_i \in \sum_i, \; i = 1, \ldots, m \tag{4.26c}$$

$$\gamma_{i\sigma_i} \left( g_{i\sigma_i}(y) - g_{i\sigma_i}(x) - t \right) = 0, \; \gamma_{i\sigma_i} \geq 0, \; \sigma_i \in \sum_i, i = 1, \ldots, m \tag{4.26d}$$

$$\bar{c}^T y \leq \bar{c}^T x, \; \gamma_0 (\bar{c}^T y - \bar{c}^T x) = 0, \; \gamma_0 \geq 0 \tag{4.26e}$$

$$\bar{A}_I y - \bar{A}_I x \leq 0, \; \lambda_I^T (\bar{A}_I y - \bar{A}_I x) = 0, \; \lambda_I \geq 0 \tag{4.26f}$$

$$\bar{A}_{II} y = \bar{A}_{II} x, \; \lambda_{II} \in \mathbb{R}^{m - |J|} \tag{4.26g}$$

$$g(y) - g_0 \leq 0, \; \mu^T \left( g(y) - g_0 \right) = 0, \; \mu \geq 0. \tag{4.26h}$$

In the non–convex case the conditions (4.26a-h) are still necessary for an optimal solution $(y^*, t^*)$ of (4.25a'-f'), provided that $(y^*, t^*)$ fulfills a certain regularity condition, see e.g. [42].

***Fundamental properties of optimal solutions of (4.25a'-f')***: For given $x \in D$, let $(y^*, t^*)$ be an optimal solution of (4.25a'-f'). According to Lemma 4.5 a) we know that $y^*$ solves then (4.15a-c,e), (4.16a) and therefore also (4.15a-e). Moreover, one of the following different cases will occur, cf. Corollary 4.4:

*Case i:* $t^* < 0$.

This yields $g_{ii}(y^*) < g_{ii}(x), i = 1, \ldots, m$, and $Q_x \succeq Q_{y^*}$. Thus, $y^* \neq x$, and $h = y^* - x$ is a feasible descent direction for $F$ at $x$, provided that $u \in \hat{C}^J(P)$ or $u \in C^J(P)$ and $F = F_u$ is not constant on $xy^*$.

*Case ii.1:* $t^* = 0$ and $g_{ii}(y^*) < g_{ii}(x)$ for an index $i, i = 1, \ldots, m$, or $\bar{c}^T y^* < \bar{c}^T x$ or $\bar{A}_i y^* < \bar{A}_i x$ for some $i \in J$.

Then $y^* \neq x$, and $h = y^* - x$ is a feasible descent direction for $F$ at $x$, provided that $u \in \hat{C}^J(P), u \in C^J(P), u \in C^{JJ}(P)$, respectively.

*Case ii.2:* $t^* = 0$ and $Q_{y^*} = Q_x, \bar{c}^T y^* = \bar{c}^T x, \bar{A}_i y^* = \bar{A}_i x, i \in J$, with $y^* \neq x$.

Here, $h = y^* - x$ is a feasible descent direction for all $u \in C^J(P)$ such that $F = F_u$ is not constant on $xy^*$.

*Case ii.3:* $t^* = 0$ and $(y^*, t^*) = (x, 0)$ is the unique optimal solution of (4.25a'-f').

According to Lemma 4.5 c) we know that (4.15a-c,e), (4.16a) has then only the trivial solution $y = x$. In this case the construction of a descent direction $h = y - x$ for $F$ by means of (4.15a-c,e), (4.16a) and Corollary 4.4 fails at $x$. Points of this type are candidates for optimal solutions of (4.9a,b), see Section 4.3.

Lemma 4.5 d) and (4.18c) yield still the auxiliary convex program

$$\min t \tag{4.27a}$$

s.t.

$$\bar{c}^T y \leq \bar{c}^T x \tag{4.27b}$$

$$\bar{A}_I y \leq \bar{A}_I x, \ A_{II} y = \bar{A}_{II} x \qquad \text{(4.27c), (4.27d)}$$

$$2(R_{ii}x - d_{ii})^T (y - x) - t \leq 0, \ i = 1, \ldots, m \tag{4.27e}$$

$$y \in D. \tag{4.27f}$$

Obviously, for given $x \in D$, each feasible solution $(y, t)$ of (4.21a'-f') or (4.25a'-f') is also feasible in (4.27a-f). If (4.25a'-f') is convex, then (4.27e) can be sharpened, see (4.16d), (4.18a), (4.25e''), as follows:

$$2\left[\left(R_{ii} - \sum_{j \neq i} \sigma_{ij} R_{ij}\right)x - \left(d_{ii} - \sum_{j \neq i} \sigma_{ij} d_{ij}\right)\right]^T (y - x) - t \leq 0,$$

$$\sigma_i \in \textstyle\sum_i, \ 1 \leq i \leq m. \tag{4.27e'}$$

## 4.3 Efficient Solutions (Points)

Instead of solving approximatively the complicated stochastic optimization problem (4.9a,b), where one has not always a unique loss function $u$, we may look for an alternative solution concept: Obvious alternatives for optimal solution of (4.9a,b) with a loss function $u \in \mathcal{C}^J_{sep}(\hat{\mathcal{C}}^J_{sep}, \mathcal{C}^{JJ}_{sep}, \hat{\mathcal{C}}^{JJ}_{sep}), u \in \mathcal{C}^J(\hat{\mathcal{C}}^J, \mathcal{C}^{JJ}, \hat{\mathcal{C}}^{JJ})$, resp., are just the points $x^0 \in D$ such that the above procedure for finding a feasible descent direction for $F = F_u$ fails at $x^0$. Based on (4.14a-e), (4.15a-e), resp., and Corollary 4.4, *efficient* solutions are defined as follows:

**Definition 4.4.** *A vector $x^0 \in D$ is called a $\mathcal{C}^J_{sep}$–efficient solution of (4.9a,b) if there is no $y \in D$ such that*

$$\bar{c}^T x^0 \geq \bar{c}^T y \tag{4.28a}$$

$$\bar{A}_I x^0 \geq \bar{A}_I y, \ \bar{A}_{II} x^0 = \bar{A}_{II} y \qquad \text{(4.28b), (4.28c)}$$

$$g_{ii}(x^0) \geq g_{ii}(y), \ i = 1, \ldots, m \tag{4.28d}$$

*and*

$$(\bar{c}^T x^0, \bar{A}_I x^0, g_{ii}(x^0), 1 \leq i \leq m) \neq (\bar{c}^T y, \bar{A}_I y, g_{ii}(y), 1 \leq i \leq m) \tag{4.28e}$$

*$x^0 \in D$ is called a strongly (or uniquely) $\mathcal{C}^J_{sep}$–efficient solution of (4.9a,b) if there is no vector $y \in D, y \neq x^0$, such that (4.28a-d) holds. Let $\mathcal{E}_{sep}, \mathcal{E}^{st}_{sep}$, resp., denote the set of all $\mathcal{C}^J_{sep}$–strongly $\mathcal{C}^J_{sep}$–efficient solutions of (4.9a,b).*

We observe that the elements of $\mathcal{E}_{sep}, \mathcal{E}^{st}_{sep}$ are the Pareto–optimal solutions [23, 121] of the vector optimization problem

$$\text{"}\min_{y \in D}\text{"} \left\{ \bar{c}^T y, \bar{A}_I y - \bar{b}_I, \bar{A}_{II} y - \bar{b}_{II}, \bar{b}_{II} - \bar{A}_{II} y, g_{ii}(y), 1 \leq i \leq m \right\}. \tag{4.28'}$$

**Definition 4.5.** *A vector $x^0 \in D$ is called a $\mathcal{C}^J$–efficient solution of (4.9a,b) if there is no $y \in D$ such that*

$$\bar{c}^T x^0 \geq \bar{c}^T y \tag{4.29a}$$

$$\bar{A}_I x^0 \geq \bar{A}_I y, \ \bar{A}_{II} x^0 = \bar{A}_{II} y \qquad \text{(4.29b), (4.29c)}$$

$$g_{i\sigma_i}(x^0) \geq g_{i\sigma_i}(y), \sigma_i \in \sum\nolimits_i, \ i = 1, \ldots, m \tag{4.29d}$$

*and*

$$(\bar{c}^T x^0, \bar{A}_I x^0, g_{ii}(x^0), 1 \leq i \leq m) \neq (\bar{c}^T y, \bar{A}_I y, g_{ii}(y), 1 \leq i \leq m). \tag{4.29e}$$

*$x^0 \in D$ is called a strongly (or uniquely) $\mathcal{C}^J$–efficient solution of (4.9a,b) if there is no $y \in D, y \neq x^0$, such that (4.29a-d) holds. Let $\mathcal{E}, \mathcal{E}^{st}$, resp., denote the set of all $\mathcal{C}^J$–, strongly $\mathcal{C}^J$–efficient solutions of (4.9a,b).*

Note that the elements of $\mathcal{E}, \mathcal{E}^{\mathrm{st}}$ are the Pareto–optimal solutions of the multi–objective program

$$\text{" }\min_{y\in D}\text{ "}\Big\{\bar{c}^T y, \bar{A}_I y - \bar{b}_I, \bar{A}_{II} y - \bar{b}_{II}, \bar{b}_{II} - \bar{A}_{II} y, g_{i\sigma_i}(y), \sigma_i \in \sum_i, 1 \le i \le m\Big\}. \tag{4.29'}$$

*Discussion of Definitions 4.4,4.5:*
Relations (4.28a d) and (4.14a-d) agree. Furthermore, (4.29a-d) and (4.15a-c), (4.16b) coincide, where (4.16b) is equivalent to (4.16a). We have the inclusions

$$\mathcal{E}^{\mathrm{st}}_{\mathrm{sep}} \subset \mathcal{E}_{\mathrm{sep}}, \mathcal{E}^{\mathrm{st}} \subset \mathcal{E}, \tag{4.30a}$$

where equality in (4.30a) holds under the following conditions:

**Lemma 4.6.** *a) If $g_{ii}$ is strictly convex for at least one index $i, 1 \le i \le m$, then $\mathcal{E}^{st}_{sep} = \mathcal{E}_{sep}$.*
*b) If $g_{i\sigma_i}$ is convex for all $\sigma_i \in \sum_i, i = 1, 2, \ldots, m$, and $g_{ii}$ is strictly convex for at least one index $i, 1 \le i \le m$, then $\mathcal{E}^{st} = \mathcal{E}$.*
*c) If for arbitrary $x, y \in D$*

$$x = y \Leftrightarrow \bar{c}^T x = \bar{c}^T y,\ \bar{A}x = \bar{A}y,\ g_{ii}(x) = g_{ii}(y),\ i = 1, \ldots, m, \tag{4.30b}$$

*then $\mathcal{E}^{st}_{sep} = \mathcal{E}_{sep}$ and $\mathcal{E}^{st} = \mathcal{E}$.*

*Proof.* a) We may suppose e.g. that $g_{11}$ is strictly convex. Let then $x^0 \in \mathcal{E}_{\mathrm{sep}}$ and assume that $x^0 \notin \mathcal{E}^{\mathrm{st}}_{\mathrm{sep}}$. Hence, there exists $y \in D, y \ne x^0$, such that (4.28a-d) holds. Since all $g_{ii}$ are convex and $g_{11}$ is strictly convex, it is easy to see that $\eta := \lambda y^0 + (1-\lambda)y$ fulfills (4.28a-e) for each $0 < \lambda < 1$. This contradiction yields $x^0 \in \mathcal{E}^{\mathrm{st}}_{\mathrm{sep}}$ and therefore $\mathcal{E}^{\mathrm{st}}_{\mathrm{sep}} = \mathcal{E}_{\mathrm{sep}}$, cf. (4.30a). Assertion b) follows in the same way. In order to show c), let $x^0 \in \mathcal{E}_{\mathrm{sep}}, x^0 \in \mathcal{E}$, resp., and assume $x^0 \notin \mathcal{E}^{\mathrm{st}}_{\mathrm{sep}}, x^0 \notin \mathcal{E}^{\mathrm{st}}$. Consider $y \in D, y \ne x^0$, such that (4.28a-d), (4.29a-d), resp., holds. Because of $x^0 \in \mathcal{E}_{\mathrm{sep}}, x^0 \in \mathcal{E}$, resp., we have

$$\Big(\bar{c}^T x^0, \bar{A}_I x^0, g_{ii}(x^0), 1 \le i \le m\Big) = \Big(\bar{c}^T y, \bar{A}_I y, g_{ii}(y), 1 \le i \le m\Big).$$

This equation, (4.28c), (4.29c), resp., and (4.30b) yield the contradiction $y = x^0$. Thus, $x^0 \in \mathcal{E}^{\mathrm{st}}_{\mathrm{sep}}, x^0 \in \mathcal{E}^{\mathrm{st}}$, resp., and therefore $\mathcal{E}^{\mathrm{st}}_{\mathrm{sep}}$ and $\mathcal{E}^{\mathrm{st}} = \mathcal{E}$.

According to Lemma 4.4 c), Lemma 4.5 c), a point $x^0 \in D$ is strongly $C^J_{\mathrm{sep}}-$, strongly $C^J$–efficient if and only if (4.21a'-f'), (4.25a'-f'), resp., with $x := x^0$, has the unique optimal solution $(y^*, t^*) = (x^0, 0)$. Since (4.29d), being equivalent to (4.16a), implies (4.28d), we find

$$\mathcal{E}_{\mathrm{sep}} \subset \mathcal{E}, \mathcal{E}^{\mathrm{st}}_{\mathrm{sep}} \subset \mathcal{E}^{\mathrm{st}}. \tag{4.30c}$$

A special subset of efficient solutions of (4.9a,b) is given by

$$\mathcal{E}_0 := \left\{x^0 \in D : (4.19\text{a-f}) \text{ has no solution } h \neq 0\right\} \tag{4.31}$$

Indeed, Lemma 4.2 and Remark 4.4 yield the following inclusion:

**Corollary 4.5.** $\mathcal{E}_0 \subset \mathcal{E}_{sep}^{st}$.

By Lemma 4.3 we have this parametric representation for $\mathcal{E}_0$:

**Corollary 4.6.** *a) If $R_{ii}$ is positive definite for at least one index $1 \leq i \leq m$, then $x \in \mathcal{E}_0$ if and only if $x \in D$ and there are parameters $\gamma_i, i = 0, 1, \ldots, m, \lambda_I, \lambda_{II}$ and $\pi^{(1)}, \pi^{(2)}$ such that (4.20a,b) holds.*

*b) Let $R_{ii}$ be singular for each $i = 1, \ldots, m$. If $\nu_x^{(2)} > 0$ for each $x \in D$, then $x \in \mathcal{E}_0$ if and only if $x \in D$ and there exist $\gamma_i, 0 \leq i \leq m, \lambda_I, \lambda_{II}, \pi^{(1)}, \pi^{(2)}$ such that (4.20a,b) holds. If $\nu_x^{(2)} = 0, x \in D$, then $x \in \mathcal{E}_0$ if and only if $x \in D$, rank $\prod_x = n$ (with the matrix $\prod_x$ defined in Lemma 4.3c)) or there are parameters $\gamma_i, 0 \leq i \leq m, \lambda_I, \lambda_{II}$ and $\pi^{(1)}$ such that (4.20a',b) holds.*

### 4.3.1 Necessary Optimality Conditions Without Gradients

As was already discussed in Section 4.2.2, Cases ii.3, efficient solutions are candidates for optimal solutions $x^*$ of (4.9a,b). This will be made more precise in the following:

**Theorem 4.5.** *If $x^* \in argmin_{x \in D} F(x)$ with any $u \in \hat{C}_{sep}^{JJ}(P), u \in \hat{C}^{JJ}(P)$. then $x^* \in \mathcal{E}_{sep}, x^* \in \mathcal{E}$, respectively.*

*Proof.* Supposing that $x^* \notin \mathcal{E}_{\text{sep}}, x^* \notin \mathcal{E}$, resp., we find $y \in D$ such that (4.28a-e), (4.29a-e) holds, where $x^0 := x^*$. Thus, (4.14a-e), (4.15a-c,e), (4.16a), resp., hold. Moreover at least one of the relations (4.14a',b',d'), (4.15a',b') and (4.16b'), resp., hold with $x := x^*$. Consequently, according to Corollary 4.4 and Remark 4.3a we have a feasible descent direction for $F$ at $x^*$. This contradiction to the optimality of $x^*$ yields the assertion.

Approximating the elements of $C_{\text{sep}}^J(P), C^J(P)$ by certain elements of $\hat{C}_{\text{sep}}^{JJ}(P), C^{JJ}(P)$, resp., we find the next condition:

**Theorem 4.6.** *If $D$ is compact or $D$ is closed and $F(x) \to +\infty$ as $\|x\| \to +\infty$, then for each $u \in C_{sep}^J(P), u \in C^J(P)$ there exists $x^* \in$ argmin $F(x)$ contained in the closed hull $\bar{\mathcal{E}}_{sep}, \bar{\mathcal{E}}$ of $\mathcal{E}_{sep}, \mathcal{E}$, respectively.*

*Proof.* Consider any $u \in C_{\text{sep}}^J(P), u \in C^J(P)$, and define $u_\rho \in \hat{C}_{\text{sep}}^{JJ}(P), u_\rho \in \hat{C}^{JJ}(P)$, resp., by

$$u_\rho(z) := u(z) + \rho \sum_{i=1}^{m} v_i(z_i), \quad z \in \mathbb{R}^m,$$

where $\rho > 0$ and $v_i = v_i(z_i)$ is defined by

$$v_i(z_i) = \exp(z_i), i \in J, \text{ and } v_i(z_i) = z_i^2 \text{ for } i \notin J.$$

If $D$ is compact, then for each $\rho > 0$ we find $x_\rho^* \in \mathrm{argmin}_{x\in D} F_\rho(x)$, where $F_\rho(x) := \bar{c}^T x + E u_\rho\Big(A(\omega)x - b(\omega)\Big)$, see (4.9b). Furthermore, since $D$ is compact and $x \to E\sum_{i=1}^{m} v_i\big(A(\omega)x - b_i(\omega)\big)$ is convex and therefore continuous on $D$, there is a constant $C > 0$ such that $\Big|F(x) - F_\rho(x)\Big| \le \rho C$ for all $x \in D$. This yields $\Big|\min_{x\in D} F(x) - \min_{x\in D} F_\rho(x)\Big| = \Big|\min_{x\in D} F(x) - F_\rho(x_\rho^*)\Big| \le \rho C$. Hence, $F_\rho(x_\rho^*) \to \min_{x\in D} F(x)$ as $\rho \downarrow 0$. Furthermore, we have $\Big|F(x_\rho^*) - F_\rho(x_\rho^*)\Big| \le \rho C, \rho > 0$, and for any accumulation point $x^*$ of $(x_\rho^*)$ we get

$$\Big|F(x^*) - \min_{x\in D} F(x)\Big| \le \Big|F(x^*) - F(x_\rho^*)\Big| + \Big|F(x_\rho^*) - F_\rho(x_\rho^*)\Big| + \Big|F_\rho(x_\rho^*) - \min_{x\in D} F(x)\Big| \le \Big|F(x^*) - F(x_\rho^*)\Big| + 2\rho C, \quad \rho > 0.$$

Consequently, since $x_\rho^* \in \mathcal{E}_{\mathrm{sep}}, x_\rho^* \in \mathcal{E}$, resp., for all $\rho > 0$, cf. Theorem 4.5, for every accumulation point $x^*$ of $(x_\rho^*)$ we find that $x^* \in \bar{\mathcal{E}}_{\mathrm{sep}}, x^* \in \bar{\mathcal{E}}$, resp., and $F(x^*) = \min_{x\in D} F(x)$. Suppose now that $D$ is closed and $F(x) \to +\infty$ as $\|x\| \to +\infty$. Since $F_\rho(x) \ge F(x), x \in \mathbb{R}^n, \rho > 0$, we also have that $F_\rho(x) \to +\infty$ for each fixed $\rho > 0$. Hence, there exist optimal solutions $x_0^* \in \mathrm{argmin}_{x\in D} F(x)$ and $x_\rho^* \in \mathrm{argmin}_{x\in D} F_\rho(x)$ for every $\rho > 0$. For all $0 < \rho \le 1$ and with an arbitrary, but fixed $\tilde{x} \in D$ we find then $x_\rho^* \in \mathcal{E}_{\mathrm{sep}}, x_\rho^* \in \mathcal{E}$, resp., and

$$F(x_\rho^*) \le F_\rho(x_\rho^*) \le F_\rho(\tilde{x}) \le F_1(\tilde{x}), \quad 0 < \rho \le 1.$$

Thus, $(x_\rho^*)$ must lie in a bounded set. Let $R := \max\Big\{\|x_0^*\|, R_0\Big\}$, where $R_0$ denotes a norm bound of $(x_\rho^*)$. If $\tilde{D} := \Big\{x \in D : \|x\| \le R\Big\}$, then $\min_{x\in D} F(x) = \min_{x\in\tilde{D}} F(x)$ and $\min_{x\in D} F_\rho(x) = \min_{x\in\tilde{D}} F_\rho(x)$. Since $\tilde{D}$ is compact, the rest of the proof follows now as in the first part.

For elements of $C_{\mathrm{sep}}^J(P,D), C^J(P,D)$, see Definition 4.3, we have the following condition:

**Theorem 4.7.** *If $x^* \in \mathrm{argmin}_{x\in D} F(x)$ with any $u \in C_{sep}^J(P,D), u \in C^J(P,D)$, then $x^* \in \mathcal{E}_{sep}^{st}, x^* \in \mathcal{E}^{st}$, respectively.*

*Proof.* Let $u \in C_{\mathrm{sep}}^J(P,D)$ and assume that $x^* \notin \mathcal{E}_{\mathrm{sep}}^{\mathrm{st}}$. According to Definition 4.4 we find then $y \in D, y \ne x^*$, such that (4.28a-d) hold with $x^0 := x^*$. Corollary 4.4a yields that $h = y - x^*$ is a feasible descent direction for $F$ at $x^*$ which is a contradiction to the optimality of $x^*$. Hence, $x^* \in \mathcal{E}_{\mathrm{sep}}^{\mathrm{st}}$. The rest follows in the same way.

### 4.3.2 Existence of Feasible Descent Directions in Non–Efficient Solutions of (4.9a,b)

The preceding Theorems 4.3, 4.4 and Corollary 4.4 yield this final result:

**Corollary 4.7.** *a) Let $x^0 \notin \mathcal{E}, x^0 \notin \mathcal{E}_{sep}$, respectively. Then there exists $y \in D, y \neq x^0$, such that (4.29a-e), (4.28a-e), resp., holds. Consequently, $F(y) \leq F(x^0)$ for all $u \in C^J(P), u \in C^J_{sep}(P)$, resp., and $F(y) < F(x^0)$ for all $u \in \hat{C}^{JJ}(P), u \in \hat{C}^{JJ}_{sep}(P)$, respectively. Hence, $h = y - x^0$ is a feasible descent direction for $F$ at $x^0$ for all $u \in \hat{C}^{JJ}(P), u \in \hat{C}^J_{sep}(P)$, and for all $u \in C^J(P), u \in C^J_{sep}(P)$, resp., such that $F$ is not constant on $x^0y$. b) Let $x^0 \notin \mathcal{E}^{st}, x^0 \notin \mathcal{E}^{st}_{sep}$, respectively. Then there is $y \in D, y \neq x^0$, such that (4.29a-d), (4.28a-d), resp., holds. Hence, $F(y) \leq F(x^0)$, and $h = y - x^0$ is a feasible descent direction for $F$ at $x^0$ for all $u \in C^J(P), u \in C^J_{sep}(P)$, such that $F$ is not constant on $x^0y$.*

## 4.4 Descent Directions in Case of Elliptically Contoured Distributions

In extension of the results in Section 4.2, here we suppose that the random $m \times (r+1)$ matrix $\big(A(\omega), b(\omega)\big)$ has an elliptically contoured probability distribution, cf. [50,108,141]. Members of this family are e.g. the multivariate normal, Cauchy, stable, Student $t$–, inverted Student $t$–distribution as also their truncations to elliptically contoured sets in $\mathbb{R}^{m(r+1)}$ and the uniform distribution on elliptically contoured sets in $\mathbb{R}^{m(r+1)}$. The parameters of this class can be described by an integer order parameter $s \geq 1$ and an $m(r+1)$ location matrix

$$\Xi = (\Theta, \theta) \tag{4.32a}$$

Moreover, there are positive semidefinite $m(r+1) \times m(r+1)$ scale matrices

$$Q(\sigma) = \Big(Q_{ij}(\sigma)\Big)_{i,j=1,\dots,m}, \ \sigma = 1, \dots, s, \tag{4.32b}$$

having $(r+1) \times (r+1)$ submatrices $Q_{ij}$ such that

$$Q_{ji} = Q_{ij}^T, \ i, j = 1, \dots, m. \tag{4.32c}$$

In the following $M \in \mathbb{R}^{m(r+1)}$ is always represented by an $m \times (r+1)$ matrix.

The characteristic function $\widehat{\mathcal{P}}_{\big(A(\cdot),b(\cdot)\big)} = \widehat{\mathcal{P}}_{\big(A(\cdot),b(\cdot)\big)}(M)$ of this class of distributions is then given by

$$\widehat{\mathcal{P}}_{\left(A(\cdot),b(\cdot)\right)}(M) := E\exp\left(i\mathrm{tr}M\left(A(\omega),b(\omega)\right)^T\right) \tag{4.33a}$$

$$= \exp(i\mathrm{tr}M\Xi^T)\exp\left(-\sum_{\sigma=1}^{s} g\left(\mathrm{vec}MQ(\sigma)(\mathrm{vec}M)^T\right)\right),$$

where

$$\mathrm{vec}M := (M_1, M_2, \ldots, M_m) \tag{4.33b}$$

denotes the $m(r+1)$–vector of the rows $M_1, \ldots, M_m$ of $M$. Furthermore, $g = g(t)$ is a certain nonnegative 1–1–function on $\mathbb{R}_+$. E.g., in the case of multivariate symmetric stable distributions of order $s$ and characteristic exponent $\alpha, 0 < \alpha \leq 2$, we have $g(t) := \frac{1}{2}t^{\alpha/2}, t \geq 0$, see [108].

Consequently, the probability distribution $\mathcal{P}_{A(\cdot)x-b(\cdot)}$ of the random $m$–vector

$$A(\omega)x - b(\omega) = \left(A(\omega), b(\omega)\right)\hat{x} \text{ with } \hat{x} := \begin{pmatrix} x \\ -1 \end{pmatrix} \tag{4.34a}$$

may be represented by the characteristic function

$$\widehat{\mathcal{P}}_{A(\cdot)x-b(\cdot)}(z) = E\exp\left(iz^T\left(A(\omega)x - b(\omega)\right)\right)$$

$$= E\exp\left(i\mathrm{tr}(z\hat{x}^T)\left(A(\omega),b(\omega)\right)^T\right) = \widehat{\mathcal{P}}_{\left(A(\cdot),b(\cdot)\right)}(z\hat{x}^T)$$

$$= \exp\left(iz^T(\Theta x - \theta)\exp\left(-\sum_{\sigma=1}^{s} g\left(z^T Q_x(\sigma)z\right)\right)\right). \tag{4.34b}$$

In (4.34b) the positive semidefinite $m \times m$ matrices $Q_x(\sigma)$ are defined by

$$Q_x(\sigma) := \left(\hat{x}^T Q_{ij}(\sigma)\hat{x}\right)_{i,j=1,\ldots,m}, \sigma = 1, \ldots, s, \tag{4.34c}$$

cf. (4.32b,c) and (4.13a-c).

Corresponding to Section 4.2, in the following we consider again the derivative–free construction of (feasible) descent directions $h = y - x$ for the objective function $F$ of the convex optimization problem (4.9a,b), hence,

$$F(x) = \bar{c}^T x + F_0(x)$$

where

$$F_0(x) = Eu\left(A(\omega)x - b(\omega)\right) \text{ with } u \in C^J(P).$$

Following the proof of Theorem 4.4, for a given vector $x$ and a vector $y \neq x$ to be determined, we consider the quotient of the characteristic functions of $A(\omega)x - b(\omega)$ and $A(\omega)y - b\omega)$, see (4.34a,b):

$$\frac{\widehat{\mathcal{P}}_{A(\cdot)x-b(\cdot)}(z)}{\widehat{\mathcal{P}}_{A(\cdot)y-b(\cdot)}(z)} = \exp\left(iz^T(\Theta x - \Theta y)\right)\varphi^{(0)}_{x,y}(z), \tag{4.35a}$$

where

$$\varphi_{x,y}^{(0)}(z) := \exp\left(-\sum_{\sigma=1}^{s}\left(g\left(z^T Q_x(\sigma) z\right) - g\left(z^T Q_y(\sigma) z\right)\right)\right). \tag{4.35b}$$

Thus, a necessary condition such that $\varphi_{x,y}^{(0)}$ is the characteristic function

$$\varphi_{x,y}^{(0)}(z) = \widehat{K}_{x,y}^{(0)}(z), z \in \mathbb{R}^m, \tag{4.35c}$$

of a symmetric probability distribution $K_{x,y}^{(0)}$ reads

$$Q_x(\sigma) \succsim Q_y(\sigma), \text{ i.e. } Q_x(\sigma) - Q_y(\sigma) \text{ is positive semidefinite}, \sigma = 1, \ldots, s \tag{4.36}$$

for arbitrary functions $g = g(t)$ having the above mentioned properties. Note that (4.35b,c) means that

$$\mathcal{P}_{A_0(\cdot)x - b_0(\cdot)} = K_{x,y}^{(0)} * \mathcal{P}_{A_0(\cdot)y - b_0(\cdot)}, \tag{4.37}$$

where $\left(A_0(\omega), b_0(\omega)\right) := \left(A(\omega), b(\omega)\right) - \Xi$.

Based on (4.35a-c) and (4.36), corresponding to Theorem 4.4 for normal distributions $\mathcal{P}_{\left(A(\cdot), b(\cdot)\right)}$, here we have the following results:

**Theorem 4.8.** *Distribution invariance. For given $x \in \mathbb{R}^r$, let $y \neq x$ fulfill the relations*

$$\bar{c}^T x \geq \bar{c}^T y \tag{4.38a}$$

$$\Theta x = \Theta y \tag{4.38b}$$

$$Q_x(\sigma) = Q_y(\sigma), \sigma = 1, \ldots, s. \tag{4.38c}$$

*Then $\mathcal{P}_{A(\cdot)x - b(\cdot)} = \mathcal{P}_{A(\cdot)y - b(\cdot)}$, $F_0(x) = F_0(y)$ for arbitrary $u \in \mathcal{C}(P) := \mathcal{C}^J(P)$ with $J = \emptyset$, and $F(x) \geq F(y)$.*

*Proof.* According to (4.34a,b), conditions (4.38b,c) imply $\widehat{\mathcal{P}}_{A(\cdot)x-b(\cdot)} = \widehat{\mathcal{P}}_{A(\cdot)y-b(\cdot)}$, hence, $\mathcal{P}_{A(\cdot)x-b(\cdot)} = \mathcal{P}_{A(\cdot)y-b(\cdot)}$, and the rest follows immediately.

**Corollary 4.8.** *Under the above assumptions $h = y - x$ is a descent direction for $F$ at $x$, provided that $F$ is not constant on $xy$.*

**Theorem 4.9.** *Suppose that (4.36) is also sufficient for (4.35c). Assume that the distribution $K_{x,y}^{(0)}$ has zero mean for all $x, y$ under consideration. For given $x \in \mathbb{R}^r$, let $y \neq x$ be a vector satisfying the relations*

$$\bar{c}^T x \geq \bar{c}^T y \quad (\bar{c}^T x > \bar{c}^T y) \tag{4.39a}$$

$$\Theta_I x \geq \Theta_I y \ (\Theta_i x > \Theta_i y \text{ for at least one } i \in J) \tag{4.39b}$$

$$\Theta_{II} x = \Theta_{II} y \tag{4.39c}$$

$$Q_x(\sigma) \succsim Q_y(\sigma), \sigma = 1, \ldots, s\Big(Q_x(\sigma) \neq Q_y(\sigma) \text{ for at least one } 1 \leq \sigma \leq s\Big), \tag{4.39d}$$

*where* $\Theta_i := (\Theta_i)_{i \in J}, \Theta_{II} := (\Theta_i)_{i \notin J}$. *Then* $F(x) \geq (>) F(y)$ *for all* $u \in \mathcal{C}^J(P), \mathcal{C}^{JJ}(P), \hat{\mathcal{C}}^J(P)$, *respectively.*

*Proof.* Under the above assumptions equation (4.37) holds and

$$F_0(x) = Eu\Big(A(\omega)x - b(\omega)\Big) = Eu\Big(\Theta x - \theta + Y_x(\omega)\Big),$$

$Y_x(\omega) := A(\omega)x - b(\omega) - (\Theta_x - \theta)$. Now the proof of Theorem 4.4 can be transfered to the present case.

## 4.5 Construction of Descent Directions by Using Quadratic Approximations of the Loss Function

Suppose that there exists a closed convex subset $Z_0 \subset \mathbb{R}^m$ such that the loss function $u$ fulfills the relations

$$d^T V D \leq d^T \nabla^2 u(z) d \leq d^T W d \text{ for all } z \in Z_0 \text{ and } d \in \mathbb{R}^m, \tag{4.40a}$$

where $V, W$ are given fixed symmetric $m \times m$ matrices. Moreover, suppose

$$A(\omega)x - b(\omega) \in Z_0 \text{ for all } x \in D \text{ and } \omega \in \Omega. \tag{4.40b}$$

Let $(\bar{A}, \bar{b}) = E\Big(A(\omega), b(\omega)\Big)$ be the mean of $\Big(A(\omega), b(\omega)\Big)$, and let

$$\begin{aligned} Q_{ij} &= \operatorname{cov}\Big(\big(A_i(\cdot), b_i(\cdot)\big), \big(A_j(\cdot), b_j(\cdot)\big)\Big) \\ &= E\Big(\big(A_i(\omega), b_i(\omega)\big) - \big(\bar{A}_i, \bar{b}_i\big)\Big)^T \Big(\big(A_j(\omega), b_j(\omega)\big) - (\bar{A}_j, \bar{b}_j)\Big) \end{aligned} \tag{4.41}$$

denote the covariance matrix of the $i$–th and $j$–th row $\Big(A_i(\cdot), b_i(\cdot)\Big), \Big(A_j(\cdot), b_j(\cdot)\Big)$ of $\Big(A(\omega), b(\omega)\Big)$. For any symmetric $m \times m$ matrix $U = (u_{ij})$ we define then the symmetric $(r+1) \times (r+1)$ matrix $\hat{U}$ by

$$\begin{aligned} \hat{U} &= E\Big(A(\omega) - \bar{A}, b(\omega) - \bar{b}\Big)^T U \Big(A(\omega) - \bar{A}, b(\omega) - \bar{b}\Big) \\ &= \sum_{i,j=1}^{m} u_{ij} Q_{ij}. \end{aligned} \tag{4.42}$$

For a given $r$–vector $x \in D$ we consider now the Taylor expansion of $u$ at

$$\bar{z}_x = \bar{A}x - \bar{b} = E\Big(A(\omega)x - b(\omega)\Big), \tag{4.43}$$

hence,

$$u(z) = u(\bar{z}_x) + \nabla u(\bar{z})^T (z - \bar{z}_x) + \frac{1}{2}(z - \bar{z})^T \nabla^2 u(w)(z - \bar{z}_x),$$

where $w = \bar{z}_x + \vartheta(z - \bar{z}_x)$ for some $0 < \vartheta < 1$. Because of (4.40b) and the convexity of $Z_0$ it is $\bar{z}_x \in Z_0$ for all $x \in D$ and therefore $w \in Z_0$ for all $z \in Z_0$. Hence, (4.40a) yields

$$\begin{aligned} u(\bar{z}_x) + \nabla u(\bar{z}_x)^T (z - \bar{z}_x) + \tfrac{1}{2}(z - \bar{z}_x)^T V (z - \bar{z}_x) &\le u(z) \\ \le u(\bar{z}_x) + \nabla u(\bar{z}_x)^T (z - \bar{z}_x) + \tfrac{1}{2}(z - \bar{z}_x)^T W (z - \bar{z}_x)& \\ \text{for all } x \in D \text{ and } z \in Z_0.& \end{aligned} \tag{4.44}$$

Putting $z := A(\omega)x - b(\omega)$ into (4.44), because of (4.40b) we find

$$\begin{aligned} & u(\bar{z}_x) + \nabla u(\bar{z}_x)^T \Big(A(\omega)x - b(\omega) - \bar{z}_x\Big) \\ & +\tfrac{1}{2}\Big(A(\omega)x - b(\omega) - \bar{z}_x\Big)^T V \Big(A(\omega)x - b(\omega) - \bar{z}_x\Big) \le u\Big(A(\omega)x - b(\omega)\Big) \le \\ & \le u(\bar{z}_x) + \nabla u(\bar{z}_x)^T \Big(A(\omega)x - b(\omega) - \bar{z}_x\Big) \\ & +\tfrac{1}{2}\Big(A(\omega)x - b(\omega) - \bar{z}_x\Big)^T W \Big(A(\omega)x - b(\omega) - \bar{z}_x\Big) \\ & \text{for all } x \in D \text{ and all } \omega \in \Omega. \end{aligned} \tag{4.45}$$

Taking now expectations on both sides of (4.45), using (4.42) and $\hat{x} = \binom{x}{-1}$ we get

$$u(\bar{z}_x) + \tfrac{1}{2}\hat{x}^T \hat{V} \hat{x} \le Eu\Big(A(\omega)x - b(\omega)\Big) \le u(\bar{z}_x) + \tfrac{1}{2}\hat{x}^T \hat{W} \hat{x} \quad \text{for all } x \in D. \tag{4.46a}$$

Furthermore, putting $z := \bar{A}y - \bar{b}$, where $y \in D$, into the second inequality of (4.44), we find

$$\begin{aligned} u(\bar{A}y - \bar{b}) &\le u(\bar{z}_x) + \nabla u(\bar{z}_x)^T \bar{A}(y - x) + \tfrac{1}{2}(y - x)^T \bar{A}^T W \bar{A}(y - x) \\ &= u(\bar{z}_x) + \Big(\bar{A}^T \nabla u(\bar{z}_x)\Big)^T (y - x) + \tfrac{1}{2}(y - x)^T (\bar{A}^T W \bar{A})(y - x) \\ &\text{for all } x \in D \text{ and } y \in D. \end{aligned} \tag{4.46b}$$

Considering now $F(y) - F(x)$ for vectors $x, y \in D$, from (4.46a) we get

$$\begin{aligned} F(y) - F(x) &= \Big(\bar{c}^T y + Eu\Big(A(\omega)y - b(\omega)\Big)\Big) - \Big(\bar{c}^T x + Eu\Big(A(\omega)x - b(\omega)\Big)\Big) \\ &\le \bar{c}^T (y - x) + \frac{1}{2}\hat{y}^T \hat{W} \hat{y} + u(\bar{A}y - \bar{b}) - \left(\frac{1}{2}\hat{x}^T \hat{V} \hat{x} + u(\bar{A}x - \bar{b})\right). \end{aligned}$$

Hence, we have the following result:

**Theorem 4.10.** *If* $D, \Big(A(\omega), b(\omega)\Big), Z_0$ *and* $V, W$ *are selected such that (4.40a,b) hold, then for all* $x, y \in D$ *it is*

$$F(y)-F(x) \leq \bar{c}^T(y-x)+\frac{1}{2}(\hat{y}^T\hat{W}\hat{y}-\hat{x}^T\hat{V}\hat{x})+u(\bar{A}y-\bar{b})-u(\bar{A}x-\bar{b}). \quad (4.47a)$$

*Using also inequality (4.46b), we get this corollary:*

**Corollary 4.9.** *Under the assumption of Theorem 4.10 it holds*

$$\begin{aligned} F(y)-F(x) \leq \bar{c}^T(y-x) + \frac{1}{2}(\hat{y}^T\hat{W}\hat{y}-\hat{x}^T\hat{V}\hat{x}) + \\ +\Big(\bar{A}^T\nabla u(\bar{A}x-\bar{b})\Big)^T(y-x) + \frac{1}{2}(y-x)^T\bar{A}^TW\bar{A}(y-x). \quad (4.47b) \end{aligned}$$

Represent the $(r+1)\times(r+1)$ matrix $\hat{W}$ by

$$\hat{W} = \begin{pmatrix} \widetilde{W} & \vdots & w \\ \cdots & \cdots & \cdots \\ w^T & \vdots & \gamma \end{pmatrix}, \quad (4.47c)$$

where $\widetilde{W}$ is a symmetric $r\times r$ matrix, $w$ is an $r$–vector and $\gamma$ is a real number. Then

$$\hat{y}^T\hat{W}\hat{y} = \hat{x}^T\hat{W}\hat{x} + (y-x)^T\widetilde{W}(y-x) + 2(\widetilde{W}x-w)^T(y-x),$$

and therefore

$$\begin{aligned} F(y)-F(x) &\leq \bar{c}^T(y-x) + \frac{1}{2}\hat{x}^T(\hat{W}-\hat{V})\hat{x} + \frac{1}{2}(y-x)^T(\widetilde{W}+\bar{A}^TW\bar{A})(y-x) \\ &\quad + \Big(\bar{A}^T\nabla u(\bar{A}x-\bar{b}) + \widetilde{W}x - w\Big)^T(y-x) \\ &= \frac{1}{2}\hat{x}^T(\hat{W}-\hat{V})\hat{x} + \Big(\bar{c}+\bar{A}^T\nabla u(\bar{A}x-\bar{b}) + \widetilde{W}x - w\Big)^T(y-x) \quad (4.47d) \\ &\quad + \frac{1}{2}(y-x)^T(\widetilde{W}+\bar{A}^TW\bar{A})(y-x). \end{aligned}$$

Thus, $F\Big(x+(y-x)\Big)-F(x)$ is estimated from above by a quadratic form in $h=y-x$.

Concerning the construction of descent directions of $F$ at $x$ we have this first immediate consequence from Theorem 4.10:

**Theorem 4.11.** *Suppose that the assumptions of Theorem 4.10 hold. If vectors $x, y \in D, y \neq x$, are related such that*

$$\bar{c}^Tx \geq (>)\bar{c}^Ty \quad (4.48a)$$

$$\bar{A}x = \bar{A}y \quad (4.48b)$$

$$\hat{x}^T\hat{V}\hat{x} \geq (>)\hat{y}^T\hat{W}\hat{y}, \quad (4.48c)$$

*then $F(x) \geq (>)F(y)$. Moreover, if $F(x) > F(y)$ or $F(x) = F(y)$ and $F$ is not constant on $xy$, then $h = y-x$ is a feasible descent direction of $F$ at $x$.*

Note. If the loss function $u$ is **monotone** with respect to the partial order on $\mathbb{R}^m$ induced by $\mathbb{R}^m_+$, then (4.48b) can be replaced by one of the weaker relations

$$\bar{A}x \geq \bar{A}y \text{ or } \bar{A}x \leq \bar{A}y. \tag{4.48b'}$$

If $u$ has the partial monotonicity property (4.10a,b), then (4.48b) can be replaced by (4.14b,b',c).
The next result follows from (4.47d).

**Theorem 4.12.** *Suppose that the assumptions of Theorem 4.10 hold. If vectors $x, y \in D, y \neq x$, are related such that*

$$(\widetilde{W} + \bar{A}^T W \bar{A})(y - x) = -\Big(\bar{c} + \bar{A}^T \nabla u(\bar{A}x - \bar{b}) + \widetilde{W}x - w\Big) \tag{4.49a}$$

$$\Big(\bar{c} + \bar{A}^T \nabla u(\bar{A}x - \bar{b}) + \widetilde{W}x - w\Big)^T (y - x) \leq (<) - \hat{x}^T(\hat{W} - \hat{V})\hat{x}, \tag{4.49b}$$

*then $F(y) \leq (<) F(x)$. Furthermore, if $F(y) < F(x)$ or $F(y) = F(x)$ and $F$ is not constant on $\overline{xy}$, then $h = y - x$ is a feasible descent direction for $F$ at $x$.*

*Proof.* If (4.49a,b) holds, then (4.47d) yields

$$\begin{aligned}
F(y) - F(x) &\leq \frac{1}{2}\hat{x}^T(\hat{W} - \hat{V})\hat{x} + \Big(\bar{c} + \bar{A}^T \nabla u(\bar{A}x - \bar{b}) + \widetilde{W}x - w\Big)^T (y - x) \\
&+ \frac{1}{2}(y - x)^T(\widetilde{W} + \bar{A}^T W \bar{A})(y - x) \\
&\leq \frac{1}{2}\hat{x}^T(\hat{W} - \hat{V})\hat{x} + \Big(\bar{c} + \bar{A}^T \nabla u(\bar{A}x - \bar{b}) + \widetilde{W}x - x\Big)^T (y - x) \\
&+ (-1)\frac{1}{2}(y - x)^T\Big(\bar{c} + \bar{A}^T \nabla u(\bar{T}x - \bar{b}) + \widetilde{W}x - w\Big) \\
&= \frac{1}{2}\Big(\hat{x}^T(\hat{W} - \hat{V})\hat{x} + \Big(\bar{c} + \bar{A}^T \nabla u(\bar{A}x - \bar{b}) + \widetilde{W}x - w\Big)^T (y - x)\Big) \\
&\leq (<) 0.
\end{aligned}$$

*Discussion of (4.49a,b)*: Suppose that

$$C = \tilde{W} + \bar{A}^T W \bar{A} \text{ is positive definite,} \tag{4.50a}$$

hence, there are positive numbers $\alpha, \beta$ such that

$$\alpha\|x\|^2 \leq x^T C x \leq \beta\|x\|^2. \tag{4.50b}$$

According to (4.49a) it is

$$y - x = -C^{-1}\Big(\bar{c} + \bar{A}^T \nabla u(\bar{A}x - \bar{b}) + \widetilde{W}x - w\Big),$$

and (4.49b) reads

$$\left(\bar{c}+\bar{A}^T\nabla u(\bar{A}x-\bar{b})+\widetilde{W}x-w\right)^T C^{-1}\left(\bar{c}+\bar{A}^T\nabla u(\bar{A}x-\bar{b})+\widetilde{W}x-w\right) \geq (>)\hat{x}^T(\hat{W}-\hat{V})\hat{x}$$

or

$$a^T C^{-1} a \geq \hat{x}^T(\hat{W}-\hat{V})\hat{x},$$

where

$$a = \bar{c} + \bar{A}^T\nabla u(\bar{A}x - \bar{b}) + \widetilde{W}x - w.$$

Because of (4.50b) it is

$$\frac{1}{\beta}\|a\|^2 \leq a^T C^{-1} a \leq \frac{1}{\alpha}\|a\|^2.$$

Thus, we find the following criterion for (4.49b):

i) If $\frac{1}{\beta}\left\|\bar{c} + \bar{A}^T\nabla u(\bar{A}x + \bar{b}) + \widetilde{W}x - w\right\|^2 \geq (>)\hat{x}^T(\hat{W} - \hat{V})\hat{x}$, then (4.49b) holds.

ii) If $\hat{x}^T(\hat{W} - \hat{V})\hat{x} > \frac{1}{\alpha}\left\|\bar{c} + \bar{A}^T\nabla u(\bar{A}x - \bar{b}) + \widetilde{W}x - w\right\|^2$, then (4.49b) is not valid.

# Part IV

# Semi-Stochastic Approximation Methods

# 5

# RSM–Based Stochastic Gradient Procedures

## 5.1 Introduction

According to the methods for converting an optimization problem with a random parameter vector $a = a(\omega)$ into a deterministic substitute problem, see Chapter 1 and Sections 2.1.1, 2.1.2, 2.2, a basic problem is the minimization of the total expected cost function $F = F(x)$ on a closed, convex feasible domain $D$ for the input or design vector $x$.

For simplification of notation, here the random total cost function $f = f\Big(a(\omega), x\Big)$ is denoted also by $f = f(\omega, x)$, hence,

$$f(\omega, x) := f\Big(a(\omega), x\Big).$$

Thus, the expected total cost minimization problem

$$\text{minimize } Ef(\omega, x) \text{ s.t. } \quad x \in D \tag{5.1}$$

must be solved. In this chapter we consider therefore the approximate solution of the mean value minimization problem (5.1) by a semi-stochastic approximation method based on the response surface methodology (RSM), see e.g. [16, 49, 96]. Hence, $f = f(\omega, x)$ denotes a function on $\mathbb{R}^r$ depending also on a random element $\omega$, and *"E"* is the expectation operator with respect to the underlying probability space $(\Omega, \mathcal{A}, P)$ containing $\omega$. Moreover, $D$ designates a closed convex subset of $\mathbb{R}^r$. We suppose that the objective function of (5.1)

$$F(x) = Ef(\omega, x), \qquad x \in \mathbb{R}^r, \tag{5.2}$$

is defined and finite for every $x \in \mathbb{R}^r$. Furthermore, we assume that the derivatives $\nabla F(x), \nabla^2 F(x), \ldots$ exist up to a certain order $\kappa$. In some cases, cf. [11] and Section 2.2, we may simply interchange differentiation and integration, hence,

$$\nabla F(x) = E\nabla_x f(\omega, x) \text{ or } \nabla F(x) = E\partial_x f(\omega, x), \tag{5.3a}$$

where $\nabla_x f, \partial_x f$ denotes the gradient, subgradient, resp., of $f$ with respect to $x$. Corresponding formulas may hold also for higher derivatives of $F$. However, in some important practical situations, especially for probability–valued functions $F(x)$, equation (5.3a) does not hold. According to Chapter 3 we know that gradient representation in integral form

$$\nabla F(x) = \int G(w,x)\mu(dw) \tag{5.3b}$$

with a certain vector-valued function $G = G(w,x)$, may be derived then, cf. also [136,145].

One of the main method for solving problem (5.1) is the stochastic approximation procedure [31,39,41,46,61,63,101]

$$X_{n+1} = p_D(X_n - \rho_n Y_n^s), \quad n = 1,2,\ldots, \tag{5.4a}$$

where $Y_n^s$ denotes an estimator of the gradient $\nabla F(x)$ at $x = X_n$. Important examples are the simple stochastic (sub)gradient or the finite difference gradient estimator

$$Y_n^s = \nabla_x f(\omega_n, X_n)\Big(Y_n^s \in \partial_x f(\omega_n, X_n)\Big), \text{ or } Y_n^s = G(w_n, X_n) \tag{5.4b}$$

$$Y_n^s = \sum_{j=1}^{r} \frac{1}{2c_n}\Big(f(\omega_{n,j}^{(1)}, X_n + c_n e_j) - f(\omega_{n,j}^{(2)}, X_n - c_n e_j)\Big)e_j \tag{5.4c}$$

with the unit vectors $e_1,\ldots,e_r$ or $\mathbb{R}^r$. Here, $\omega_n(w_n), n = 1,2,\ldots,\omega_{n,j}^{(k)}$, $k = 1,2$, $j = 1,2,\ldots,r$, $n = 1,2,\ldots$, resp., are sequences of independent realizations of the random element $\omega(w)$. Using formula (5.4b), (5.4c), resp., algorithm (5.4a) represents a Robbins-Monro (RM)-, a Kiefer-Wolfowitz (KW)-type stochastic approximation procedure. Moreover, $p_D$ designates the projection of $\mathbb{R}^r$ onto $D$, and $\rho_n > 0, n = 1,2,\ldots$, denotes the sequence of step sizes. A standard condition for the selection of $(\rho_n)$ is given [29–31] by

$$\sum_{n=1}^{\infty} \rho_n = +\infty, \qquad \sum_{n=1}^{\infty} \rho_n^2 < +\infty, \tag{5.4d}$$

and corresponding conditions for $(c_n)$ can be found in [149]. Due to its stochastic nature, algorithms of the type (5.4a) have only a very small asymptotic convergence rate in general cf. [149].

Considerable improvements of (5.4a) can be obtained [78] by

I) replacing $-Y_n^s$ at certain iteration points $X_n, n \in \mathbb{N}_1$, by an improved step direction $h_n$ of $F$ at $X_n$, see [70,71,75,77,81],
II) step size control, cf. [76,80,106].

Using (I), we get the following **hybrid stochastic approximation procedure**

$$X_{n+1} = \begin{cases} p_D(X_n + \rho_n h_n) & \text{for } n \in \mathbb{N}_1 \\ p_D(X_n - \rho_n Y_n^s) & \text{for } n \in \mathbb{N}_2 \end{cases} \tag{5.5}$$

where $\mathbb{N}_1, \mathbb{N}_2$ is a certain partition of the set $\mathbb{N}$ of integers.

Important types of improved step directions $h_n$ are:

* (feasible) deterministic descent directions $h_n$ existing at nonstationary/nonefficient points $X_n$ in problems (5.1) having certain functional properties, see Section 4 and [72–75,84]
* $h_n = -Y_n^e$, where $Y_n^e$ is a more exact estimator of $\nabla F(X_n)$, as e.g. the arithmetic mean of a certain number $M = M_n$ of simple stochastic gradients $Y_{n,i}^s = \nabla_x f(\omega_{n,i}, X_n), Y_{n,i}^s \in \partial_x f(\omega_{n,i}, X_n)$, resp., $i = 1, \ldots, M$, of the type (5.4b), cf. [77], or the estimator $Y_n^e$ defined by the extended central difference schemes given by Fabian [34,36].

By an appropriate selection of the partition $\mathbb{N}_1, \mathbb{N}_2$ we may guarantee that the stochastic approximation method is accelerated sufficiently, without using the improved and therefore more expensive step direction $h_n$ too often. Several numerical experiments show that improved step directions $h_n$ are most effective during an initial phase of the algorithm.

Having several generators of deterministic descent directions $h_n$ in special situations, as e.g. in stochastic linear programs with appropriate parameter distributions, see Section 4, in the following a general method for the computation of more exact gradient estimators $Y_n^e$ at an arbitrary iteration point $X_n = x_0$ is described. This method, which includes as a special case the well known finite difference schemes, used e.g. in the $KW$-type procedures [31,36,149], is based on the following fundamental principle of numerical differentiation, see e.g. [21,43,45]:

Starting from certain supporting points $x^{(i)}, i = 1, \ldots, p$, lying in a neighborhood $S$ of $x_0$ and the corresponding function values $y^{(i)} = F\left(x^{(i)}\right), i = 1, \ldots p$, by interpolation, first an interpolation function (e.g. a polynomial of degree $p-1$) $\hat{F} = \hat{F}(x)$ is computed. Then, the $k$th order numerical derivatives $\hat{\nabla}^k F$ of $F$ at $x_0$ are defined by the analytical derivatives

$$\hat{\nabla}^k F(x_0) := \nabla^k \hat{F}(x_0), \qquad k = 0, 1, \ldots, \kappa$$

of the interpolation function $\hat{F}$ at $x_0$. Clearly, in order to guarantee a certain accuracy of the higher numerical derivatives $\hat{\nabla}^k F(x_0)$, a sufficiently large number $p$ of interpolation points $x^{(i)}, i = 1, \ldots, p$ , is needed.

## 5.2 Gradient Estimation Using the Response Surface Methodology (RSM)

We consider a subdomain $S$ of the feasible domain $D$ such that

$$x_0 \in S \subset D, \tag{5.6a}$$

where $x_0$ is the point, e.g. $x_0 := X_n$, at which an estimate $\hat{\nabla}F(x_0)$ of $\nabla F$ must be computed. Corresponding to the principle of numerical differentiation described above, in order to estimate first the objective function $F(x)$ on $S$, estimates $y^{(i)}, i = 1,\dots,p$, of the function values $F\left(x^{(i)}\right), i = 1,\dots,p$, are determined at so-called design points

$$x^{(i)} = x^{(i)}(x_0) \in S, \qquad i = 1,2,\dots,p, \tag{5.6b}$$

chosen by the decision maker [16,56,96]. Simple estimators are

$$y^{(i)} := f\left(\omega^{(i)}, x^{(i)}\right) \qquad i = 1,2,\dots,p, \tag{5.6c}$$

where $\omega^{(1)}, \omega^{(2)}, \dots, \omega^{(p)}$ are independent realizations of the random element $\omega$.

The objective function $F$ is then approximated - on $S$ - mostly by a polynomial response surface model $\hat{F}(x) = \hat{F}(x|\beta_j, j \in J)$. In practice, usually first and/or second order models are used. The coefficients $\beta_j, j \in J$, are determined by means of regression analysis (mainly least squares estimation), see [22,49,56]. Having $\hat{F}(x)$, the RSM-gradient estimator $\hat{\nabla}F(x_0)$ at $x_0$ is defined by the gradient (with respect to $x$)

$$\hat{\nabla}F(x_0) := \nabla\hat{F}(x_0) \tag{5.6d}$$

of $\hat{F}$ at $x_0$. Using higher order polynomial response surface models $\hat{F}$, the estimates $\hat{\nabla}^k F(x_0)$ of higher order derivatives of $F$ at $x_0$ can be defined in the same way:

$$\hat{\nabla}^k F(x_0) := \nabla^k \hat{F}(x_0), \qquad k = 1,\dots,\kappa. \tag{5.6e}$$

*Remark 5.1.* The above gradient estimation procedure (5.6a-e) is especially useful if the gradient representation (5.3a) or (5.3b) does not hold (or is not known explicitly) for every $x \in \mathbb{R}^r$, or the numerical evaluation of $\nabla_x f(\omega, x), \partial_x f(\omega, x), G = G(w, x)$, resp., is too complicated.

The following very important example is taken from structural reliability [26,33,142]: A decisive role is played, see Sections 2.1, 3.1, by the failure probability $F(x) = 1 - P\left(y_l \le y(a(\omega), x) \le y_u\right)$ that, for a given design vector $x$ (e.g. structural dimensions), the structure does not fulfill the basic behavioral constraints $y_l \le y\left(a(\omega), x\right) \le y_u$. Here, $y_l, y_u$ are given lower and upper bounds, $y = y\left(a(\omega), x\right)$ is an $m$-vector of displacement and/or stress components of the structure, and $a = a(\omega)$ denotes the random vector of structural parameters, e.g. load parameters, yield stresses, elastic moduli. Obviously,

$$F(x) = Ef(\omega, x) \text{ with } f(\omega, x) := 1 - 1_B\left(y\left(a(\omega), x\right)\right),$$

where $1_B$ designates the characteristic function of $B = \{y \in \mathbb{R}^m : y_l \leq y \leq y_u\}$. Here, equation (5.3a) does not hold, and the numerical evaluation of differentiation formulas of the type (5.3b), see Chapter 3, may be very complicated - if they are available at all.

*Remark 5.2.* In addition to Remark 5.1, the RSM-estimator has still the following advantages:

a) A priori knowledge on $F$ can be incorporated easily in the selection of an appropriate response surface model. As an important example we mention here the use of so-called reciprocal variables $1/x_k$ or more general intermediate variables $t_k = t_k(x_k)$ besides the decision variables $x_k, k = 1, \ldots, r$, in structural optimization, see e.g. [5, 37].
b) The present method does not only yield estimates of the gradient (and some higher derivatives) of $F$ at $x_0$, but also estimates $\hat{F}(x_0)$ of the function value $F(x_0)$. However, besides gradient estimations, estimates of function values are the main indicators used to evaluate the performance of an algorithm, see [39]. Indeed, having these indicators, we may check whether the process runs into the "right direction", i.e. decreases (increases, resp.) the mean performance criterion $F(x)$, see [95]. Hence, appropriate stopping criterions can be formulated.

*Remark 5.3.* In many practical situations in a certain neighborhood of an optimal solution $x^*$ of (5.1), the objective function $F$ is nearly quadratic. Hence, $F$ may be described then by means of second order polynomial models $\hat{F}$ with high accuracy. Based on second order models $\hat{F}$ of $F$, estimates $\hat{x}^*$ of an optimal solution $x^*$ of (5.1) may be obtained by solving analytically the approximate program

$$\min \hat{F}(x) \text{ s.t. } \quad x \in D.$$

Furthermore, Newton-type step directions $h_n := -\nabla^2 \hat{F}(X_n)^{-1} \nabla \hat{F}(X_n)$, to be used in procedure (5.5), can be calculated easily.

*Remark 5.4.* The above remarks show that RSM has many very interesting features and advantages. But we have also to mention that for problems (5.1) with large dimensions $r$, the computational expenses to find a full second order model $\hat{F}$ or more accurate gradient estimates may become very large. However, an improved search direction $h_n$ is required in the semi-stochastic approximation procedure only at iteration points $X_n$ with $n \in \mathbb{N}_1$, cf. Section 5.1. Furthermore, within RSM there are several suggestions to reduce the estimation expenses. E.g., by the group screening few variables $x_{k_l}, l = 1, \ldots, r_1 (< r)$, can be identified. On the other hand, by decomposition - e.g. in structural optimization - a large optimization problem that might be intractable because of the large number of unknowns and constraints combined with a computationally expensive evaluation of the objective function $F$ can be converted into a set of much smaller and separate, but coordinated subproblems, see e.g. [132]. Hence, in many cases the optimization problem (5.1) can be reduced again to a moderate size.

### 5.2.1 The Two Phases of RSM

In Phase 1 of RSM, i.e., if the process $(X_n)$ is still far away from an optimal point $x^*$ of (5.1), then $F$ is estimated on $S$ by the linear empirical model

$$\hat{F}(x) = \beta_0 + \beta_I^T (x - x_0), \tag{5.7a}$$

where

$$\beta^T = (\beta_0, \beta_I^T) = (\beta_0, \beta_1, \ldots, \beta_r)^T \tag{5.7b}$$

is the $(r+1)$-vector of unknown coefficients of the linear model (5.7a). Having estimates $y^{(i)}$ of the function values $F\left(x^{(i)}\right)$ at the so-called design points $x^{(i)}, i = 1, \ldots, p$, in $S$, by least squares estimation (LSQ) the following estimate $\hat{\beta}$ of $\beta$ is obtained:

$$\hat{\beta} = (W^T W)^{-1} W^T y. \tag{5.8a}$$

Here, the $p \times (r+1)$ matrix $W$ and the $p$-vector $y$ are defined by

$$W = \begin{pmatrix} 1 & d^{(1)^T} \\ 1 & d^{(2)^T} \\ \vdots & \vdots \\ 1 & d^{(p)^T} \end{pmatrix}, \quad y = \begin{pmatrix} y^{(1)} \\ y^{(2)} \\ \vdots \\ y^{(p)} \end{pmatrix} \tag{5.8b}$$

with $d := x - x_0$ and therefore

$$d^{(i)} = x^{(i)} - x_0, \quad i = 1, 2, \ldots, p. \tag{5.8c}$$

In order to guarantee the regularity of $W^T W$, we require that

$$d^{(2)} - d^{(1)}, \ldots, d^{(p)} - d^{(1)}, p \geq r+1, \text{ span the } \mathbb{R}^r. \tag{5.8d}$$

Having $\hat{\beta}$ by (5.8a), the gradient $\nabla F$ of $F$ can be estimated on $S$ by the gradient of (5.7a), hence,

$$\hat{\nabla} F(x) = \hat{\beta}_I \text{ for all } x \in S. \tag{5.9}$$

The accuracy of the estimator (5.9) on $S$, cf. (5.6a), is studied in the following way: Denote by

$$\varepsilon^{(i)} = \varepsilon^{(i)}\left(\omega, x^{(i)}\right), \quad i = 1, 2, \ldots, p, \tag{5.10a}$$

the stochastic error of the estimate $y^{(i)}$ of $F\left(x^{(i)}\right)$, see e.g. (5.6c). By means of first order Taylor expansion of $F$ at $x_0$ we obtain, cf. (5.8c),

$$y^{(i)} = F\left(x^{(i)}\right) + \varepsilon^{(i)} = F(x_0) + \nabla F(x_0)^T d^{(i)} + R_1^{(i)} + \varepsilon^{(i)}, \tag{5.10b}$$

where

$$R_1^{(i)} = \frac{1}{2} d^{(i)^T} \nabla^2 F\left(x_0 + \vartheta_i d^{(i)}\right) d^{(i)}, \quad 0 < \vartheta_i < 1, \tag{5.10c}$$

denotes the remainder of the first order Taylor expansion. Obviously, if $S$ is a convex subset of the feasible domain $D$ of (5.1), and the Hessian $\nabla^2 F(x)$ of the objective function $F$ of (5.1) is bounded on $S$, then

$$\begin{aligned} \left|R_1^{(i)}\right| &\leq \frac{1}{2}\left\|d^{(i)}\right\| \sup_{0 \leq \vartheta \leq 1}\left\|\nabla^2 F\left(x_0 + \vartheta d^{(i)}\right)\right\| \\ &\leq \frac{1}{2}\left\|d^{(i)}\right\|^2 \sup_{x \in S}\left\|\nabla^2 F(x)\right\|. \end{aligned} \tag{5.10d}$$

Thus, representation (5.10b) yields the data model

$$y = W\beta + R_1 + \varepsilon, \tag{5.11a}$$

where $W, y$ are given by (5.8b), and the vectors $\beta, R_1$ and $\varepsilon$ are defined by

$$\beta = \begin{pmatrix} F(x_0) \\ \nabla F(x_0) \end{pmatrix}, \quad R_1 = \begin{pmatrix} R_1^{(1)} \\ R_1^{(2)} \\ \vdots \\ R_1^{(p)} \end{pmatrix}, \quad \varepsilon = \begin{pmatrix} \varepsilon^{(1)} \\ \varepsilon^{(2)} \\ \vdots \\ \varepsilon^{(p)} \end{pmatrix}. \tag{5.11b}$$

Concerning the $p$-vector $\varepsilon$ of random errors $\varepsilon^{(i)}, i = 1, \ldots, p$, we make the following (standard) assumptions

$$E(\varepsilon|x_0) = 0, \ \mathrm{cov}(\varepsilon|x_0) = E(\varepsilon\varepsilon^T|x_0) = Q := (\sigma_i^2 \delta_{ij}), \qquad (5.11c),(5.11d)$$

where $\delta_{ij}, i, j = 1, \ldots, p$, denotes the Kronecker symbol, and

$$\sigma_i^2 = \sigma_i^2(x_0, d^{(1)}, d^{(2)}, \ldots, d^{(p)}), \quad i = 1, 2, \ldots, p. \tag{5.11e}$$

Putting now the data model (5.11a) into (5.8a), we get

$$\hat{\beta} = \begin{pmatrix} F(x_0) \\ \nabla F(x_0) \end{pmatrix} + (W^T W)^{-1} W^T (R_1 + \varepsilon). \tag{5.12a}$$

Hence, for the estimator $\hat{\beta}_I$ of $\nabla F(x_0)$, cf. (5.9), we find

$$\begin{aligned} \hat{\beta}_I = \nabla F(x_0) + \left(\frac{1}{p}\sum_{i=1}^{p}\left(d^{(i)} - \overline{d}\right)\left(d^{(i)} - \overline{d}\right)'\right)^{-1} \frac{1}{p}\Big(d^{(1)} - \overline{d}, \ldots, \\ d^{(p)} - \overline{d}\Big)(R_1 + \varepsilon), \end{aligned} \tag{5.12b}$$

where $\overline{d}$ is defined by

$$\overline{d} = \frac{1}{p}\sum_{i=1}^{p} d^{(i)}, \quad d^{(i)} = x^{(i)} - x_0. \tag{5.12c}$$

Clearly, (5.12b,c) represents the accuracy of the estimator (5.9) on $S$.

*Note.* According to (5.12a), the accuracy of $\hat{\beta}$ depends on the deterministic model error $R_1$, the stochastic observation error $\varepsilon$ and also on the choice of the design points $x^{(1)}, \ldots, x^{(p)}$, see e.g. [16,123].

In Phase 2 of RSM, i.e. if the process $(X_n)$ is approaching a certain neighborhood of a minimal point $x^*$ of (5.1), see Remark 5.3, then the linear model (5.7a) of $F$ is no more adequate, in general. It is replaced therefore by the more general quadratic empirical model

$$\hat{F}(x) = \beta_0 + \beta_I^T(x - x_0) + (x - x_0)^T B(x - x_0), \quad x \in S. \tag{5.13a}$$

The unknown parameters $\beta_0, \beta_i, i = 1, \ldots, r$, in the $r$-vector $\beta_I$, and $\beta_{ij}, i, j = 1, \ldots, r$, in the symmetric $r \times r$ matrix $B$, are determined again by least squares estimation. The LSQ-estimate

$$\hat{\beta} = (\hat{\beta}_0, \hat{\beta}_I, \hat{B})$$

of the $\frac{1}{2}(r^2 + 3r + 2)$-vector

$$\begin{aligned}\beta^T = (\beta_0, \beta_1, \beta_2, \ldots, \beta_r, \beta_{11}, \beta_{12}, \ldots, \beta_{1r}, \beta_{22}, \beta_{23}, \ldots,\\ \beta_{2r}, \ldots, \beta_{r-1r-1}, \beta_{r-1r}, \beta_{rr}),\end{aligned} \tag{5.13b}$$

of unknown parameters in model (5.13a), denoted for simplicity also by

$$\beta = (\beta_0, \beta_1, B), \tag{5.13b'}$$

is given - in correspondence with formula (5.8a) - again by

$$\hat{\beta} = (W^T W)^{-1} W^T y. \tag{5.14a}$$

Here the $p \times \frac{1}{2}(r^2 + 3r + 2)$ matrix $W$ and the $p$-vector $y$ are defined by

$$W = \begin{pmatrix} 1 & d^{(1)^T} & z^{(1)^T} \\ 1 & d^{(2)^T} & z^{(2)^T} \\ \vdots & \vdots & \vdots \\ 1 & d^{(p)^T} & z^{(p)^T} \end{pmatrix}, \quad y = \begin{pmatrix} y^{(1)} \\ y^{(2)} \\ \vdots \\ y^{(p)} \end{pmatrix}, \tag{5.14b}$$

where $d^{(i)} = x^{(i)} - x_0$ see (5.8c), and

$$z^{(i)} = \left(d_1^{(i)2}, 2d_1^{(i)}d_2^{(i)}, \ldots, 2d_1^{(i)}d_r^{(i)}, d_2^{(i)2}, 2d_2^{(i)}d_3^{(i)}, \ldots, 2d_2^{(i)}d_r^{(i)}, \ldots, d_r^{(i)2}\right)^T. \tag{5.14c}$$

Corresponding to the first phase, the existence of $(W'W)^{-1}$ is guaranteed - for the full quadratic model - by the assumption

$$\begin{pmatrix} d^{(2)} - d^{(1)} \\ z^{(2)} - z^{(1)} \end{pmatrix}, \ldots, \begin{pmatrix} d^{(p)} - d^{(1)} \\ z^{(p)} - z^{(1)} \end{pmatrix} \text{ span the } \mathbb{R}^{(r^2+3r+2)/2}, \tag{5.14d}$$

see [16], [96], hence,

$$p \geq \frac{(r+1)(r+2)}{2} + 1.$$

Having $\hat{\beta} = (\hat{\beta}_0, \hat{\beta}_I, \hat{B})$, the gradient of $F$ is then estimated on $S$ by the gradient of (5.13a), thus

$$\hat{\nabla} F(x) = \hat{\beta}_I + 2\hat{B}(x - x_0) \quad \text{for} \quad x \in S \tag{5.15a}$$

and therefore

$$\hat{\nabla} F(x_0) = \hat{\beta}_I. \tag{5.15b}$$

*Note.* In practice, in the beginning of Phase 2 only the pure quadratic terms $\beta_{ii}(x_i - x_{0i})^2, i = 1, \ldots, r$, may be added (successively) to the linear model (5.7a).

For the consideration of the accuracy of the estimator (5.15b) we use now second order Taylor expansion. Corresponding to (5.10b), we obtain

$$y^{(i)} = F(x^{(i)}) + \varepsilon^{(i)} = F(x_0) + \nabla F(x_0)' d^{(i)} + \frac{1}{2} d^{(i)'} \nabla^2 F(x_0) d^{(i)} + R_2^{(i)} + \varepsilon^{(i)}, \tag{5.16a}$$

where

$$R_2^{(i)} = \frac{1}{3!} \nabla^3 F\left(x_0 + \vartheta_i d^{(i)}\right) \cdot d^{(i)3}, 0 < \vartheta_i < 1, \tag{5.16b}$$

denotes the remainder of the second order Taylor expansion. Clearly, if $S$ is a convex subset of $D$ and the multilinear form $\nabla^3 F(x)$ of all 3rd order derivatives of $F$ is bounded on $S$, then

$$\begin{aligned} |R_2^{(i)}| &\leq \frac{1}{6} \|d^{(i)}\|^3 \sup_{0 \leq \vartheta \leq 1} \|\nabla^3 F(x_0 + \vartheta d^{(i)})\| \\ &\leq \frac{1}{6} \|d^{(i)}\|^3 \sup_{x \in S} \|\nabla^3 F(x)\|. \end{aligned} \tag{5.16c}$$

Representation (5.16a) yields now the data model

$$y = W\beta + R_2 + \varepsilon. \tag{5.17a}$$

Here, $W, y$ are given by (5.14b), and the vectors $\beta, R_2, \varepsilon$ are defined, cf. (5.13b), (5.13b') for the notation, by

$$\beta = \left(F(x_0), \nabla F(x_0), \frac{1}{2} \nabla^2 F(x_0)\right), \quad R_2 = \begin{pmatrix} R_2^{(1)} \\ R_2^{(2)} \\ \vdots \\ R_2^{(p)} \end{pmatrix}, \quad \varepsilon = \begin{pmatrix} \varepsilon^{(1)} \\ \varepsilon^{(2)} \\ \vdots \\ \varepsilon^{(p)} \end{pmatrix}. \tag{5.17b}$$

Corresponding to (5.11c-e) we also suppose in Phase 2 that

$$E(\varepsilon|x_0) = 0, \quad \mathrm{cov}(\varepsilon|x_0) = E(\varepsilon\varepsilon^T|x_0) = Q := (\sigma_i^2\delta_{ij}) \quad (5.17c), (5.17d)$$

$$\sigma_i^2 = \sigma_i^2\Big(x_0, d^{(1)}, d^{(2)}, \dots, d^{(p)}\Big), \quad i = 1, \dots, p. \quad (5.17e)$$

Putting the data model (5.17a) into (5.14a), we get - see (5.13b), (5.13b') concerning the notation -

$$\hat{\beta} = \Big(F(x_0), \nabla F(x_0), \frac{1}{2}\nabla^2 F(x_0)\Big) + (W^TW)^{-1}W^TR_2 + (W^TW)^{-1}W^T\varepsilon. \quad (5.18a)$$

Especially, for the estimator $\hat{\beta}_I$ of $\nabla F(x_0)$, see (5.15a,b), we find

$$\hat{\beta}_I = \nabla F(x_0) + \frac{1}{p}U^{-1}\Big\{(d^{(1)} - \overline{d}, \dots, d^{(p)} - \overline{d}) \quad (5.18b)$$

$$- \Big(\frac{1}{p}\sum_{i=1}^{p}(d^{(i)} - \overline{d})(z^{(i)} - \overline{z})^T\Big) Z^{-1}(z^{(1)} - \overline{z}, \dots, z^{(p)} - \overline{z})\Big\}(R_2 + \varepsilon).$$

The vectors $d^{(i)}, \overline{d}, z^{(i)}$ are given by (5.8c),(5.12c),(5.14c), resp., and $\overline{z}, Z, U$ are defined by

$$\overline{z} = \frac{1}{p}\sum_{i=1}^{p} z^{(i)}, \quad Z = \frac{1}{p}\sum_{i=1}^{p}(z^{(i)} - \overline{z})(z^{(i)} - \overline{z})^T, \quad (5.18c), (5.18d)$$

$$U = \frac{1}{p}\sum_{i=1}^{p}(d^{(i)} - \overline{d})(d^{(i)} - \overline{d})^T - \Big(\frac{1}{p}\sum_{i=1}^{p}(d^{(i)} - \overline{d})(z^{(i)} - \overline{z})^T\Big) Z^{-1} \quad (5.18e)$$

$$\times \Big(\frac{1}{p}\sum_{i=1}^{p}(z^{(i)} - \overline{z})(d^{(i)} - \overline{d})^T\Big).$$

### 5.2.2 The Mean Square Error of the Gradient Estimator

According to (5.9) and (5.12b), (5.15b) and (5.18b), resp., in both phases for the gradient estimator $\hat{\nabla} F(x_0)$ we find

$$\hat{\nabla} F(x_0) = \nabla F(x_0) + H(R + \varepsilon) \quad (5.19)$$

with $R = R_1, R = R_2$. Furthermore, $H$ designates the $r \times p$ matrix arising in the representation (5.12b),(5.18b), resp., of $\hat{\nabla} F(x_0) = \hat{\beta}_1$.

The mean square error $V = V(x_0)$ of the estimator $\hat{\nabla} F(x_0)$

$$V := E\Big(\|\hat{\nabla} F(x_0) - \nabla F(x_0)\|^2 |x_0\Big) \quad (5.20a)$$

is given, cf. see (5.11c),(5.17c), resp., by the sum

$$V = \|e^{\text{det}}\|^2 + E\Big(\|e^{\text{stoch}}\|^2 | x_0\Big) \tag{5.20b}$$

of the squared bias $\|e^{\text{det}}\|^2 := \|HR\|^2$ and the variance of $\hat{\nabla} F(x_0)$

$$E\Big(\|e^{\text{stoch}}\|^2 | x_0\Big) := E\Big(\|H\varepsilon\|^2 | x_0\Big).$$

Since $\|w\|^2 = \text{tr}(ww^T) = \|ww^T\|$ for a vector $w$, where $\text{tr}(M), \|M\|$, resp., denotes the trace, the operator norm of a square matrix $M$, we find

$$\|e^{\text{det}}\|^2 = \|HRR^TH^T\| = \sup_{\|w\|=1} w^T(HRR^TH^T)w \leq \|R\|^2\|HH^T\|. \tag{5.20c}$$

Moreover, using that $\text{tr}(M)$ is equal to the sum of the eigenvalues of $M$, we get

$$\begin{aligned} E(\|e^{\text{stoch}}\|^2 | x_0) &= E\Big(tr(H\varepsilon\varepsilon^T H^T) | x_0\Big) \\ &= tr(HQH^T) \leq r \sup_{\|w\|=1} (H^T w)^T (\sigma_i^2 \delta_{ij})(H^T w) \\ &\leq r \left(\max_{1 \leq i \leq p} \sigma_i^2\right) \sup_{\|w\|=1} \|H^T w\|^2 = r \left(\max_{1 \leq i \leq p} \sigma_i^2\right) \|HH^T\|, \end{aligned} \tag{5.20d}$$

cf. (5.11d),(5.17d), respectively. Consequently, inequalities (5.20c),(5.20d) yield in the first and in the second phase

$$V \leq \left(\|R\|^2 + r \max_{1 \leq i \leq p} \sigma_i^2\right) \|HH^T\|. \tag{5.20e}$$

Here, the matrix product $HH^T$ can be computed explicitly: In Phase 1, due to (5.12b) we have

$$H = H_0^{-1} \frac{1}{p} (d^{(1)} - \overline{d}, d^{(2)} - \overline{d}, \ldots, d^{(p)} - \overline{d}), \tag{5.21a}$$

where $d^{(i)}, \overline{d}$ are given by (5.8c),(5.12c), resp., and

$$H_0 = \frac{1}{p} \sum_{i=1}^{p} (d^{(i)} - \overline{d})(d^{(i)} - \overline{d})^T. \tag{5.21b}$$

By an easy calculation, from (5.21a,b) we get

$$HH^T = \frac{1}{p} H_0^{-1} H_0 H_0^{-1} = \frac{1}{p} H_0^{-1}. \tag{5.21c}$$

Assume now that the function value $F\Big(x^{(i)}\Big)$ is estimated $\nu$ times at

$$x^{(i)} = x_0 + d^{(i)} \text{ for each } i = 1, \ldots, p_0. \tag{5.22a}$$

Hence, $p = \nu p_0$ and

$$d^{(kp_0+i)} = d^{(i)}, x^{(kp_0+i)} = x^{(i)}, i = 1, 2, \ldots, p_0, k = 1, 2, \ldots, \nu - 1. \tag{5.22b}$$

Putting

$$H_0^{(0)} := \frac{1}{p_0} \sum_{i=1}^{p_0} \left(d^{(i)} - \overline{d}^{(0)}\right)\left(d^{(i)} - \overline{d}^{(0)}\right)^T, \quad \overline{d}^{(0)} := \frac{1}{p_0} \sum_{i=1}^{p_0} d^{(i)}, \tag{5.23a}$$

in case (5.22a,b) we easily find

$$HH^T = \frac{1}{\nu}\left(\frac{1}{p_0} H_0^{(0)-1}\right). \tag{5.23b}$$

In Phase 2 matrix $H$ reads, see (5.18b),

$$\begin{aligned} H = H_0^{-1}\frac{1}{p}\Bigg\{&\left(d^{(1)} - \overline{d}, \ldots, d^{(p)} - \overline{d}\right) \\ &- \left(\frac{1}{p}\sum_{i=1}^{p}\left(d^{(i)} - \overline{d}\right)\left(z^{(i)} - \overline{z}\right)^T\right) Z^{-1}\left(z^{(1)} - \overline{z}, \ldots, z^{(p)} - \overline{z}\right)\Bigg\}, \end{aligned} \tag{5.24a}$$

where $d^{(i)}, \overline{d}, z^{(i)}, \overline{z}, Z$ are given by (5.8c), (5.12c), (5.14c), (5.18c), (5.18d), resp., and $H_0$ is defined, cf. (5.18e), here by

$$H_0 := U. \tag{5.24b}$$

Corresponding to (5.21c), from (5.24a,b), also in Phase 2 we get

$$HH^T = \frac{1}{p} H_0^{-1}. \tag{5.24c}$$

In case (5.22a,b) of estimating $F$ $\nu$ times at $p_0$ different points, one can easily see that an equation corresponding to (5.23b) holds also in Phase 2.

Consequently, from (5.20e), and (5.21c), (5.24c), resp., for the mean square estimation error $V$ we obtain in both phases

$$V \leq \left(\|R\|^2 + r \max_{1 \leq i \leq p} \sigma_i^2\right) \frac{1}{p} \|H_0^{-1}\|, \tag{5.25a}$$

where $H_0$ is given by (5.21b), (5.24b), respectively. In case (5.22a,b) we get

$$V \leq \left(\|R\|^2 + r \max_{1 \leq i \leq p} \sigma_i^2\right) \frac{1}{\nu p_0} \|H_0^{(0)-1}\|, \tag{5.25b}$$

provided that

$$\sigma^2_{(kp_0+i)} = \sigma_i^2, \quad i = 1,2,\ldots,p_0, \quad k = 1,\ldots,\nu-1. \tag{5.26}$$

Finally, we have to consider $\|R^2\|$ : First, in Phase 1 according to (5.10d) and (5.11b) we find with certain numbers $\vartheta_i, 0 < \vartheta_i < 1, i = 1,\ldots,p$,

$$\begin{aligned}\|R_1\|^2 &\le \frac{1}{4}\sum_{i=1}^{p} \|d^{(i)}\|^4 \left\|\nabla^2 F\left(x_0 + \vartheta_i d^{(i)}\right)\right\|^2 \\ &\le \frac{1}{4}\sum_{i=1}^{p} \|d^{(i)}\|^4 \sup_{0\le\vartheta\le 1} \left\|\nabla^2 F\left(x_0 + \vartheta d^{(i)}\right)\right\|^2 \\ &\le \frac{1}{4}\left(\sum_{i=1}^{p} \|d^{(i)}\|^4\right) \sup_{x\in S}\left\|\nabla^2 F(x)\right\|^2,\end{aligned} \tag{5.27a}$$

where the last inequality holds if $S$, cf. (5.6a,b) is convex, and $\left\|\nabla^2 F(x)\right\|$ is bounded on $S$. In Phase 2, with (5.16b), (5.17b), we find

$$\begin{aligned}\|R_2\|^2 &\le \frac{1}{36}\sum_{i=1}^{p} \|d^{(i)}\|^6 \left\|\nabla^3 F\left(x_0 + \vartheta_i d^{(i)}\right)\right\|^2 \\ &\le \frac{1}{36}\sum_{i=1}^{p} \|d^{(i)}\|^6 \sup_{0\le\vartheta\le 1} \left\|\nabla^3 F\left(x_0 + \vartheta d^{(i)}\right)\right\|^2 \\ &\le \frac{1}{36}\left(\sum_{i=1}^{p} \|d^{(i)}\|^6\right) \sup_{x\in S}\|\nabla^3 F(x)\|^2,\end{aligned} \tag{5.27b}$$

where the last inequality holds under corresponding assumptions as used above. Summarizing (5.27a), (5.27b), we have

$$\|R_l\|^2 \le \frac{1}{((l+1)!)^2}\sum_{i=1}^{p} \|d^{(i)}\|^{2(l+1)} \sup_{0\le\vartheta\le 1} \left\|\nabla^{l+1} F(x_0 + \vartheta d^{(i)})\right\|^2, \quad l = 1,2. \tag{5.27c}$$

Putting (5.27c) into (5.25a), for $l = 1,2$ we find

$$\begin{aligned}V \le \Bigg(&\frac{1}{((l+1)!)^2}\frac{1}{p}\sum_{i=1}^{p} \|d^{(i)}\|^{2(l+1)} \sup_{0\le\vartheta\le 1} \left\|\nabla^{l+1} F(x_0 + \vartheta d^{(i)})\right\|^2 \\ &+ \frac{r}{p}\max_{1\le i\le p}\sigma_i^2\Bigg)\|H_0^{-1}\|.\end{aligned} \tag{5.28}$$

We consider now the case

$$d^{(i)} = \mu \tilde{d}^{(i)}, \quad i = 1,2,\ldots,p, \quad 0 < \mu \le 1, \tag{5.29a}$$

where $\tilde{d}^{(i)}, i = 1,\ldots,p$, are given $r$-vectors. Obviously

$$H_0 = \mu^2 \tilde{H}_0, \tag{5.29b}$$

where the $r \times r$ matrix $\tilde{H}_0$ follows from (5.21b), (5.24b), resp., by replacing $d^{(i)}$ by $\tilde{d}^{(i)}, i = 1, \ldots, p$. Using (5.29a), from (5.28) we get for $l = 1, 2$

$$V \leq \left( \frac{\mu^{2l}}{\left((l+1)!\right)^2} \frac{1}{p} \sum_{i=1}^{p} \|\tilde{d}^{(i)}\|^{2(l+1)} \sup_{0 \leq \vartheta \leq 1} \left\| \nabla^{l+1} F(x_0 + \vartheta \tilde{d}^{(i)}) \right\|^2 \right.$$
$$\left. + \frac{r\mu^{-2}}{p} \max_{l \leq i \leq p} \sigma_i^2 \right) \|\tilde{H}_0^{-1}\|. \tag{5.30}$$

## 5.3 Estimation of the Mean Square (Mean Functional) Error

Considering now the convergence behavior of the hybrid stochastic approximation procedure (5.5) working with the RSM-gradient estimator $Y_n^e$, we need the following notations.

Let $\mathbb{N}_1, \mathbb{N}_2$ be a given partition of $\mathbb{N}$, and suppose that $X_{n+1} = p_D(X_n - \rho_n Y_n^s)$ with a simple gradient estimator $Y_n^s$, see (5.4b), at each iteration point $X_n$ with $n \in \mathbb{N}_2$. Furthermore, define for each $X_n$ with $n \in \mathbb{N}_1$ the improved step direction $h_n = -Y_n^e$ by the following scheme:

Following Section 5.2, at $x_0 = X_n, n \in \mathbb{N}_1$, we select $r$-vectors $d^{(i)}, i = 1, \ldots, p$. In the most general case it is

$$d^{(i)} = d^{(i,n)} = d^{(i,n)}(X_n), \quad i = 1, 2, \ldots, p = p(n), \tag{5.31a}$$

i.e. the increments $d^{(1)}, \ldots, d^{(p)}$ and $p$ may depend on the stage index $n$ and the iteration point $X_n$. In the most simple case we have

$$d^{(i,n)} = d^{(i,0)}, \quad i = 1, 2, \ldots, p = \bar{p} \text{ for all } n = 1, 2, \ldots, \tag{5.31b}$$

where $d^{(1,0)}, \ldots, d^{(\bar{p},0)}$ are $\bar{p}$ given, fixed $r$-vectors.

The unknown objective function $F$ is then estimated at

$$x^{(i)} = x^{(i,n)} = X_n + d^{(i,n)}, \quad i = 1, 2, \ldots, p(n), \tag{5.31c}$$

i.e., we have the estimates

$$y^{(i)} = y^{(i,n)} = F(x^{(i,n)}) + \varepsilon^{(i,n)}, \quad i = 1, 2, \ldots, p(n). \tag{5.31d}$$

According to (5.8b),(5.14b), resp., we define

$$y_n = \begin{pmatrix} y^{(1,n)} \\ y^{(2,n)} \\ \vdots \\ y^{(p,n)} \end{pmatrix} \text{ with } p = p(n) \tag{5.31e}$$

$$W_n = \begin{pmatrix} 1 & d^{(1,n)^T} \\ 1 & d^{(2,n)^T} \\ \vdots & \vdots \\ 1 & d^{(p,n)^T} \end{pmatrix}, W_n = \begin{pmatrix} 1 & d^{(1,n)^T} & z^{(1,n)^T} \\ 1 & d^{(2,n)^T} & z^{(2,n)^T} \\ \vdots & \vdots & \vdots \\ 1 & d^{(p,n)^T} & z^{(p,n)^T} \end{pmatrix} \quad (5.31f)$$

for Phase 1, Phase 2, resp., where

$$z^{(i,n)} = \Big(d_1^{(i,n)^2}, 2d_1^{(i,n)}d_2^{(i,n)}, \ldots, 2d_1^{(i,n)}d_r^{(i,n)}, d_2^{(i,n)^2},$$

$$2d_2^{(i,n)}d_3^{(i,n)}, \ldots, d_r^{(i,n)^2}\Big)^T, i = 1, 2, \ldots, p, \quad (5.31g)$$

cf. (5.14c). Note that in case (5.31b) the matrices $W_n$ are fixed.

Finally, the RSM-step directions $h_n$ are defined by

$$h_n = -Y_n^e = -\hat{\nabla} F(X_n) := -\hat{\beta}(n)_I \text{ with } \hat{\beta}(n) = (W_n^T W_n)^{-1} W_n^T y_n, \quad (5.32)$$

see (5.8a),(5.9) for Phase 1 and (5.14a), (5.15b) for Phase 2. Of course, we suppose that all matrices $W_n^T W_n, n = 1, 2, \ldots$, are regular, see (5.8d),(5.14d).

### 5.3.1 The Argument Case

Here we consider the behavior of the hybrid stochastic approximation method (5.5) by means of the mean square errors

$$b_n = E\|X_n - x^*\|^2, n \geq 1, \text{ with an optimal solution } x^* \text{ of } (5.1). \quad (5.33)$$

Following [80], in the present case we suppose

$$k_0\|x - x^*\|^2 \leq (x - x^*)^T\Big(\nabla F(x) - \nabla F(x^*)\Big) \leq k_1\|x - x^*\|^2 \quad (5.34a)$$

$$k_0^2\|x - x^*\|^2 \leq \|\nabla F(x) - \nabla F(x^*)\|^2 \leq k_1^2\|x - x^*\|^2 \quad (5.34b)$$

for all $x \in D$ where $k_0, k_1$ are some constants such that

$$k_1 \geq k_0 > 0. \quad (5.34c)$$

Moreover, for the mean square estimation error

$$V_n^s(X_n) = E\left(\left\|Y_n^s - \nabla F(X_n)\right\|^2 |X_n\right) \quad (5.35a)$$

of the simple stochastic gradient $Y_n^s$ according to (5.4b), if (5.3a) holds true, we suppose that

$$V_n^s(X_n) \leq \sigma_1^{s2} + C_1^s\|X_n - x^*\|^2 \text{ with constants } \sigma_1^2, C_1^s \geq 0. \quad (5.35b)$$

For $n \in \mathbb{N}_2$ with $Y_n^s$ given by (5.4b) we obtain [80]

$$b_{n+1} \leq \left(1 - 2k_0\rho_n + (k_1^2 + C_1^s)\rho_n^2\right)b_n + \sigma_1^{s2}\rho_n^2. \tag{5.36}$$

For $n \in \mathbb{N}_1$ we have, see (5.32), $X_{n+1} = p_D(X_n - \rho_n Y_n^e)$, and therefore [80]

$$\begin{aligned} E\Big(\|X_{n+1} - x^*\|^2 | X_n\Big) \leq \|X_n - x^*\|^2 \\ - 2\rho_n E\Big((X_n - x^*)^T\Big(Y_n^e - \nabla F(x^*)\Big)|X_n\Big) \\ + \rho_n^2\Big(E\Big(\|Y_n^e - \nabla F(X_n)\|^2|X_n\Big) \\ + 2E\Big(\Big(Y_n^e - \nabla F(X_n)\Big)^T\Big(\nabla F(X_n) - \nabla F(x^*)\Big)|X_n\Big) \\ + \|\nabla F(X_n) - \nabla F(x^*)\|^2\Big). \end{aligned} \tag{5.37}$$

According to (5.19) in Phase 1 and Phase 2 we get

$$E(Y_n^e|X_n) = \nabla F(X_n) + E\Big(H_n(R_n + \varepsilon_n)|X_n\Big). \tag{5.38a}$$

In this equation, cf. (5.10c) and (5.11b), (5.16b) and (5.17b), resp., (5.31a-g),

$$R_n = \begin{pmatrix} R_l^1(X_n, d^{(1,n)}) \\ R_l^2(X_n, d^{(2,n)}) \\ \vdots \\ R_l^p(X_n, d^{(p(n),n)}) \end{pmatrix}, l = 1, 2, \varepsilon_n = \begin{pmatrix} \varepsilon^{(1,n)} \\ \varepsilon^{(2,n)} \\ \vdots \\ \varepsilon^{(p(n),n)} \end{pmatrix} \tag{5.38b}$$

is the vector of remainders resulting from using first, second order empirical models for $F$, the vector of stochastic errors in estimating $F(x)$ at $X_n + d^{(i,n)}, i = 1, 2, \ldots, p(n)$, respectively. Furthermore, $H_n$ is the $r \times p$ matrix given by (5.21a),(5.24a), resp., if there we put $d^{(i)} := d^{(i,n)}, z^{(i)} := z^{(i,n)}, i = 1, 2, \ldots, p = p(n)$. Hence,

$$E(Y_n^e|X_n) = \nabla F(X_n) + e_n^{\text{det}},$$

where, cf. (5.20a-c), $e_n^{\text{det}} = H_n R_n$ is the bias resulting from using a first or second order empirical model for $F$. Consequently

$$\begin{aligned} & E\Big((X_n - x^*)^T\Big(Y_n^e - \nabla F(x^*)\Big)|X_n\Big) \\ &= (X_n - x^*)^T\Big(\nabla F(X_n) - \nabla F(x^*) + e_n^{\text{det}}\Big) \\ &\geq k_0\|X_n - x^*\|^2 + (X_n - x^*)^T e_n^{\text{det}} \end{aligned} \tag{5.39a}$$

see (5.34a). Moreover,

$$E\Big(\Big(Y_n^e - \nabla F(X_n)\Big)^T\Big(\nabla F(X_n) - \nabla F(x^*)\Big)|X_n\Big) = e_n^{\text{det}^T}\Big(\nabla F(X_n) - \nabla F(x^*)\Big), \tag{5.39b}$$

and (5.34b) yields

$$\left\|\nabla F(X_n) - \nabla F(x^*)\right\|^2 \leq k_1^2 \|X_n - x^*\|^2. \tag{5.39c}$$

Finally, (5.25a) implies that

$$\begin{aligned} V_n^e(X_n) &:= E\left(\left\|Y_n^e - \nabla F(X_n)\right\|^2 | X_n\right) \\ &\leq \left(\|R_n\|^2 + r \max_{1\leq i\leq p(n)} \sigma_i(X_n)\right) \frac{1}{p(n)} \|H_{n0}^{-1}\|, \end{aligned} \tag{5.39d}$$

where $H_{n0}$ is given by (5.21b),(5.24b), respectively. From (5.37) and (5.39a-d) we now have

$$\begin{aligned} E\Big(\|X_n - x^*\|^2 | X_n\Big) &\leq (1 - 2k_0\rho_n + k_1^2\rho_n^2)\|X_n - x^*\|^2 \\ &\quad -2\rho_n e_n^{\det T}(X_n - x^*) + 2\rho_n^2 e_n^{\det T}\Big(\nabla F(X_n) - \nabla F(x^*)\Big) \\ &\quad +\rho_n^2\left(\|R_n\|^2 + r \max_{1\leq i\leq p(n)} \sigma_i(X_n)^2\right)\frac{1}{p(n)}\|H_{n0}^{-1}\|. \end{aligned} \tag{5.40}$$

Estimating next $e_n^{\det T}(X_n - x^*)$ and $e_n^{\det T}\Big(\nabla F(X_n) - \nabla F(x^*)\Big)$, we get, see (5.20c),(5.21c), (5.24c),(5.34b)

$$\left|e_n^{\det T}(X_n - x^*)\right| \leq \sqrt{\|H_{n0}^{-1}\|}\frac{1}{\sqrt{p(n)}}\|R_n\| \cdot \|X_n - x^*\|, \tag{5.41a}$$

$$\left|e_n^{\det T}\Big(\nabla F(X_n) - \nabla F(x^*)\Big)\right| \leq k_1\sqrt{\|H_{n0}^{-1}\|}\frac{1}{\sqrt{p(n)}}\|R_n\| \cdot \|X_n - x^*\|. \tag{5.41b}$$

According to (5.27c) we find

$$\begin{aligned} \frac{1}{\sqrt{p(n)}}\|R_n\| \leq \frac{1}{(l_n+1)!} &\left(\frac{1}{p(n)} \sum_{i=1}^{p(n)} \|d^{(i,n)}\|^{2(l_n+1)}\right. \\ &\left. \times \sup_{0\leq\vartheta\leq 1} \left\|\nabla^{l_n+1} F(X_n + \vartheta d^{(i,n)})\right\|^2\right)^{1/2}. \end{aligned} \tag{5.41c}$$

Here, $(l_n)_{n\in\mathbb{N}_1}$ is a sequence with $l_n \in \{1,2\}$ for all $n \in \mathbb{N}_1$ such that

$$\{n \in \mathbb{N}_1 : \text{ algorithm (5.5) is in Phase } \lambda\} = \{n \in \mathbb{N}_1 : l_n = \lambda\}, \quad \lambda = 1, 2. \tag{5.41d}$$

*Note.* For the transition from Phase 1 to Phase 2 statistical $F$ tests may be applied. However, since a sequence of tests must be applied, the significance level is not easy to calculate.

Furthermore, suppose that representation (5.29a) holds, i.e.,

$$d^{(i,n)} = \mu_n \tilde{d}^{(i,n)}, \quad 0 < \mu_n \le 1, \quad i = 1,2,\ldots,p(n),$$

where $\tilde{d}^{(i,n)}, i = 1,\ldots,p(n)$, are given $r$-vectors. Then $H_{n0} = \mu_n^2 \widetilde{H}_{n0}$, cf. (5.29b), and (5.41c) yields

$$\sqrt{\|H_{n0}^{-1}\|}\frac{1}{\sqrt{p(n)}}\|R_n\| \le \sqrt{\|\widetilde{H}_{n0}^{-1}\|}\frac{\mu_n^{l_n}}{(l_n+1)!}\left(\frac{1}{p(n)}\sum_{i=1}^{p(n)}\|\tilde{d}^{(i,n)}\|^{2(l_n+1)}\right.$$
$$\left.\times \sup_{0\le\vartheta\le 1}\left\|\nabla^{l_n+1}F(X_n+\vartheta\tilde{d}^{(i,n)})\right\|^2\right)^{1/2}. \quad (5.41e)$$

Now we assume that

$$\sup_{0\le\vartheta\le 1}\left\|\nabla^{l_n+1}F(X_n+\vartheta\tilde{d}^{(i,n)})\right\|^2 \le \gamma_n^2 + \Gamma_n^2\|X_n - x^*\|^2,\ 1\le i\le p(n),\ n\in \mathbb{N}_1, \quad (5.42)$$

where $\gamma_n \ge 0, \Gamma_n \ge 0, n \in \mathbb{N}_1$, are certain coefficients. Then (5.41e) yields

$$\sqrt{\|H_{n0}^{-1}\|}\frac{1}{\sqrt{p(n)}}\|R_n\| \quad (5.43)$$
$$\le \sqrt{\|\widetilde{H}_{n0}^{-1}\|}\frac{\mu_n^{l_n}}{(l_n+1)!}\left(\frac{1}{p(n)}\sum_{i=1}^{p(n)}\|\tilde{d}^{(i,n)}\|^{2(l_n+1)}\right)^{1/2}\Big(\gamma_n + \Gamma_n\|X_n - x^*\|\Big).$$

Putting (5.41a),(5.41b),(5.41e) and (5.43) into (5.40), with (5.42) we get

$$E\Big(\|X_{n+1}-x^*\|^2|X_n\Big) \le \big(1 - 2[k_0 - \Gamma_n\mu_n^{l_n}T_n]\rho_n$$
$$+ (k_1 + \Gamma_n\mu_n^{l_n}T_n)^2\rho_n^2\big)\|X_n - x^*\|^2$$
$$+ \rho_n^2\left(r\frac{\mu_n^{-2}}{p(n)}\|\widetilde{H}_{n0}^{-1}\|\max_{1\le i\le p(n)}\sigma_i(X_n)^2 + \gamma_n^2(\mu_n^{l_n}T_n)^2\right)$$
$$+ 2\gamma_n(\rho_n + k_1\rho_n^2)\mu_n^{l_n}T_n\|X_n - x^*\|, \quad n \in \mathbb{N}_1. \quad (5.44a)$$

Here, $T_n$ is given by

$$T_n = \frac{1}{(l_n+1)!}\sqrt{\|\widetilde{H}_{n0}^{-1}\|}\left(\frac{1}{p(n)}\sum_{i=1}^{p(n)}\|\tilde{d}^{(i,n)}\|^{2(l_n+1)}\right)^{1/2}. \quad (5.44b)$$

In the following we suppose, cf. (5.31a)-(5.31d), that

$$E\left(\left(y^{(i,n)} - F(x^{(i,n)})\right)^2|X_n\right) =: \sigma_i(X_n)^2 \le \sigma_1^{F2} + C_1^F\|X_n - x^*\|^2,$$
$$1 \le i \le p(n), \quad (5.45)$$

with constants $\sigma_1^F, C_1^F \ge 0$. Taking expectations on both sides of (5.44a), we find the following result:

**Theorem 5.1.** *Assume that (5.34a-c), (5.42), (5.45) hold, and use representation (5.29a). Then*

$$
\begin{aligned}
b_{n+1} \leq \Big(1 - 2[k_0 - \Gamma_n \mu_n^{l_n} T_n]\rho_n + \Big[(k_1 + \Gamma_n \mu_n^{l_n} T_n)^2 \\
+ r C_1^F \frac{\mu_n^{-2}}{p(n)} \|\widetilde{H}_{n0}^{-1}\|\Big] \rho_n^2\Big) b_n \\
+ 2\gamma_n (\rho_n + k_1 \rho_n^2) \mu_n^{l_n} T_n E\|X_n - x^*\| \\
+ \left( r\sigma_1^{F2} \frac{\mu_n^{-2}}{p(n)} \|\widetilde{H}_{n0}^{-1}\| + \gamma_n^2 (\mu_n^{l_n} T_n)^2 \right) \rho_n^2 \text{ for all } n \in \mathbb{N}_1, \quad (5.46)
\end{aligned}
$$

*where $T_n$ is given by (5.44b).*

*Remark 5.5.* Since the KW-gradient estimator (5.4c) can be represented as a special (Phase 1-)RSM-gradient estimator, for the estimator $Y_n^s$ according to (5.4c) we obtain again an error recursion of type (5.46). Indeed, we have only to set $l_n = 1$ and to use appropriate (KW-)increments $d^{(i,n)} := d^{(i,0)}, i = 1, \ldots, \overline{p}, n \geq 1$.

### 5.3.2 The Criterial Case

Here the performance of (5.5) is evaluated by means of the mean functional errors

$$b_n = E\Big(F(X_n) - F^*\Big), n \geq 1, \text{ with the minimum value } F^* \text{ of (5.1).}$$

Under corresponding assumptions, see [80], also in the criterial case an inequality of the type (5.46) is obtained.

## 5.4 Convergence Behavior of Hybrid Stochastic Approximation Methods

In the following we compare the hybrid stochastic approximation method (5.5) with the standard stochastic gradient procedure (5.4a), i.e. $X_{n+1}^s := p_D(X_n - \rho_n Y_n^s), n = 1, 2, \ldots$ . For this purpose we suppose that (5.3a) holds and $Y_n^s$ is given by (5.4b).

### 5.4.1 Asymptotically Correct Response Surface Model

Suppose that $\gamma_n = 0$ for all $n \in \mathbb{N}_1$. Hence, according to (5.42), the sequence of response surface models $\hat{F}_n$ is asymptotically correct. In this case inequalities (5.36) and (5.46) yield

$$b_{n+1} \le \begin{cases} (1 - 2K^e_{0,n}\rho_n + K^{e2}_{1,n}\rho_n^2)b_n + \sigma^{e2}_{1,n}\rho_n^2 & \text{for } n \in \mathbb{N}_1 \\ \left(1 - 2k_0\rho_n + (k_1^2 + C_1^s)\rho_n^2\right)b_n + \sigma_1^{s2}\rho_n^2 & \text{for } n \in \mathbb{N}_2, \end{cases} \tag{5.47a}$$

where $k_0, k_1^2 + C_1^s, \sigma_1^{s2}$ are the same constants as in (5.36) and

$$K^e_{0,n} = k_0 - \Gamma_n \mu_n^{l_n} T_n \tag{5.47b}$$

$$K^{e2}_{1,n} = (k_1 + \Gamma_n \mu_n^{l_n} T_n)^2 + \frac{rC_1^F}{\mu_n^2 p(n)} \|\widetilde{H}_{n0}^{-1}\| \tag{5.47c}$$

$$\sigma^{e2}_{1,n} = \frac{r\sigma_1^{F2}}{\mu_n^2 p(n)} \|\widetilde{H}_{n0}^{-1}\|. \tag{5.47d}$$

A weak assumption is that

$$K^e_{0,n} \ge K^e_0 > 0 \text{ for all } n \in \mathbb{N}_1 \tag{5.48a}$$

$$K^{e2}_{1,n} \le K^{e2}_1 < +\infty \text{ for all } n \in \mathbb{N}_1 \tag{5.48b}$$

$$\sigma^{e2}_{1,n} \le \sigma^{e2}_1 < +\infty \text{ for all } n \in \mathbb{N}_1 \tag{5.48c}$$

with constants $K^e_0 > 0, K^{e2}_1 < +\infty$ and $\sigma^{e2}_1 < +\infty$; note that $k_0 \ge K^e_0$, see (5.47b).

*Remark 5.6.* Suppose that $(\Gamma_n)_{n\in\mathbb{N}_1}$, cf. (5.42), is bounded, i.e., $0 \le \Gamma_n \le \Gamma_0 < +\infty$ for all $n \in \mathbb{N}_1$ with a constant $\Gamma_0$. The assumptions (5.48a-c) can be guaranteed e.g. as follows:

a) If, for $n \in \mathbb{N}_1$, $F(x)$ is estimated $\nu(n)$ times at each one of $p_0$ different points $X_n + d^{(i)}, i = 1, 2, \ldots, p_0$, see (5.22a,b), then $\widetilde{H}_{n0} = \widetilde{H}_0^{(0)}$ with a fixed matrix $\widetilde{H}_0^{(0)}$ for all $n \in \mathbb{N}_1$. Since $l_n \in \{1, 2\}$ for all $n \in \mathbb{N}_1$, we find, cf. (5.44b), $T_n \le T_0$ for all $n \in \mathbb{N}_1$ with a constant $T_0 < +\infty$. Because of $0 < \mu_n \le 1, p(n) = \nu(n)p_0$, (5.47b-d) yield

$$K^e_{0,n} \ge k_0 - \Gamma_0 T_0 \mu_n$$

$$K^{e2}_{1,n} \le (k_1 + \Gamma_0 T_0)^2 + \frac{rC_1^F}{\mu_n^2} \|\widetilde{H}_0^{(0)-1}\|, \quad \sigma^{e2}_{1,n} \le \frac{r\sigma_1^{F2}}{\mu_n^2} \|\widetilde{H}_0^{(0)-1}\|.$$

Thus, conditions (5.48a-c) hold for arbitrary $\nu(n), n \in \mathbb{N}_1$, if

$$\frac{k_0 - K^e_0}{\Gamma_0 T_0} \ge \mu_n \ge \mu_0 > 0 \text{ for all } n \in \mathbb{N}_1 \text{ with}$$

$$0 < K^e_0 < k_0, 0 < \mu_0 < \frac{k_0 - K^e_0}{\Gamma_0 T_0}. \tag{5.48d}$$

b) In case (5.31b), i.e. if $\tilde{d}^{(i,n)} = \tilde{d}^{(i,0)}, i = 1, 2, \ldots, \overline{p}, n \ge 1$, then $\tilde{H}_{n0} = \tilde{H}_0$ for all $n \in \mathbb{N}_1$ with a fixed matrix $\tilde{H}_0$ and therefore $T_n \le T_0 < +\infty$ for all $n \in \mathbb{N}_1$, with fixed $T_0$, since $l_n \in \{1, 2\}, n \in \mathbb{N}_1$. Thus, for $K^e_{0,n}, K^{e2}_{1,n}, \sigma^{e2}_{1,n}$, resp., we obtain the same estimates as above if $\widetilde{H}_0^{(0)}$ is replaced by $\tilde{H}_0$. Hence, also in this case, conditions (5.48a-c) hold for an arbitrary $\overline{p}$ if $(\mu_n)_{n\in\mathbb{N}_1}$ is selected according to (5.48d).

According to (5.36), for the mean square errors of (5.4a,b)

$$b_n^s := E\|X_n^s - x^*\|^2, \quad n = 1, 2, \ldots,$$

we have the "upper" recurrence relation

$$b_{n+1}^s \leq \left(1 - 2k_0\rho_n + (k_1^2 + C_1^s)\rho_n^2\right) b_n^s + \sigma_1^{s2}\rho_n^2, \quad n = 1, 2, \ldots . \tag{5.49}$$

In addition to (5.35b) assume

$$V_n^s(X_n) = E\left(\|Y_n^s - \nabla F(X_n)\|^2 | X_n\right) \geq \sigma_0^{s2} + C_0^s\|X_n - x^*\|^2 \tag{5.50}$$

with coefficients $0 < \sigma_0^s \leq \sigma_1^s, 0 \leq C_0^s \leq C_1^s$. Using (5.34a,b), we find, see [76], [80], for $(b_n^s)$ also the "lower" recursion

$$b_{n+1}^s \geq \left(1 - 2k_1\rho_n + (k_0^2 + C_0^s)\rho_n^2\right) b_n^s + \sigma_0^{s2}\rho_n^2 \text{ for } n \geq n_0, \tag{5.51a}$$

provided that e.g.

$$X_n^s - \rho_n Y_n^s \in D \text{ a.s. for all } n \geq n_0. \tag{5.51b}$$

Having the lower recursion (5.51a) for $(b_n^s)$, we can compare algorithms (5.4a,b) and (5.5) by means of the quotients

$$\frac{b_n}{b_n^s} \leq \frac{b_n}{\underline{B}_n^s} \text{ for } n \geq n_0, \tag{5.52a}$$

where the sequence $(\underline{B}_n^s)$ of lower bounds $\underline{B}_n^s \leq b_n^s, n \geq n_0$, is defined by

$$\underline{B}_{n+1}^s = \left(1 - 2k_1\rho_n + (k_0^2 + C_0^s)\rho_n^2\right)\underline{B}_n^s + \sigma_0^{s2}\rho_n^2, n \geq n_0, \text{ with } \underline{B}_{n_0} \leq b_{n_0}^s. \tag{5.52b}$$

In all other cases we consider

$$\frac{b_n}{\overline{B}_n^s}, n = 1, 2, \ldots, \tag{5.52c}$$

where the sequence of upper bounds $\overline{B}_n^s \geq b_n^s, n = 1, 2, \ldots$, defined by

$$\overline{B}_{n+1}^s = \left(1 - 2k_0\rho_n + (k_1^2 + C_1^2)\rho_n^2\right)\overline{B}_n^s + \sigma_1^{s2}\rho_n^2, \quad n = 1, 2, \ldots, \overline{B}_1^s \geq b_1^s, \tag{5.52d}$$

represents the "worst case" of (5.4a,b). For the standard step sizes

$$\rho_n = \frac{c}{n+q} \text{ with constants } c > 0, q \in \mathbb{N} \cup \{0\} \tag{5.53}$$

and with $k_1 \geq k_0 > 0$ we have, see [149],

$$\lim_{n\to\infty} n \cdot \underline{B}_n^s = \sigma_0^{s2} \frac{c^2}{2k_1 c - 1} \text{ for all } c > \frac{1}{2k_1}, \tag{5.54a}$$

$$\lim_{n\to\infty} n \cdot \overline{B}_n^s = \sigma_1^{s2} \frac{c^2}{2k_0 c - 1} \geq \sigma_0^{s2} \frac{c^2}{2k_1 c - 1} \text{ for all } c > \frac{1}{2k_0}. \tag{5.54b}$$

For the standard step sizes (5.53) the asymptotic behavior of $(b_n/b_n^s)$, $(b_n/\overline{B}_n^s)$, resp., can be determined by means of the methods developed in the following Section 5.5:

**Theorem 5.2.** *Suppose that for the RSM-based stochastic approximation procedure (5.5) the upper error recursion (5.47a) holds, where (5.48a-c) is fulfilled, and $(\rho_n)$ is defined by (5.53). Moreover, suppose that groups of M iterations with the RSM-gradient estimator $Y_n^e$ and N iterations with the simple gradient estimator $Y_n^s$ are taken by turns.*

*i) If the lower estimates (5.51a) of $(b_n^s)$ hold, then*

$$\limsup_{n\to\infty} \frac{b_n}{b_n^s} \leq \frac{\sigma_1^{e2} M + \sigma_1^{s2} N}{\sigma_0^{s2}(M+N)} \cdot \frac{2k_1 c - 1}{2K_0^e c - 1} \text{ for all } c > \frac{1}{2K_0^e} \tag{5.55a}$$

*ii) If $(b_n)$ is compared with $(\overline{B}_n^s)$, then*

$$\limsup_{n\to\infty} \frac{b_n}{\overline{B}_n^s} \leq \frac{\sigma_1^{e2} M + \sigma_1^{s2} N}{\sigma_1^{s2}(M+N)} \cdot \frac{2k_0 c - 1}{2K_0^e c - 1} \text{ for all } c > \frac{1}{2K_0^e}. \tag{5.55b}$$

*Note.* It is easy to see that the right hand side of (5.55a,b), resp., takes any value below 1, provided that $\sigma_1^{e2}$ and the rate $N/(N+M)$ of steps with a "simple" gradient estimator $Y_n^s$ are sufficiently small.

Instead of applying the standard step size rule (5.53) and having then Theorem 5.2, we may use the optimal step sizes developed in [80]. Here estimates of the coefficients appearing in the optimal step size rules may be obtained from second order response surface models, see also the next Chapter 6.

### 5.4.2 Biased Response Surface Model

Assume now that $\gamma_n > 0$ for all $n \in \mathbb{N}_1$. For $n \in \mathbb{N}_1$, according to Theorem 5.1 we have first to estimate $E\|x_n - x^*\|$: For any given sequence $(\delta_n)_{n\in\mathbb{N}_1}$ of positive numbers, we know [149] that

$$E\|X_n - x^*\| \leq \delta_n + \frac{1}{\delta_n} b_n \text{ for all } n \in \mathbb{N}_1. \tag{5.56}$$

Taking, see (5.47b), (5.48a),

$$\delta_n := \frac{4\gamma_n(1 + k_1\rho_n)T_n}{k_0 - \Gamma_n \mu_n^{l_n} T_n} \cdot \mu_n^{l_n}, \qquad n \in \mathbb{N}_1, \tag{5.57}$$

according to (5.46), (5.56) and (5.57) we obtain, see (5.47b-d),

$$b_{n+1} \le \left(1 - \frac{3}{2}K_{0,n}^e\rho_n + K_{1,n}^{e2}\rho_n^2\right)b_n + \frac{8\Big(\gamma_n(1+k_1\rho_n)T_n\Big)^2(\mu_n^{l_n})^2}{K_{0,n}^e}\rho_n$$
$$+\Big(\sigma_{1,n}^{e2} + \gamma_n^2(\mu_n^{l_n}T_n)^2\Big)\rho_n^2 \text{ for all } n \in \mathbb{N}_1. \tag{5.58}$$

Inequalities (5.36) and (5.58) may be represented jointly by

$$b_{n+1} \le u_n b_n + v_n, \qquad n = 1, 2, \ldots \tag{5.59}$$

where the coefficients $u_n, v_n$ can be taken easily from (5.36) and (5.58).

We want now to select $(\rho_n), (\mu_n)_{n\in\mathbb{N}_1}$ and $\Big(p(n)\Big)_{n\in\mathbb{N}_1}$ such that

$$u_n \le 1 - A\rho_n \text{ for } n \ge n_0 \text{ with } n_0 \in \mathbb{N} \text{ and fixed } A > 0. \tag{5.60}$$

According to (5.36), (5.58) we have to select $A > 0$ such that

$$-2k_0 + (k_1^2 + C_1^s)\rho_n \le -A \text{ for } n \in \mathbb{N}_2, n \ge n_0 \tag{5.61a}$$

$$-\frac{3}{2}(k_0 - \Gamma_n\mu_n^{l_n}T_n) + (k_1 + \Gamma_n\mu_n^{l_n}T_n)^2\rho_n + rC_1^F\|\tilde{H}_{n0}^{-1}\|\frac{\rho_n}{\mu_n^2 p(n)} \le -A$$
$$\text{for } n \in \mathbb{N}_1, n \ge n_0. \tag{5.61b}$$

Assuming that with fixed numbers $K_0^e > 0, K_{11}^2 \ge 0$ and $\tilde{K}_{12}^2$ it is

$$K_{0,n}^e = k_0 - \Gamma_n\mu_n^{l_n}T_n \ge K_0^e \quad \text{for all } n \in \mathbb{N}_1 \tag{5.62a}$$
$$(k_1 + \Gamma_n\mu_n^{l_n}T_n)^2 \le K_{11}^2 < +\infty \quad \text{for all } n \in \mathbb{N}_1 \tag{5.62b}$$
$$rC_1^F\|\tilde{H}_{n0}^{-1}\| \le \tilde{K}_{12}^2 < +\infty \quad \text{for all } n \in \mathbb{N}_1, \tag{5.62c}$$

see (5.47b-d), (5.48a-c) and Remark 5.6, we find that (5.61b) is implied by

$$K_{11}^2\rho_n + \tilde{K}_{12}^2\frac{\rho_n}{\mu_n^2 p(n)} \le \frac{3}{2}K_0^e - A \text{ for } n \in \mathbb{N}_1, n \ge n_0. \tag{5.61b'}$$

Thus, selecting $A > 0$ such that

$$0 < A < \frac{3}{2}K_0^e\left(< \frac{3}{2}k_0 < 2k_0\right), \tag{5.63a}$$

we find that (5.61a,b) and therefore also (5.60) hold, provided that

$$\rho_n \to 0, \; n \to \infty \tag{5.63b}$$
$$\frac{\rho_n}{\mu_n^2 p(n)} \to 0, \; n \to \infty \text{ with } n \in \mathbb{N}_1. \tag{5.63c}$$

Note that (5.63c) holds for any sequence $\Big(p(n)\Big)_{n\in\mathbb{N}_1}$ if

$$\frac{\rho_n}{\mu_n^2} \to 0, n \to \infty \text{ with } n \in \mathbb{N}_1. \tag{5.63c'}$$

Having (5.62a-c) and (5.63a-c), inequalities (5.59) and (5.60) yield

$$b_{n+1} \le (1 - A\rho_n)b_n + v_n \text{ for all } n \ge n_0.$$

Finally, supposing still that

$$\sum_{n=1}^{\infty} \rho_n = +\infty \text{ and } \sum_{n=1}^{\infty} v_n < +\infty, \qquad (5.63\text{d}),\ (5.63\text{e})$$

the above recursion implies $b_n \to 0$ as $n \to \infty$. Condition (5.63e) demands, with respect to $\mathbb{N}_2$, that

$$\sum_{n \in \mathbb{N}_2} \rho_n^2 < +\infty. \qquad (5.63\text{f})$$

Concerning $\mathbb{N}_1$ we assume that

$$\gamma_n \le \gamma_0, T_n \le T_0, \|\tilde{H}_{n0}^{-1}\| \le h_0 \text{ for all } n \in \mathbb{N}_1 \qquad (5.64)$$

with certain constants $c_0 > 0, T_0 > 0, h_0 > 0$. Under these assumptions condition (5.63e) holds if – besides (5.63f) – we still demand that

$$\sum_{n \in \mathbb{N}_1} (\mu_n^{l_n})^2 \rho_n < +\infty, \qquad \sum_{n \in \mathbb{N}_1} \frac{\rho_n^2}{\mu_n^2 p(n)} < +\infty \qquad (5.65\text{a}),\ (5.65\text{b})$$

$$\sum_{n \in \mathbb{N}_1} (\mu_n^{l_n})^2 \rho_n^2 < +\infty, \qquad (5.65\text{c})$$

cf. (5.63b). Since $0 < \mu_n \le 1, n \in \mathbb{N}_1$, (5.63f) and (5.65c) hold if $\sum_{n=1}^{\infty} \rho_n^2 < +\infty$.

In summary, in the present case we have the following result:

**Theorem 5.3.** *Suppose that, besides the general assumptions guaranteeing inequalities (5.36) and (5.46),* $(\gamma_n)_{n\in\mathbb{N}_1}, (\Gamma_n)_{n\in\mathbb{N}_1}, (T_n)_{n\in\mathbb{N}_1}$ *and* $\left(\|\tilde{H}_{n0}^{-1}\|\right)_{n\in\mathbb{N}_1}$ *are bounded. If* $(\rho_n), (\mu_n)_{n\in\mathbb{N}_1}, \left(p(n)\right)_{n\in\mathbb{N}_1}$ *are selected such that*

*i)* $0 < \mu_n < \frac{k_0 - K_0^e}{\Gamma_0 T_0}$ *for all* $n \in \mathbb{N}_1$, *where* $0 < K_0^e < k_0$ *and* $\Gamma_0 > 0, T_0 > 0$ *are upper bounds of* $(\Gamma_n)_{n\in\mathbb{N}_1}, (T_n)_{n\in\mathbb{N}_1}$, *and*
*ii) conditions (5.63b-d,f) and (5.65a-c) hold, then* $E\|X_n - x^*\|^2 \to 0$ *as* $n \to \infty$.

In the literature on stochastic approximation [149] many results are available on the convergence rate of numerical sequences $(\beta_n)$ defined by recursions of the type

$$\beta_{n+1} \le \left(1 - \frac{c_n}{n}\right)\beta_n + \frac{d_n}{n^{\pi+1}}, \qquad n = 1, 2, \ldots,$$

with $\liminf_{n\to\infty} c_n = c > \pi$ and $\limsup_{n\to\infty} d_n = d > 0$. Hence, the above recursions on $(b_n)$ yield still the following result, cf. (5.54a,b):

**Theorem 5.4.** *Let the sequence of stepsizes* $(\varrho_n)$ *be given by (5.53). Assume that the general assumptions of Theorem 5.3 are fulfilled.*

*i) If* $p(n) = 0(n)$ *and* $\mu_n = 0(n^{-\frac{1}{2}})$, *then* $b_n = 0(n^{-1})$, *provided that* $c > \frac{3}{2K_0^e}$.

*ii) If* $\left(p(n)\right)_{n\in\mathbb{N}_1}$ *is bounded and* $\mu_n = 0(n^{-\frac{1}{4}})$, *then* $b_n = 0(n^{-\frac{1}{2}})$ *for all* $c > \frac{3}{2K_0^e}$.

## 5.5 Convergence Rates of Hybrid Stochastic Approximation Procedures

In order to compare the speed of convergence of the pure stochastic algorithm (5.4a,b) and the hybrid stochastic approximation algorithm (5.5), we consider the mean square error

$$b_n = E\|X_n - x^*\|^2,$$

where $x^*$ is an optimal solution of (5.1).

Some preliminaries are needed for the derivation of a recursion for the error sequence $(b_n)$. First, we note that the projection operator $p_D$ from $\mathbb{R}^r$ on $D$ is defined by

$$\left\|x - p_D(x)\right\| = \inf_{y\in D} \|x - y\| \text{ for all } x \in \mathbb{R}^r.$$

It is known [15] that

$$\left\|p_D(x) - p_D(y)\right\| \le \|x - y\| \text{ for all } x, y \in \mathbb{R}^r. \tag{5.66}$$

As mentioned in Section 5.1, see also Section 4, many stochastic optimization problems (5.1) have the following property:

**Property (DD).** At certain "nonstationary" or "nonefficient" points $x \in D$ there exits a feasible descent direction (DD) $h = h(x)$ of $F$ which can be obtained with much less computational expenses than the gradient $\nabla F(x)$. Moreover, $h = h(x)$ is stable with respect to changes of the loss function contained in $f = f(\omega, x)$, see [68, 73, 74].

Consider then the mapping $A : \mathbb{R}^r \to \mathbb{R}^r$ defined by

$$A(x) = \begin{cases} -h(x), & \text{if (DD) holds at } x \\ \nabla F(x), & \text{else.} \end{cases} \tag{5.67}$$

In many cases [68, 73, 74] $h(x)$ can be represented by $h(x) = y - x$ with a certain vector $y \in D$. Hence, in this case we get

$$p_D(X_n + \varrho_n h_n) = X_n + \varrho_n h_n, \text{ if } 0 < \varrho_n \le 1,$$

and the projection onto $D$ can be omitted then.

We suppose now

$$p_D\Big(x^* - \varrho A(x^*)\Big) = x^* \text{ for all } \varrho > 0. \tag{5.68}$$

for every optimal solution $x^*$ of (5.1). This assumption is justified, since for a convex, differentiable objective function $F$, the optimal solutions $x^*$ of (5.1) can be characterized [15] by $p_D\Big(x^* - \varrho \nabla F(x^*)\Big) = x^*$ for all $\varrho > 0$, and in $x^*$ there is no feasible descent direction $h(x^*)$ on the other hand.

The hybrid algorithm (5.5) can be represented then by

$$X_{n+1} = p_D\left(X_n - \varrho_n\Big(A(X_n) + \xi_n\Big)\right), \quad n = 1, 2, \ldots, \tag{5.69a}$$

where the noise $r$-vector $\xi_n$ is defined as follows:

i) Hybrid methods with deterministic descent directions $h(X_n)$ and standard gradient estimators $Y_n^s$

$$\xi_n := \begin{cases} 0, & \text{if } n \in \mathbb{N}_1 \\ Y_n^s - \nabla F(X_n), & \text{if } n \in \mathbb{N}_2, \end{cases} \tag{5.69b}$$

where $A(x)$ is defined by (5.67).

ii) Hybrid methods with RSM-gradient estimators $Y_n^e$ and simple gradient estimators $Y_n^s$

$$\xi_n := \begin{cases} Y_n^e - \nabla F(X_n), & \text{if } n \in \mathbb{N}_1 \\ Y_n^s - \nabla F(X_n), & \text{if } n \in \mathbb{N}_2, \end{cases} \tag{5.69c}$$

where $A(x) := \nabla F(x)$.

While for hybrid methods with mixed RSM-/standard gradient estimators $Y_n^e, Y_n^s$ a recurrence relation for the error sequence $(b_n)$ was given in Theorem 5.1 and in Section 5.4.1, for case (i) we have this result:

**Lemma 5.1 (Mixed feasible descent directions $\mathbf{A(X_n)}$ and standard gradient estimators $\mathbf{Y_n^s}$).** *Let $x^*$ be an optimal solution. Assume that the optimality condition (5.68) holds and the following conditions are fulfilled*

$$\Big\|A(x) - A(x^*)\Big\|^2 \le \gamma_1 + \gamma_2 \|x - x^*\|^2, \tag{5.70}$$

$$\Big(A(x) - A(x^*)\Big)^T (x - x^*) \ge \alpha \|x - x^*\|^2 \tag{5.71}$$

*for every $x \in D$, where $\gamma_1 \ge 0, \gamma_2 > 0$ and $\alpha > 0$ are some constants. Moreover, for $n \in \mathbb{N}_2$ assume that*

$$E(Y_n^s | X_n) = \nabla F(X_n) \text{a.s. (almost sure)} \tag{5.72}$$

$$E\|Y_n^s\|^2 < +\infty, \tag{5.73}$$

*Then* $b_n = E\|X_n - x^*\|^2$ *satisfies the recursion*

$$b_{n+1} \le (1 - 2\alpha\varrho_n + \gamma_2\varrho_n^2)b_n + \beta_n\varrho_n^2, \quad n = 1, 2, \ldots, \tag{5.74a}$$

*where* $\beta_n$ *is defined by*

$$\beta_n = \gamma_1 + E\|\xi_n\|^2. \tag{5.74b}$$

*Proof.* Using (5.66) and (5.68)-(5.73), we find

$$\begin{aligned}
\|X_{n+1} - x^*\|^2 &= \Big\|p_D\Big(X_n - \varrho_n\big(A(X_n) + \xi_n\big)\Big) - p_D\Big(x^* - \varrho_n A(x^*)\Big)\Big\|^2 \\
&\le \Big\|(X_n - x^*) - \varrho_n\Big(A(X_n) - A(x^*)\Big) - \varrho_n\xi_n\Big\|^2 \\
&\le \|X_n - x^*\|^2 - 2\varrho_n\Big(A(X_n) - A(x^*)\Big)^T(X_n - x^*) \\
&\quad + \varrho_n^2\Big\|A(X_n) - A(x^*)\|^2 \\
&\quad - 2\varrho_n\xi_n^T\Big(X_n - x^* - \varrho_n\Big(A(X_n) - A(x^*)\Big)\Big) + \varrho_n^2\|\xi_n\|^2 \\
&\le \|X_n - x^*\|^2 - 2\alpha\varrho_n\|X_n - x^*\|^2 + \varrho_n^2\Big(\gamma_1 + \gamma_2\|X_n - x^*\|^2\Big) \\
&\quad - 2\varrho\xi_n^T\Big(X_n - x^* - \varrho_n\Big(A(X_n) - A(x^*)\Big)\Big) + \varrho_n^2\|\xi_n\|^2.
\end{aligned}$$

From (5.69b) and (5.72) we obtain $E(\xi_n|X_n) = 0$ a.s. for every $n$. Hence, from the above inequality we a.s. have

$$\begin{aligned}
E\Big(\|X_{n+1} - x^*\|^2\,|X_n\Big) &\le (1 - 2\alpha\varrho_n + \gamma_2\varrho_n^2)\|X_n - x^*\|^2 \\
&\quad + \gamma_1\varrho_n^2 + \varrho_n^2 E\Big(\|\xi_n\|^2\,|X_n\Big).
\end{aligned}$$

Taking expectations on both sides, we get the asserted recursion (5.74a,b).

We want to discuss in short the assumptions in the above lemma. First we remark that (5.70) holds if

$$\Big\|A(x) - A(x^*)\Big\| \le k_1 + k_2\|x - x^*\| \tag{5.75}$$

for some constants $k_1 \ge 0, k_2 > 0$. Indeed, (5.75) implies (5.70) with $\gamma_1 = k_1^2 + k_1k_2, \gamma_2 = k_1k_2 + k_2^2$.

If inequality (5.71) is satisfied for $A(x) = \nabla F(x)$, then $F$ is called strongly convex at $x^*$. For $A(x) = -h(x)$ being a feasible descent direction, condition (5.71) reads

$$h(x)'(x^* - x) \ge \alpha\|x - x^*\|^2 \text{ for } x \in D.$$

Since $h_{opt}(x) = x^* - x$ is the best descent direction in $X$, the above inequality may be regarded as a condition on the quality of $h(x)$ as a descent direction. The remaining conditions (5.72), (5.73) state that the error $\xi_n$ in a stochastic step has mean zero and finite second order moments.

Considering now the convergence behavior of (5.74a,b), in the literature, see e.g. [119, 122, 146, 149], the following type of results is available.

**Lemma 5.2.** *Let $(b_n)$ be a sequence of nonnegative numbers such that*

$$b_{n+1} \le \left(1 - \frac{\mu_n}{n}\right) b_n + \frac{\tau}{n^{\pi+1}}, \; n \ge n_0 \tag{5.76}$$

*for some positive integer $n_0$. If $\lim_{n\to\infty} \mu_n = \mu$ and $\tau > 0$, then*

$$b_n = 0(n^{-\pi}), \; \text{if } \mu > \pi > 0, \tag{5.77a}$$

$$b_n = 0\left(\frac{\log n}{n^{\mu}}\right), \; \text{if } \mu = \pi > 0, \tag{5.77b}$$

$$b_n = 0(n^{-\mu}), \; \text{if } \pi > \mu > 0. \tag{5.77c}$$

If $(b_n)$ satisfies the recursion

$$b_{n+1} = \left(1 - \frac{\mu_n}{n}\right) b_n + \frac{\tau_n}{n^{\pi+1}}, \; n > n_0, \tag{5.78}$$

where $\lim_{n\to\infty} \mu_n = \mu > \pi > 0$ and $\lim_{n\to\infty} \tau_n = \tau \ge 0$, then

$$\lim_{n\to\infty} n^{\pi} b_n = \frac{\tau}{\mu - \pi}. \tag{5.79}$$

Applying Lemma 5.2 to (5.74a,b), this result is obtained:

**Corollary 5.1.** *Suppose that the assumptions in Lemma 5.1 are fulfilled and, in addition, let*

$$E\|\xi_n\|^2 \le \sigma^2 < +\infty \; \text{for } n \in \mathbb{N}_2 \, . \tag{5.80}$$

*If the step size $\varrho_n$ is chosen according to $\varrho_n = \dfrac{c}{n+q}$, where $c > 0$ and $q \in \mathbb{N} \cup \{0\}$, then $\lim_{n\to\infty} E\left\|X_n(\omega) - x^*\right\|^2 = 0$, where the asymptotic rate of convergence is given by (5.77a-c) with $\pi = 1$ and $\mu = 2\alpha c$.*

*Proof.* Having (5.80) and $\varrho_n = \dfrac{c}{n+q}$, it is easy to see that (5.74a,b) implies (5.76) for $\mu_n = 2\alpha c \dfrac{n}{n+q} - \gamma_2 c^2 \dfrac{n}{(n+q)^2}$, $\mu = 2\alpha c, \tau = c^2(\gamma_1 + \sigma^2)$ and $\pi = 1$. The corollary follows now from Lemma 5.2.

Assuming now that

$$\lim_{n\to\infty} \varrho_n = 0,$$

there is an integer $n_0 > 0$ such that

$$0 < 1 - 2\alpha\varrho_n + \gamma_2\varrho_n^2 < 1 - a\varrho_n, n \ge n_0, \tag{5.81}$$

where $a$ is a fixed positive number such that $0 < a < 2\alpha$. Furthermore, assume that (5.80) holds.

Then (5.74a,b) and (5.81) yield the recursion

$$b_{n+1} \leq (1 - a\varrho_n)b_n + \bar{\beta}_n \varrho_n^2, \quad n \geq n_0, \tag{5.82}$$

where $\bar{\beta}_n$ is given by

$$\bar{\beta}_n = \begin{cases} \gamma_1, & \text{if } n \in \mathbb{N}_1 \\ \gamma_1 + \sigma^2, & \text{if } n \in \mathbb{N}_2. \end{cases} \tag{5.83}$$

Next we want to consider the "worst case" in (5.82), i.e. the recursion

$$B_{n+1} = (1 - a\varrho_n)B_n + \bar{\beta}_n \varrho_n^2, \quad n \geq n_0. \tag{5.84a}$$

If $b_n$ and $B_n$ satisfy the recursions (5.82) and (5.84a), respectively, and if $B_{n_0} \geq b_{n_0}$, then by a simple induction argument we find that

$$E\left\|X_n(\omega) - x^*\right\|^2 = b_n \leq B_n \quad \text{for all} \quad n \geq n_0. \tag{5.84b}$$

Corresponding to Corollary 5.1, for the upper bound $B_n$ of the mean square error $b_n$ we have this

**Corollary 5.2.** *Suppose that all assumptions in Lemma 5.1 are fulfilled and (5.80) holds. If* $\varrho_n = \frac{c}{n+q}$ *with* $c > 0$ *and* $q \in \mathbb{N} \cup \{0\}$, *then for* $(B_n)$ *we have the asymptotic formulas*

$$B_n = 0(n^{-1}), \text{ if } ac > 1,$$

$$B_n = 0\left(\frac{\log n}{n}\right), \text{ if } ac = 1,$$

$$B_n = 0(n^{-ac}), \text{ if } ac < 1$$

*as* $n \to \infty$. *Furthermore, if* $ac > 1$ *and* $\mathbb{N}_1 = \emptyset$, *i.e. in the pure stochastic case, we have*

$$\lim_{n\to\infty} nB_n = (\gamma_1 + \sigma^2)\frac{c^2}{ac-1}.$$

*Proof.* The assertion follows by applying Lemma 5.2 to (5.84a,b).

Since by means of Corollaries 5.1 and 5.2 we can not distinguish between the convergence behavior of the pure stochastic and the hybrid stochastic approximation algorithm, we have to consider recursion (5.84a,b) in more detail.

It turns out that the improvement of the speed of convergence resulting from the use of both stochastic and deterministic directions depends on the rate of stochastic and deterministic steps taken in (5.69a-c).

Note. For simplicity of notation, an iteration step $X_n \to X_{n+1}$ using an improved step direction, a standard stochastic gradient is called a "deterministic step", a "stochastic step", respectively.

### 5.5.1 Fixed Rate of Stochastic and Deterministic Steps

Let us assume that the rate of stochastic and deterministic steps taken in (5.69a-c) is fixed and given by $\frac{M}{N}$, i.e. $M$ deterministic and $N$ stochastic steps are taken by turns beginning with $M$ deterministic steps.

Then $B_n$ satisfies recursion (5.84a,b) with $\bar{\beta}_n$ given by (5.83), i.e.

$$\bar{\beta}_n = \begin{cases} \gamma_1 & \text{for } n \in \mathbb{N}_1 \\ \gamma_1 + \sigma^2 & \text{for } n \in \mathbb{N}_2 \end{cases},$$

where $\mathbb{N}_1, \mathbb{N}_2$ are defined by

$$\mathbb{N}_1 = \Big\{ n \in \mathbb{N} : (m-1)(M+N)+1 \leq n \leq mM + (m-1)N \text{ for some } m \in \mathbb{N} \Big\},$$

$$\mathbb{N}_2 = \mathbb{N} \setminus \mathbb{N}_1 = \Big\{ n \in \mathbb{N} : \tilde{m}M + (\tilde{m}-1)N + 1 \leq n \leq \tilde{m}(M+N) \text{ for some } \tilde{m} \in \mathbb{N} \Big\}.$$

We want to compare $B_n$ with the upper bounds $B_n^s$ corresponding to the purely stochastic algorithm (5.4a,b), i.e. $B_n^s$ shall satisfy (5.84a,b) with

$$\bar{\beta}_n = \gamma_1 + \sigma^2 \quad \text{for all} \quad n \geq n_0.$$

We assume that the step size $\varrho_0$ is chosen according to

$$\varrho_n = \frac{c}{n+q}$$

with a fixed $q \in \mathbb{N} \cup \{0\}$ and $c > 0$ such that $ac \geq 1$.

Then the recursions to be considered are

$$B_{n+1} = \left(1 - \tfrac{\vartheta}{n+q}\right) B_n + \tfrac{\tilde{\beta}_n}{(n+q)^2},\ n \geq n_0, \tag{5.85}$$

$$B_{n+1}^s = \left(1 - \tfrac{\vartheta}{n+q}\right) B_n^s + \tfrac{B}{(n+q)^2},\ n \geq n_0, \tag{5.86}$$

where

$$\tilde{\beta}_n = \begin{cases} A \text{ for } n \in \mathbb{N}_1 \\ B \text{ for } n \in \mathbb{N}_2, \end{cases}$$

$$A = c^2\gamma_1, B = c^2(\gamma_1 + \sigma^2) \text{ and } \vartheta = ac.$$

We need two lemmata (see [149] for the first and [116] for the second).

**Lemma 5.3.** *Let $\vartheta \geq 1$ be a real number and*

$$C_{k,n} = \prod_{m=k+1}^{n} \left(1 - \frac{\vartheta}{m}\right) \quad \textit{for } k, n \in \mathbb{N}, k < n.$$

*Then for every* $\varepsilon > 0$ *there is a* $k_\varepsilon \in \mathbb{N}$ *such that*

$$(1-\varepsilon)\left(\frac{k}{n}\right)^{\vartheta} \leq C_{k,n} \leq (1+\varepsilon)\left(\frac{k}{n}\right)^{\vartheta}$$

*for all* $n > k \geq k_\varepsilon$.

**Lemma 5.4.** *Let the numbers* $y_n$ *satisfy the recursion*

$$y_{n+1} = U_n y_n + V_n \text{ for } n \geq \nu,$$

*where* $(U_n), (V_n)$ *are given sequences of real numbers such that* $U_n \neq 0$ *for* $n \geq \nu$. *Then*

$$y_{n+1} = \left(y_\nu + \sum_{j=\nu}^{n} \frac{V_j}{\sum_{m=\nu}^{j} U_m}\right) \prod_{m=\nu}^{n} U_m \text{ for } n \geq \nu.$$

We are going to prove the following result:

**Theorem 5.5.** *Let the sequence* $(B_n)$ *and* $(B_n^s)$ *be defined by (5.85) and (5.86), respectively. If* $\vartheta \geq 1$, *then*

$$\lim_{n\to\infty} \frac{B_n}{B_n^s} = Q := \frac{AM+BN}{B(M+N)} = \frac{\gamma_1 M + (\gamma_1+\sigma^2)N}{(\gamma_1+\sigma^2)(M+N)}. \tag{5.87}$$

*Proof.* Using the notation of Lemma 5.3 , we have

$$\prod_{m=\nu}^{j}\left(1-\frac{\vartheta}{m+q}\right) = C_{\nu+q-1,j+q} \text{ for } j \geq \nu \geq 1,$$

and as a simple consequence of (5.85), (5.86) and Lemma 5.4 we note

$$\frac{B_{n+1}}{B_{n+1}^s} = \frac{B_\nu + \sum\limits_{j=\nu}^{n} \frac{\tilde{\beta}_j}{(j+q)^2 C_{\nu+q-1,j+q}}}{B_\nu^s + \sum\limits_{j=\nu}^{n} \frac{B}{(j+q)^2 C_{\nu+q-1,j+q}}},\ n > \nu$$

for every fixed integer $\nu > n_0$. By Lemma 5.3 for every $0 < \varepsilon < 1$ there is an integer $p = \nu - 1, \nu > n_0$, such that

$$(1-\varepsilon)\left(\frac{p+q}{j+q}\right)^{\vartheta} \leq C_{p+q,j+q} \leq (1+\varepsilon)\left(\frac{p+q}{j+q}\right)^{\vartheta} \text{ for } j > p.$$

Consequently, for $n \geq \nu$ we find the inequality

$$\frac{B_\nu + \frac{1}{(1+\varepsilon)(p+q)^\vartheta}\sum\limits_{j=\nu}^{n}\tilde{\beta}_j(j+q)^{\vartheta-2}}{B_\nu^s + \frac{B}{(1-\varepsilon)(p+q)^\vartheta}\sum\limits_{j=\nu}^{n}(j+q)^{\vartheta-2}} \le \frac{B_{n+1}}{B_{n+1}^s} \tag{5.88}$$

$$\le \frac{B_\nu + \frac{1}{(1-\varepsilon)(p+q)^\vartheta}\sum\limits_{j=\nu}^{n}\tilde{\beta}_j(j+q)^{\vartheta-2}}{B_\nu^s + \frac{B}{(1+\varepsilon)(p+q)^\vartheta}\sum\limits_{j=\nu}^{n}(j+q)^{\vartheta-2}}.$$

Since $\vartheta \ge 1$, the series $\sum\limits_{j=\nu}(j+q)^{\vartheta-2}$ diverges as $n \to \infty$. So the numbers $B_\nu$ and $B_\nu^s$ have no influence on the limit of both sides of the above inequalities (5.88) as $n \to \infty$. Suppose now that we are able to show the equation

$$\lim_{n\to\infty}\frac{S(\nu,n)}{T(\nu,n)} = \frac{AM+BN}{M+N} = L, \tag{5.89}$$

where

$$S(\nu,n) = \sum_{j=\nu}^{n}\tilde{\beta}_j(j+q)^{\vartheta-2},$$

$$T(\nu,n) = \sum_{j=\nu}^{n}(j+q)^{\vartheta-2}.$$

Taking the limit in (5.88) for $n \to \infty$, for every $\varepsilon > 0$ we obtain

$$\frac{1-\varepsilon}{1+\varepsilon}\cdot\frac{L}{B} \le \liminf_{j\to\infty}\frac{B_{n+1}}{B_{n+1}^s} \le \limsup_{n\to\infty}\frac{B_{n+1}}{B_{n+1}^s} \le \frac{1+\varepsilon}{1-\varepsilon}\cdot\frac{L}{B}.$$

But this can be true only if $\lim\limits_{n\to\infty}\frac{B_{n+1}}{B_{n+1}^s}$ exists and

$$\lim_{n\to\infty}\frac{B_{n+1}}{B_{n+1}^s} = \frac{L}{B} = \frac{AM+BN}{B(M+N)}$$

which proves (5.87). Hence, it remains to show that the limit (5.89) exists and is equal to $L$. For this purpose we may assume without loss of generality that $\nu = m_0(M+N)+1$ for some integer $m_0$. Furthermore, we assert that from the existence of

$$\lim_{m\to\infty}\frac{S\big(\nu, m(M+N)\big)}{T\big(\nu, m(M+N)\big)} \tag{5.90}$$

the existence of the limit (5.89) follows. To see this, let $n \ge \nu$ be given and let $m = m_n$ be the uniquely determined integer such that

$$m(M+N)+1 \leq n < (m+1)\,(M+N).$$

Since

$$S\Big(\nu,(m+1)(M+N)\Big) = S(\nu,n) + S\Big(n+1,(m+1)(M+N)\Big),$$

for $m(M+N)+1 \leq n < (m+1)(M+N)$ a simple computation yields

$$\frac{S\Big(\nu,(m+1)(M+N)\Big)}{T\Big(\nu,(m+1)(M+N)\Big)} - \frac{S(\nu,n)}{T(\nu,n)} \tag{5.91}$$

$$= \frac{S\Big(n+1,(m+1)(M+N)\Big)}{T\Big(\nu,(m+1)(M+N)\Big)} - \frac{T\Big(n+1,(m+1)(M+N)\Big)\cdot S(\nu,n)}{T\Big(\nu,(m+1)(M+N)\Big)\cdot T(\nu,n)}.$$

Clearly, from the definition of $S(\nu,n), T(\nu,n)$ and $\tilde{\beta}_j \leq B$ for all $j \in \mathbb{N}$ follows that

$$\frac{S\Big(n+1,(m+1)(M+N)\Big)}{T\Big(\nu,(m+1)(M+N)\Big)} \leq B\frac{T\Big(m(M+N)+1,(m+1)(M+N)\Big)}{T\Big(\nu,(m+1)(M+N)\Big)} \tag{5.92a}$$

and

$$\frac{T\Big(n+1,(m+1)(M+N)\Big)\cdot S(\nu,n)}{T\Big(\nu,(m+1)(M+N)\Big)\cdot T(\nu,n)} \leq B\frac{T\Big(m(M+N)+1,(m+1)(M+N)\Big)}{T\Big(\nu,(m+1)(M+N)\Big)}. \tag{5.92b}$$

Assume now that

$$\lim_{n\to\infty} \frac{T\Big(m(M+N)+1,(m+1)(M+N)\Big)}{T\Big(\nu,(m+1)(M+N)\Big)} = 0. \tag{5.93}$$

Since $n \to \infty$ implies that $m = m_n \to \infty$, from (5.91), the inequalities (5.92a,b) and assumption (5.93) we obtain

$$\lim_{n\to\infty} \frac{S\Big(\nu,(m_n+1)(M+N)\Big)}{T\Big(\nu,(m_n+1)(M+N)\Big)} - \frac{S(\nu,n)}{T(\nu,n)} = 0.$$

Therefore, if the limit (5.90) exists, then also the limit (5.89) exists and both limits are equal.

For $1 \leq \vartheta \leq 2$ the limit (5.93) follows from

$$T\Big(m(M+N)+1,(m+1)(M+N)\Big) \leq M+N$$

and the divergence of $T\Big(\nu, (m+1)(M+N)\Big)$ as $m \to \infty$.

For $\vartheta > 2$ we have

$$T\Big(m(M+N)+1, (m+1)(M+N)\Big) \leq (M+N)\Big((m+1)(M+N)+q\Big)^{\vartheta-2}$$

and $\Big($with $\nu = m_0(M+N)+1\Big)$

$$\begin{aligned} T\Big(\nu, (m+1)(M+N)\Big) &= \sum_{k=m_0}^{m} \sum_{j=k(M+N)+1}^{(k+1)(M+N)} (j+q)^{\vartheta-2} \\ &\geq (M+N) \sum_{k=m_0}^{m} \Big(k(M+N)+q\Big)^{\vartheta-2}. \end{aligned}$$

Consider the monotonously increasing function

$$u(x) = \Big(x(M+N)+q\Big)^{\vartheta-2}, \ x \geq 0.$$

Because of

$$\int_k^{k+1} u(x-1)\, dx \leq \int_k^{k+1} u(k+1-1) dx = u(k),$$

we have that

$$T\Big(\nu, (m+1)(M+N)\Big) \geq (M+N) \sum_{k=m_0}^{m} u(k)$$

$$\geq (M+N) \sum_{k=m_0}^{m} \int_k^{k+1} u(x-1) dx = (M+N) \int_{m_0}^{m+1} u(x-1) dx$$

$$= \frac{1}{\vartheta-1}\left(\Big(m(M+N)+q\Big)^{\vartheta-1} - \Big((m_0-1)(M+N)+q\Big)^{\vartheta-1}\right).$$

Therefore, for $\vartheta > 2$ we find

$$\frac{T\Big(m(M+N)+1, (m+1)(M+N)\Big)}{T\Big(\nu, (m+1)(M+N)\Big)}$$

$$\leq (\vartheta-1)(M+N)\left\{\Big(m(M+N)+q\Big)\left(\frac{m(M+N)+q}{(m+1)(M+N)+q}\right)^{\vartheta-2}\right.$$

$$\left. - \frac{\Big((m_0-1)(M+N)+q\Big)^{\vartheta-1}}{\Big((m+1)(M+N)+q\Big)^{\vartheta-2}}\right\}^{-1}.$$

This yields then the limit (5.93).

It remains to show that (5.90) and consequently also (5.89) holds. Again with $\nu = m_0(M+N)+1$ we have

$$S\Big(\nu, m(M+N)\Big) = \sum_{k=m_0}^{m-1} \sum_{j=k(M+N)+1}^{(k+1)(M+N)} \tilde{\beta}_j (j+q)^{\vartheta-2} \tag{5.94}$$

$$= \sum_{k=m_0}^{m-1} \left( A \sum_{j=k(M+N)+1}^{(k+1)M+kN} (j+q)^{\vartheta-2} + B \sum_{j=(k+1)M+kN+1}^{(k+1)(M+N)} (j+q)^{\vartheta-2} \right)$$

$$= \sum_{k=m_0}^{m-1} \sigma_k,$$

where $\sigma_k$ is defined by

$$\sigma_k = A \sum_{i=1}^{M} \Big(k(M+N)+i+q\Big)^{\vartheta-2} + B \sum_{i=M+1}^{M+N} \Big(k(M+N)+i+q\Big)^{\vartheta-2}.$$

In the same way we find

$$T\Big(\nu, m(M+N)\Big) = \sum_{k=m_0}^{m-1} \tau_k, \tag{5.95}$$

where $\tau_k$ is given by

$$\tau_k = \sum_{i=1}^{M+N} \Big(k(M+N)+i+q\Big)^{\vartheta-2}.$$

Define now the functions

$$f(x) = A \sum_{i=1}^{M} \Big(x(M+N)+i+q\Big)^{\vartheta-2} + B \sum_{i=M+1}^{M+N} \Big(x(M+N)+i+q\Big)^{\vartheta-2}$$

and

$$g(x) = \sum_{i=1}^{M+N} \Big(x(M+N)+i+q\Big)^{\vartheta-2}.$$

Then $\sigma_k = f(k)$ and $\tau_k = g(k)$.

For $1 \leq \vartheta \leq 2$ the functions $f$ and $g$ are monotonously nonincreasing. Hence,

$$\int_k^{k+1} f(x)dx \leq f(k) = \sigma_k \leq \int_k^{k+1} f(x-1)dx$$

and therefore

$$\int_{m_0}^{m} f(x)dx \le \sum_{k=m_0}^{m-1} \sigma_k \le \int_{m_0}^{m} f(x-1)dx.$$

Analogously we find

$$\int_{m_0}^{m} g(x)dx \le \sum_{k=m_0}^{m-1} \tau_k \le \int_{m_0}^{m} g(x-1)dx.$$

By (5.94) and (5.95), in the case $1 \le \vartheta \le 2$ we have now

$$\frac{\int_{m_0}^{m} f(x)dx}{\int_{m_0}^{m} g(x-1)dx} \le \frac{S\Big(\nu, m(M+N)\Big)}{T\Big(\nu, m(M+N)\Big)} \le \frac{\int_{m_0}^{m} f(x-1)dx}{\int_{m_0}^{m} g(x)dx}. \tag{5.96a}$$

If $\vartheta > 2$, then $f$ and $g$ are monotonously increasing and therefore

$$\int_{k}^{k+1} f(x)dx \ge f(k) = \sigma_k \ge \int_{k}^{k+1} f(x-1)dx,$$

hence,

$$\int_{m_0}^{m} f(x)dx \ge \sum_{k=m_0}^{m-1} \sigma_k \ge \int_{m_0}^{m} f(x-1)dx,$$

and in the same way we get

$$\int_{m_0}^{m} g(x)dx \ge \sum_{k=m_0}^{m-1} \tau_k \ge \int_{m_0}^{m} g(x-1)dx.$$

Consequently, if $\vartheta > 2$, then by (5.94) and ( 5.95) we have

$$\frac{\int_{m_0}^{m} f(x-1)dx}{\int_{m_0}^{m} g(x)dx} \le \frac{S\Big(\nu, m(M+N)\Big)}{T\Big(\nu, m(M+N)\Big)} \le \frac{\int_{m_0}^{m} f(x)dx}{\int_{m_0}^{m} g(x-1)dx} \tag{5.96b}$$

Hence, in both cases we have the same type of upper and lower bounds for $\frac{S\Big(\nu, m(M+N)\Big)}{T\Big(\nu, m(M+N)\Big)}$.

We have to discuss two cases:

i) Let $\vartheta = 1$. In this case we find

$$\int_{m_0}^{m} f(x-1)dx = \frac{A}{M+N}\sum_{i=1}^{M}\Big(\log\big((m-1)(M+N)+i+q\big)$$
$$-\log\big((m_0-1)(M+N)+i+q\big)\Big)$$
$$+\frac{B}{M+N}\sum_{i=M+1}^{M+N}\Big(\log\big((m-1)(M+N)+i+q\big)$$
$$-\log\big((m_0-1)(M+N)+i+q\big)\Big) = \frac{A\cdot M}{M+N}\log(m-1)$$
$$+\frac{A}{M+N}\sum_{i=1}^{M}\left(\log\left(M+N+\frac{i+q}{m-1}\right)-\log\big((m_0-1)(M+N)+i+q\big)\right)$$
$$+\frac{B\cdot N}{M+N}\log(m-1)+\frac{B}{M+N}\sum_{i=M+1}^{M+N}\left(\log\left(M+N+\frac{i+q}{m-1}\right)\right.$$
$$\left.-\log\big((m_0-1)(M+N)+i+q\big)\right)$$

and in the same way

$$\int_{m_0}^{m} g(x)dx = \log m + \frac{1}{M+N}\sum_{i=1}^{M+N}\left(\log\left(M+N+\frac{i+q}{m}\right)\right.$$
$$\left.-\log\big(m_0(M+N)+i+q\big)\right).$$

Since in the above integral representations all terms besides $\frac{AM}{M+N}\log(m-1)$, $\frac{BN}{M+N}\log(m-1)$ and $\log m$ are bounded as $m\to\infty$, we have now

$$\lim_{m\to\infty}\frac{\int_{m_0}^{m} f(x-1)dx}{\int_{m_0}^{m} g(x)dx} = \lim_{m\to\infty}\left(\frac{AM}{M+N}\cdot\frac{\log(m-1)}{\log m}+\frac{BN}{M+N}\right.$$
$$\left.\times\frac{\log(m-1)}{\log m}\right) = \frac{AM+BN}{M+N}.$$

Analogously

$$\lim_{m\to\infty} \frac{\int\limits_{m_0}^{m} f(x)dx}{\int\limits_{m_0}^{m} g(x-1)dx} = \frac{AM+BN}{M+N},$$

and the theorem is proved for $\vartheta = 1$.

ii) Let now $\vartheta > 1$. Here we find

$$\int\limits_{m_0}^{m} f(x-1)dx =$$

$$= \frac{A}{(M+N)(\vartheta-1)} \sum_{i=1}^{M} \left( \left( (m-1)(M+N)+i+q \right)^{\vartheta-1} - \left( (m_0-1)(M+N)+i+q \right)^{\vartheta-1} \right)$$

$$+ \frac{B}{(M+N)(\vartheta-1)} \sum_{i=M+1}^{M+N} \left( \left( (m-1)(M+N)+i+q \right)^{\vartheta-1} - \left( (m_0-1)(M+N)+i+q \right)^{\vartheta-1} \right)$$

$$= \frac{Am^{\vartheta-1}}{(M+N)(\vartheta-1)} \sum_{i=1}^{M} \left( \left( \left(1-\frac{1}{m}\right)(M+N)+\frac{i+q}{m} \right)^{\vartheta-1} - \left( \left(\frac{m_0-1}{m}\right)(M+N)+\frac{i+q}{m} \right)^{\vartheta-1} \right)$$

$$+ \frac{Bm^{\vartheta-1}}{(M+N)(\vartheta-1)} \sum_{i=M+1}^{M+N} \left( \left( \left(1-\frac{1}{m}\right)(M+N)+\frac{i+q}{m} \right)^{\vartheta-1} - \left( \left(\frac{m_0-1}{m}\right)(M+N)+\frac{i+q}{m} \right)^{\vartheta-1} \right)$$

as well as

$$\int\limits_{m_0}^{m} g(x)dx = \frac{m^{\vartheta-1}}{(M+N)(\vartheta-1)} \sum_{i=1}^{M+N} \left( \left( M+N+\frac{i+q}{m} \right)^{\vartheta-1} - \left( \frac{m_0}{m}(M+N)+\frac{i+q}{m} \right)^{\vartheta-1} \right)$$

and therefore

$$\lim_{m\to\infty} \frac{\int\limits_{m_0}^{m} f(x-1)dx}{\int\limits_{m_0}^{m} g(x)dx} = \frac{AM(M+N)^{\vartheta-1} + BN(M+N)^{\vartheta-1}}{(M+N)(M+N)^{\vartheta-1}} = \frac{AM+BN}{M+N}.$$

Analogously

$$\lim_{m\to\infty} \frac{\int\limits_{m_0}^{m} f(x)dx}{\int\limits_{m_0}^{m} g(x-1)dx} = \frac{AM+BN}{M+N},$$

which completes now the proof of our theorem.

**Discussion of Theorem 5.5**

If $M$ deterministic an $N$ stochastic steps are taken by turns and the step size $\varrho_n$ is chosen such that $\varrho_n = \dfrac{c}{n+q}$ with $q \in \mathbb{N}$ and $c \geq \dfrac{1}{a}$, where $0 < a < 2\alpha$, then by Theorem 5.5 we have the asymptotic representation

$$B_n \approx \left(1 - \frac{\sigma^2}{\gamma_1 + \sigma^2} w\right) B_n^s \text{ as } n \to \infty,$$

where $w = \dfrac{M}{M+N}$ is the percentage of deterministic steps in one complete turn of $M+N$ deterministic and stochastic steps. Hence, here we find that asymptotically the semi–stochastic procedure is $\left(1 - \frac{\sigma^2}{\gamma_1+\sigma^2} w\right)^{-1}$ times faster than the pure stochastic approximation procedure. This conclusion is studied in more detail in Section 5.5.2.

### Comparison of Hybrid Stochastic Approximation Procedures

By the above results we are also able to compare two hybrid stochastic approximation procedures characterized by the parameters $A, B, M, N, \vartheta$ and $\tilde{A}, \tilde{B}, \tilde{M}, \tilde{N}, \tilde{\vartheta}$, respectively. Corresponding to these two tuples of parameters $A, B, M, N, \vartheta$ and $\tilde{A}, \tilde{B}, \tilde{M}, \tilde{N}, \tilde{\vartheta}$, respectively, let $B_n$ and $\tilde{B}_n$, respectively, be the upper bounds of the mean square errors of the hybrid procedures defined by the recursion (5.85). Analogously, let $B_n^s$ and $\tilde{B}_n^s$, respectively, be the solution of recursion (5.86) corresponding to the above tuples of parameters.

If $\vartheta \geq 1$ and $\tilde{\vartheta} \geq 1$, then by Theorem 5.5 we know that

$$\frac{B_n}{B_n^s} \to \frac{AM+BN}{B(M+N)} \text{ as } n \to \infty$$

and

$$\frac{\tilde{B}_n}{\tilde{B}_n^s} \to \frac{\tilde{A}\tilde{M} + \tilde{B}\tilde{N}}{\tilde{B}(\tilde{M} + \tilde{N})} \text{ as } n \to \infty.$$

Moreover, by Corollary 5.2 we know that for $\vartheta > 1, \tilde{\vartheta} > 1$

$$\frac{B_n^s}{\tilde{B}_n^s} = \frac{n \cdot B_n^s}{n \cdot \tilde{B}_n^s} \to \frac{\frac{B}{\vartheta - 1}}{\frac{\tilde{B}}{\tilde{\vartheta} - 1}} = \frac{B}{\tilde{B}} \cdot \frac{\tilde{\vartheta} - 1}{\vartheta - 1} \text{ as } n \to \infty.$$

Consequently, we find this

**Corollary 5.3.** *For any two hybrid stochastic approximation procedures characterized by the parameters $A, B, M, N, \vartheta$ and $\tilde{A}, \tilde{B}, \tilde{M}, \tilde{N}, \tilde{\vartheta}$, resp., where $\vartheta > 1$ and $\tilde{\vartheta} > 1$, it holds*

$$\lim_{n \to \infty} \frac{B_n}{\tilde{B}_n} = \frac{\frac{AM+BN}{M+N}}{\frac{\tilde{A}\tilde{M}+\tilde{B}\tilde{N}}{\tilde{M}+\tilde{N}}} \cdot \frac{\tilde{\vartheta} - 1}{\vartheta - 1}.$$

*Proof.* For $\frac{B_n}{\tilde{B}_n}$ we may write

$$\lim_{n \to \infty} \frac{B_n}{\tilde{B}_n} = \frac{B_n}{B_n^s} \cdot \frac{B_n^s}{\tilde{B}_n^s} \cdot \frac{1}{\frac{\tilde{B}_n}{\tilde{B}_n^s}}.$$

Hence, by the above considerations we get

$$\lim_{n \to \infty} \frac{B_n}{\tilde{B}_n} = \lim_{n \to \infty} \frac{B_n}{B_n^s} \cdot \lim_{n \to \infty} \frac{B_n^s}{\tilde{B}_n^s} \cdot \frac{1}{\lim_{n \to \infty} \frac{\tilde{B}_n}{\tilde{B}_n^s}} = \frac{AM + BN}{B(M + N)} \cdot \frac{B}{\tilde{B}} \cdot \frac{\tilde{\vartheta} - 1}{\vartheta - 1} \cdot \frac{\tilde{B}(\tilde{M} + \tilde{N})}{\tilde{A}\tilde{M} + \tilde{B}\tilde{N}}$$

which proves our corollary.

*Note.* If $\tilde{M} = 0, \tilde{B} = B$ and $\tilde{\vartheta} = \vartheta$, i.e. if $\tilde{B}_n = B_n^s$, then from Corollary 5.2 we gain back Theorem 5.5.

### 5.5.2 Lower Bounds for the Mean Square Error

The results obtained above may be used now to find also lower bounds for the mean square error of our hybrid procedure, provided that some further hypotheses are fulfilled.

To this end assume that for an optimal solution $x^*$ of (5.1)

$$\begin{aligned} &\left\| p_D \Big( X_n - \varrho \Big( A(X_n) + \xi_n \Big) \Big) - p_D \Big( x^* - \varrho_n A(x^*) \Big) \right\| \qquad (5.97) \\ &= \left\| X_n - \varrho_n \Big( A(X_n) + \xi_n \Big) - \Big( x^* - \varrho A(x^*) \Big) \right\| \text{ for all } n \in \mathbb{N}. \end{aligned}$$

This holds e.g. if $\nabla F(x^*) = 0$ and

$$X_n - \varrho_n\Big(A(X_n) + \xi_n\Big) \in D \text{ for all } n \in \mathbb{N},$$

see (5.67) and (5.69a).

Now, if (5.97) holds, then we find

$$\begin{aligned}
\|X_{n+1} - x^*\|^2 &= \Big\|(X_n - x^*) - \varrho_n\Big(A(X_n) - A(x^*)\Big) - \varrho_n\xi_n\Big\|^2 \qquad (5.98)\\
&= \|X_n - x^*\|^2 - 2\varrho_n\Big(A(X_n) - A(x^*)\Big)^T(X_n - x^*)\\
&\quad +\varrho_n^2\Big\|A(X_n) - A(x^*)\Big\|^2 - 2\varrho_n\xi_n^T\Big(X_n - x^*\\
&\quad -\varrho_n\Big(A(X_n) - A(x^*)\Big)\Big) + \varrho_n^2\|\xi_n^2\|^2.
\end{aligned}$$

Assume now that for all $x \in D$ the inequalities

$$\Big\|A(x) - A(x^*)\Big\|^2 \geq \delta_1 + \delta_2\|x - x^*\|^2, x \neq x^*, \tag{5.99}$$

$$\Big(A(x) - A(x^*)\Big)^T(x - x^*) \leq \beta\|x - x^*\|^2 \tag{5.100}$$

hold, and for all $n \in \mathbb{N}$ the inequality

$$E\|\xi_n\|^2 \geq \eta^2 > 0 \tag{5.101}$$

is fulfilled, where $\delta_1 \geq 0, \delta_2 > 0, \beta > 0$ and $\eta > 0$ are given constants.

Note that these inequalities are opposite to (5.70), (5.71) and (5.80), respectively. Furthermore, (5.99), (5.100) and (5.101) together with (5.70), (5.71) and (5.80) imply that

$$\delta_1 \leq \gamma_1, \delta_2 \leq \gamma_2, \alpha \leq \beta \text{ and } \eta^2 \leq \sigma^2.$$

If again (5.72) is valid, i.e. if $E(\xi_n|X_n) = 0$ a.s. for every $n \in \mathbb{N}_2$, then, using equality (5.98), for $b_n = E\|X_n - x^*\|^2$ we find in the same way as in the proof of Lemma 5.1 the following recursion being opposite to (5.74a,b) and (5.82)

$$\begin{aligned}
b_{n+1} &\geq (1 - 2\beta\varrho_n + \delta_2\varrho_n^2)b_n + \varrho_n^2\Big(\delta_1 + E\|\xi_n\|^2\Big) \qquad (5.102a)\\
&\geq (1 - 2\beta\varrho_n)b_n + \underline{\beta}_n\varrho_n^2, \quad n = 1, 2, \ldots,
\end{aligned}$$

where

$$\underline{\beta}_n = \begin{cases} \delta_1, & \text{if } n \in \mathbb{N}_1 \\ \delta_1 + \eta^2, & \text{if } n \in \mathbb{N} \end{cases} \tag{5.102b}$$

In contrast to the "worst case" in (5.82), which is represented by the recursion (5.84a,b), here we consider the "best case" in (5.102a), i.e. the recursion

$$\underline{B}_{n+1} = (1 - 2\beta\varrho_n)\underline{B}_n\varrho_n^2 + \underline{\beta}_n\varrho_n^2, \ n = 1, 2, \ldots. \tag{5.103}$$

Assuming again that $\lim_{n\to\infty} \varrho_n = 0$, there is an integer $n_0$ such that (5.81) holds, i.e.

$$0 < 1 - 2\alpha\varrho_n + \gamma_2\varrho_n^2 < 1 - a\varrho_n \text{ for all } n \geq n_0,$$

where $0 < a < 2\alpha$, as well as

$$1 - 2\beta\varrho_n > 0 \text{ for all } n \geq n_0.$$

Let $B_{n_0} = b_{n_0}$ and $\underline{B}_{n_0} = b_{n_0}$.

Then, according to Section 5.5, inequality (5.84a), we know that $b_n \leq B_n$ for all $n \geq n_0$, and by a similar simple induction argument we find

$$\underline{B}_n \leq b_n \leq B_n \text{ for all } n \geq n_0. \tag{5.104a}$$

Since the pure stochastic procedure is characterized simply by $\mathbb{N}_1 = \emptyset$, we also have

$$\underline{B}_n^s \leq b_n^s \leq B_n^s, \text{ for all } n \geq n_0, \tag{5.104b}$$

where $b_n^s$ is the mean square error of the pure stochastic algorithm. Furthermore, $B_n^s$ and $\underline{B}_n^s$ are the solutions of (5.84a) and (5.103), resp., with $\bar{\beta}_n = \gamma_1 + \sigma^2$ and $\underline{\beta}_n = \delta_1 + \eta^2$ for all $n \in \mathbb{N}$, where, of course, $B_{n_0}^s = b_{n_0}^s$ and $\underline{B}_{n_0}^s = b_{n_0}^s$.

Thus, the inequalities (5.104a,b) yield the estimates

$$\frac{\underline{B}_n}{B_n^s} \leq \frac{b_n}{b_n^s} \leq \frac{B_n}{\underline{B}_n^s} \text{ for all } n \geq n_0. \tag{5.105}$$

Let now the step size $\varrho_n$ be defined by

$$\varrho_n = \frac{c}{n+q}, \quad n = 1, 2, \ldots$$

where $c > 0$ and $q \in \mathbb{N} \cup \{0\}$ are fixed numbers.

The (hybrid) stochastic algorithms having the lower and upper mean square error estimates $(\underline{B}_n)$ and $(B_n^s)$ are characterized – see Section 5.5.1 – by the parameters

$$A = c^2\delta_1, \quad B = c^2(\delta_1 + \eta^2), \quad M, N, \vartheta = 2\beta c,$$

and

$$\tilde{B} = c^2(\gamma_1 + \sigma^2), \quad \tilde{M} = 0, \tilde{\vartheta} = ac, \text{ respectively.}$$

Hence, if $c > 0$ is selected such that $ac > 1$ and $2\beta c > 1$, then Corollary 5.3 yields

$$\lim_{n\to\infty} \frac{\underline{B}_n}{B_n^s} = \frac{\frac{c^2\delta_1 M + c^2(\delta_1+\eta^2)N}{M+N}}{c^2(\gamma_1 + \sigma^2)} \cdot \frac{ac-1}{2\beta c - 1}. \tag{5.106a}$$

Furthermore, the sequences $(B_n), (B_n^s)$, resp., are estimates of the mean square error of the (hybrid) stochastic algorithms represented by the parameters

$$A = c^2\gamma_1,\ B = c^2(\gamma_1 + \sigma^2),\ M, N, \vartheta = ac$$

and

$$\tilde{B} = c^2(\delta_1 + \eta^2),\ \tilde{M} = 0,\ \tilde{\vartheta} = 2\beta c, \text{ respectively.}$$

Thus, if $ac > 1$ and $2\beta c > 1$, then Corollary 5.3 yields

$$\lim_{n\to\infty} \frac{B_n}{B_n^s} = \frac{\frac{c^2\gamma_1 M + c^2(\gamma_1+\sigma^2)N}{M+N}}{c^2(\delta_1 + \eta^2)} \cdot \frac{2\beta c - 1}{ac - 1}. \tag{5.106b}$$

Consequently, we have now this next result:

**Theorem 5.6.** *Suppose that the minimization problem (5.1) fulfills the inequalities (5.70), (5.71) and (5.99), (5.100). Furthermore, assume that the noise conditions (5.72), (5.80) and (5.101) hold. If* $\varrho_n = \dfrac{c}{n+q}$ *with* $q \in \mathbb{N} \cup \{0\}$ *and* $c > 0$ *such that* $ac > 1$, *where* $0 < a < 2\alpha$, *and if (5.97) is fulfilled, then*

$$Q_1 \le \liminf_{n\to\infty} \frac{b_n}{b_n^s} \le \limsup_{n\to\infty} \frac{b_n}{b_n^s} \le Q_2, \tag{5.107}$$

*where*

$$Q_1 = \frac{\delta_1 M + (\delta_1 + \eta^2)N}{(M+N)(\gamma_1 + \sigma^2)} \cdot \frac{ac-1}{2\beta c - 1}, \tag{5.108a}$$

$$Q_2 = \frac{\gamma_1 M + (\gamma_1 + \sigma^2)N}{(M+N)(\delta_1 + \eta^2)} \cdot \frac{2\beta c - 1}{ac - 1}. \tag{5.108b}$$

*Proof.* Since $0 < a < 2\alpha$ and $\alpha \le \beta$, we have $ac < 2\beta c$ for every $c > 0$ and therefore $2\beta c - 1 > ac - 1 > 0$. The assertion follows now from (5.106a) and (5.106b).

**Discussion of Theorem 5.6**

a) Obviously, (5.107) yields $Q_1 \le Q_2$, and, using formulas (5.108a), (5.108b), we find $Q_1 < Q_2$.
Moreover, by means of Theorem 5.5 and because of $\delta_1 \le \gamma_1, \alpha \le \beta, \eta^2 \le \sigma^2$ as also $1 < ac < 2\alpha c \le 2\beta c$, we obtain

$$Q_1 \le \frac{\gamma_1 M + (\gamma_1 + \sigma^2)N}{(M+N)(\gamma_1 + \sigma^2)} \cdot \frac{ac-1}{2\beta c - 1} = \frac{ac-1}{2\beta c - 1} \cdot \lim_{n\to\infty} \frac{B_n}{B_n^s} < \lim_{n\to\infty} \frac{B_n}{B_n^s} \tag{5.109}$$

as also

$$Q_2 \ge \frac{\delta_1 M + (\delta_1 + \eta^2)N}{(M+N)(\delta_1 + \eta^2)} \cdot \frac{2\beta c - 1}{ac - 1} = \frac{2\beta c - 1}{ac - 1} \cdot \lim_{n\to\infty} \frac{B_n}{B_n^s} > \lim_{n\to\infty} \frac{B_n}{B_n^s}. \tag{5.110}$$

b) According to (5.107) we have

$$Q_1 b_n^s \le b_n \le Q_2 b_n^s \text{ as } n \to \infty.$$

This means that asymptotically the hybrid stochastic approximation procedure is

$$\text{at most } \frac{1}{Q_1} \text{ times faster and at least } \frac{1}{Q_2} \text{ times faster}$$

than the pure stochastic procedure.
According to (5.109) we have $\frac{1}{Q_1} > 1$ for the best case. Moreover, $Q_2 < 1$ holds if and only if

$$\frac{\gamma_1 M + (\gamma_1 + \sigma^2)N}{(M+N)(\delta_1 + \eta^2)} < f := \frac{ac-1}{2\beta c - 1} (< 1). \tag{5.111}$$

Since always

$$\gamma_1 + \sigma^2 - f(\delta_1 + \eta^2) > \gamma_1 + \sigma^2 - \delta_1 - \eta^2 \geq 0,$$

condition (5.111) is equivalent to

$$\frac{N}{M} < \frac{f(\delta_1 + \eta^2) - \gamma_1}{\gamma_1 + \sigma^2 - f(\delta_1 + \eta^2)} \tag{5.112a}$$

which can be fulfilled if and only if

$$\delta_1 + \eta^2 > \frac{\gamma_1}{f} = \gamma_1 \frac{2\beta c - 1}{ac - 1} \tag{5.112b}$$

c) If $c$ is selected according to $c > \frac{1}{a_0}$, where $a_0$ is a fixed number such that $0 < a_0 < 2\alpha$, then $ac \geq a_0 c > 1$ for all $a$ with $a_0 \leq a < 2\alpha$. Hence, if $c > \frac{1}{a_0}$, then Theorem 5.6 holds for all numbers $a$ with $a_0 \leq a < 2\alpha$. Consequently, if $c > \frac{1}{a_0}$, where $0 < a_0 < 2\alpha$, then the relations (5.107) with (5.108a) and (5.108b), (5.109) and (5.110) as also (5.111), (5.112a) and (5.112b) hold also if $a$ is replaced by $2\alpha$.

### 5.5.3 Decreasing Rate of Stochastic Steps

A further improvement of the speed of convergence is obtained if we have a decreasing rate of stochastic steps taken in (5.69a-c). In order to handle this case, we consider the sequence

$$c_n = n^\lambda E\|X_n - x^*\|^2, \quad n = 1, 2, \ldots$$

for some $\lambda$ with $0 \leq \lambda < \bar{\lambda}$, where $\bar{\lambda}$ is a given fixed positive number. Then, under the same conditions and using the same methods as in Lemma 5.1, corresponding to (5.74a,b) we find

$$c_{n+1} \leq \left(1 + \frac{1}{n}\right)^\lambda (1 - 2a\varrho_n + \gamma_2 \varrho_n^2) c_n + (n+1)^\lambda \beta_n \varrho_n^2, \quad n = 1, 2, \ldots, \tag{5.113a}$$

where again

$$\beta_n = \gamma_1 + E\|\xi_n\|^2 \tag{5.113b}$$

and $\xi_n$ is defined by (5.69b).

We claim that for $n \geq n_1, n_1 \in \mathbb{N}$, sufficiently large,

$$\left(1+\frac{1}{n}\right)^{\lambda}(1-2\alpha\varrho_n+\gamma_2\varrho_n^2) \leq 1, \tag{5.114}$$

provided that the step size $\varrho_n$ is suitably chosen.

Indeed, if $\varrho_n \to 0$ as $n \to \infty$, then there is an integer $n_0$ such that (see (5.81))

$$0 < 1-2\alpha\varrho_n+\gamma_2\varrho_n^2 < 1-a\varrho_n, \quad n \geq n_0$$

for any number $a$ such that $0 < a < 2\alpha$.

According to (5.114) and (5.81) we have to choose now $\varrho_n$ such that

$$\left(1+\frac{1}{n}\right)^{\lambda}(1-a\varrho_n) \leq 1, \quad n \geq n_1 \tag{5.115}$$

with an integer $n_1 \geq n_0$. Write

$$\left(1+\frac{1}{n}\right)^{\lambda} = 1+\frac{p_n(\lambda)}{n}$$

with $p_n(\lambda) = n\left(\left(1+\frac{1}{n}\right)^{\lambda}-1\right)$. Hence, condition (5.115) has the form

$$1-a\varrho_n \leq \frac{n}{n+p_n(\lambda)}, \quad n \geq n_1.$$

Since $0 \leq \lambda < \bar{\lambda}$ and therefore $p_n(\lambda) \leq p_n(\bar{\lambda})$ for all $n \in \mathbb{N}$, the latter is true if

$$1-a\varrho_n \leq \frac{n}{n+p_n(\bar{\lambda})}, \quad n \geq n_1.$$

Hence, for the step size $\varrho_n$ we find the condition

$$\varrho_n \geq \frac{1}{a}\frac{p_n(\bar{\lambda})}{n+p_n(\bar{\lambda})} = \frac{1}{n}\frac{1}{a}\frac{p_n(\bar{\lambda})}{1+\frac{p_n(\bar{\lambda})}{n}} \tag{5.116}$$

for all $n$ sufficiently large. Writing

$$p_n(\bar{\lambda}) = \frac{\left(1+\frac{1}{n}\right)^{\bar{\lambda}}-1}{\frac{1}{n}},$$

we see that

$$\lim_{n\to\infty} p_n(\bar{\lambda}) = \frac{d}{dx}(1+x)^{\bar{\lambda}}|_{x=0} = \bar{\lambda}.$$

Therefore

$$\lim_{n\to\infty} \frac{p_n(\bar{\lambda})}{1+\frac{p_n(\bar{\lambda})}{n}} = \bar{\lambda},$$

which implies that

$$\vartheta_1 \le \frac{p_n(\bar{\lambda})}{1+\frac{p_n(\bar{\lambda})}{n}} \le \vartheta_2, \quad n = 1,2,\ldots, \tag{5.117}$$

with some constants $\vartheta_1, \vartheta_2$ such that $0 < \vartheta_1 < \bar{\lambda} \le \vartheta_2$. If $\bar{\lambda} \in \mathbb{N}$, then an easy consideration shows that

$$\frac{p_n(\bar{\lambda})}{1+\frac{p_n(\bar{\lambda})}{n}}, \quad n = 1,2,\ldots,$$

in a monotonously increasing sequence. Hence, for every $\bar{\lambda} \in \mathbb{N}$ the bounds $\vartheta_1, \vartheta_2$ can be selected according to $\vartheta_1 = \frac{2^{\bar{\lambda}}-1}{2^{\bar{\lambda}}}$ and $\vartheta_2 = \bar{\lambda}$. Consequently, if $\varrho_n$ is defined by

$$\varrho_n = \frac{1}{n}\frac{1}{a}\frac{p_n(\bar{\lambda})}{1+\frac{p_n(\bar{\lambda})}{n}} \tag{5.118a}$$

or $\varrho_n$ is given by

$$\varrho_n = \frac{c}{n+q} = \frac{1}{n}\frac{c}{1+\frac{q}{n}} \text{ with } q \in \mathbb{N} \text{ and } c > \frac{\vartheta_2}{a}, \tag{5.118b}$$

then there is an integer $n_1 \ge n_0$ such that (5.116) holds for $n \ge n_1$.

Let $\vartheta = \vartheta_2, \vartheta = ac$, respectively. Then in both cases we have $\varrho_n \le \frac{\vartheta}{a} \cdot \frac{1}{n}$, and from (5.113a,b) - (5.115) follows

$$\begin{aligned} c_{n+1} &\le c_n + (n+1)^{\lambda}\beta_n\varrho_n^2 \le c_n + (n+1)^{\lambda}\beta_n\left(\frac{\vartheta}{a}\right)^2\frac{1}{n^2} \\ &\le c_n + \left(\frac{n+1}{n}\right)^{\bar{\lambda}}\beta_n\left(\frac{\vartheta}{a}\right)^2 n^{\lambda-2} \le c_n + 2^{\bar{\lambda}}\beta_n\left(\frac{\vartheta}{a}\right)^2 n^{\lambda-2} \end{aligned}$$

for $n \ge n_1$ and therefore

$$c_{N+1} - c_{n_1} = \sum_{n=n_1}^{N}(c_{n+1}-c_n) \le 2^{\bar{\lambda}}\left(\frac{\vartheta}{a}\right)^2\sum_{n=n_1}^{N}\beta_n n^{\lambda-2}, \; N \ge n_1. \tag{5.119}$$

Suppose now that also (5.80) holds, hence $E\|\xi_n\|^2 \le \sigma^2$ for all $n \in \mathbb{N}_2$. Therefore $\beta_n \le \bar{\beta}_n$, where (see (5.83)) $\bar{\beta}_n = \gamma_1$ for $n \in \mathbb{N}_1$ and $\bar{\beta}_n = \gamma_1 + \sigma^2$ for $n \in \mathbb{N}_2$. Inequality (5.119) yields now

$$c_{N+1} - c_{n_1} \leq 2^{\bar{\lambda}} \left(\frac{\vartheta}{a}\right)^2 \left( \sum_{\substack{n=n_1 \\ n \in \mathbb{N}_1}}^{N} \frac{\gamma_1}{n^{2-\lambda}} + \sum_{\substack{n=n_1 \\ n \in \mathbb{N}_2}}^{N} \frac{\gamma_1 + \sigma^2}{n^{2-\lambda}} \right).$$

Choosing $\bar{\lambda} \geq 2$, we find this result:

**Theorem 5.7.** *Let the assumptions of Lemma 5.1 be fulfilled, and suppose that (5.80) holds. Moreover, let the step size* $\varrho_n$ *be defined by (5.118a) or (5.118b). If* $\gamma_1 = 0$ *and* $\mathbb{N}_1 = \mathbb{N} \backslash \mathbb{N}_2$ *with*

$$\mathbb{N}_2 = \{n_k = k^p : k = 1, 2, \ldots\} \tag{5.120}$$

*for some fixed integer* $p = 1, 2, \ldots,$ *then*

$$E\|X_n - x^*\|^2 = 0(n^{-\lambda}) \text{ as } n \to \infty$$

*for all* $\lambda$ *such* $0 \leq \lambda < 2 - \frac{1}{p}$.

*Proof.* Under the given hypotheses we have

$$\begin{aligned} c_{N+1} - c_{n_1} &\leq 2^{\bar{\lambda}} \left(\frac{\vartheta}{a}\right)^2 \sigma^2 \sum_{n_1 \leq k^p \leq N} \frac{1}{(k^p)^{2-\lambda}} \\ &\leq 2^{\bar{\lambda}} \left(\frac{\vartheta}{a}\right)^2 \sigma^2 \sum_{k=1}^{\infty} \frac{1}{k^{p(2-\lambda)}}, \quad N \geq n_1, \end{aligned}$$

where $c_n = n^{\lambda} E\|X_n - x^*\|^2$ with $0 \leq \lambda < \bar{\lambda}$ and a fixed $\bar{\lambda} \geq 2$. Since $2 - \frac{1}{p} < \bar{\lambda}$, and $0 \leq \lambda < 2 - \frac{1}{p}$ is equivalent to $p(2-\lambda) > 1$, the above series converges. Hence, the sequence $(c_n)$ is bounded which yields the assertion of our theorem.

**Discussion of Theorem 5.7**

The condition $\gamma_1 = 0$ means that the operator $A(x)$ defined by (5.67) must satisfy (see (5.70)) the LIPSCHITZ condition

$$\left\| A(x) - A(x^*) \right\| \leq \sqrt{\gamma_2} \|x - x^*\|.$$

Definition (5.120) of $\mathbb{N}_2$ obviously means that we have a decreasing rate $n^{\frac{1}{p}-1}$ of stochastic steps taken in (5.69a,b).

According to Lemma 5.2, for a pure stochastic approximation procedure we know that $E\|X_n - x^*\|^2 = 0(n^{-\kappa})$ with $0 < \kappa \leq 1$. Hence, if stochastic and deterministic steps are taken in (5.69a,b) as described by (5.120), then the speed of convergence may be increased from $\mathcal{O}(n^{-\kappa})$ with $0 < \kappa \leq 1$ in the pure stochastic case to the order of $\mathcal{O}(n^{-\lambda})$ with $1 < \lambda < 2 - \frac{1}{p}$ in the hybrid stochastic case. If $p = 1, p \to \infty$, respectively, then we obtain the pure stochastic, pure deterministic case, respectively.

Denoting by $s_n$ the number of stochastic steps taken in (5.69a,b) up to the $n$-th stage, in the case (5.120) we have

$$s_n \approx n^{1/p}$$

and therefore $\frac{s_n}{n^2} \approx n^{1/p-2}$. Hence, the estimate of the convergence rate given by Theorem 5.7 can also be represented in the form

$$E\|X_n - x^*\|^2 \approx \mathcal{O}\left(\frac{s_n}{n^2}\right) = \mathcal{O}\left(\frac{r_n}{n}\right)$$

if $r_n = n^{1/p-1}$ is the rate of stochastic steps taken in (5.69a,b), cf. the example in [71].

# 6

# Stochastic Approximation Methods with Changing Error Variances

## 6.1 Introduction

As already discussed in the previous Chapter 5, hybrid stochastic gradient procedures are very important tools for the iterative solution of stochastic optimization problems as discussed in Chapters 1–4:

$$\min F(x) \text{ s.t. } x \in D$$

with the expectation value function $F(x) = Ef\Big(a(\omega), x\Big)$ having gradient $G(x) = \nabla F(x)$. Moreover, quite similar stochastic approximation procedures may be used for the solution of systems of nonlinear equations

$$G(x) = 0$$

involving a vectorial so–called "regression function" $G = G(x)$ on $\mathbb{R}^\nu$. These algorithms generating a sequence $(X_n)$ of estimators $X_n$ of a solution $x^*$ of one of the above mentioned related problems are based on estimators

$$Y_n = G(X_n) + Z_n, n = 0, 1, 2, \ldots$$

of the gradient $G$ of the objective function $F$ or the regression function $G$ at an iteration point $X_n$. The resulting iterative procedures of the type

$$X_{(n+1)} := p_D(X_n - R_n Y_n), n = 0, 1, 2, \ldots ,$$

cf. (5.5), are well studied in the literature. Here, $p_D$ denotes the projection of $\mathbb{R}^\nu$ onto the feasible domain $D$ known to contain an (optimal) solution, and $R_n, n = 0, 1, 2, \ldots$ is a sequence of scalar or matrix valued step sizes.

In the extensive literature on stochastic approximation procedures, cf. [34–36, 53, 54, 114, 149], various sufficient conditions can be found for the underlying optimization problem, for the regression function, resp., the sequence of estimation errors $(Z_n)$ and the sequence of step sizes $R_n, n = 0, 1, 2, \ldots$, such that

$$X_n \to x^*, n \to \infty,$$

in some probabilistic sense (in the (quadratic) mean, almost sure (a.s.), in distribution). Furthermore, numerous results concerning the asymptotic distribution of the random algorithm $(X_n)$ and the adaptive selection of the step sizes for increasing the convergence behavior of $(X_n)$ are available.

As shown in Chapter 5, the convergence behavior of stochastic gradient procedures may be improved considerably by using

* more exact gradient estimators $Y_n$ at certain iteration stages $n \in N_1$ and/or
* deterministic (feasible) descent directions $h_n = h(X_n)$ available at certain iteration points $X_n$.

While the convergence of hybrid stochastic gradient procedures was examined in Chapter 5 in terms of the sequence of mean square errors

$$b_n = E\|X_n - x^*\|^2, n = 0, 1, 2, \ldots,$$

in the present Chapter 6, for stochastic approximation algorithms with changing error variances further probabilistic convergence properties are studied in detail: Convergence almost sure (a.s.), in the mean, in distribution. Moreover, the asymptotic distribution is determined, and the adaptive selection of scalar and matrix valued step sizes is considered for the case of changing error variances. The results presented in this chapter are taken from the doctoral thesis of E. Plöchinger [104].

*Remark 6.1.* Based on the RSM–gradient estimators considered in Chapter 5, implementations of the present stochastic approximation procedure, especially the Hessian approximation given in Section 6.6, are developed in the thesis of J. Reinhart [110], see also [111], for stochastic structural optimization problems.

## 6.2 Solution of Optimality Conditions

For a given set $\emptyset \neq D \subset \mathbb{R}^\nu$ and a measurable function $G : D \longrightarrow \mathbb{R}^\nu$, consider the following problem

P: Find a vector $x^* \in D$ such that

$$x^* = p_D\left(x^* - r \cdot G(x^*)\right) \quad \text{for all } r > 0\,. \tag{6.1}$$

Here, $D$ is a convex und compact set, and for a vector $y \in \mathbb{R}^\nu$, let $p_D(y)$ denote the projection of $y$ onto $D$, determined uniquely by the relation

$$p_D(y) \in D \quad \text{and} \quad \left\|y - p_D(y)\right\| = \min_{x \in D} \|y - x\|\,.$$

Problems of this type have to be solved very often. E.g., in the special case that $G = \nabla F$, where $F : D \longrightarrow \mathbb{R}$ is a convex und continuously differentiable function, according to [15] condition (6.1) is equivalent to

$$F(x^*) = \min_{x \in D} F(x) .$$

Furthermore, for a vector $x^* \in \mathring{D}$ (:= interior of $D$), condition (6.1) is equivalent with the system of (nonlinear) equations

$$G(x^*) = 0 .$$

In many practical situations, the function $G(x)$ can not be determined directly, but we may assume that for every $x \in D$ a stochastic approximate (estimate) $Y$ of $G(x)$ is available. Hence, for the numerical solution of problem $(P)$ we consider here the following **stochastic approximation method**:

Select a vector $X_1 \in D$, and for $n = 1, 2, \ldots$ determine recursively

$$X_{n+1} := p_D\,(X_n - R_n \cdot Y_n) \tag{6.2a}$$

$$\text{with } R_n := \rho_n \cdot M_n \quad \text{and} \quad Y_n := G(X_n) + Z_n . \tag{6.2b}$$

The **matrix step size** $R_n = \rho_n M_n$, where $\rho_n$ is a positive number and $M_n$ is a real $\nu \times \nu$ matrix, can be chosen by the user in order to control the algorithm (6.2a,b). $Y_n$ denotes the *search direction*, and $Z_n$ is the *estimation error* occuring at the computation of $G(X_n)$ in step $n$. Furthermore, $\rho_n, M_n$ and $Z_n$ are random variables defined on an appropriate probability space $(\Omega, \mathcal{A}, \mathcal{P})$ .

## 6.3 General Assumptions and Notations

In order to derive different convergence properties of the sequence $(X_n)$ defined by (6.2a,b), in this part we suppose that the following assumptions, denoted by (U), are fulfilled. Note that these conditions are the standard assumptions which can be found in similar way in all relevant work on stochastic approximation:

a) $D \subset \mathbb{R}^\nu$ is a nonempty, convex and compact stet with diameter $\delta_0$ .
b) Let $G : D_0 \longrightarrow \mathbb{R}^\nu$ denote a vector function, where $D_0 \supset \{x + e : x \in D \text{ and } \|e\| < \varepsilon_0\}$ with a positive number $\varepsilon_0$ .
c) There exist positive numbers $L_0$ and $\mu_0$, as well as functions $H : D \longrightarrow \mathbb{R}^\nu \times \mathbb{R}^\nu$ and $\delta : D \times \mathbb{R}^\nu \longrightarrow \mathbb{R}^\nu$ such that

$$G(x + e) = G(x) + H(x) \cdot e + \delta(x; e) , \tag{U1a}$$

$$\|\delta(x; e)\| \le L_0 \cdot \|e\|^{1+\mu_0} , \tag{U1b}$$

$$H(x) \text{ is continuous in } x \tag{U1c}$$

for all $x \in D$ and $\|e\| < \varepsilon_0$ .

d) There is a unique $x^* \in D$ such that

$$x^* = p_D\big(x^* - r \cdot G(x^*)\big) \quad \text{for all } r > 0 \,. \tag{U2}$$

e) For $x^* \in \partial D$ (:= boundary of $D$) a scalar step size is selected, i.e.

$$R_n = \rho_n \cdot M_n = r_n \cdot I \quad \text{with } r_n \in \mathbb{R}_+ \,. \tag{U3}$$

f) Concerning the selection of the matrices $M_n$ in the step sizes $R_n = \rho_n \cdot M_n$, suppose that

$$\sup_n \|M_n\| < \infty \text{ a.s. (almost sure)} \,, \tag{U4a}$$

and there is a partition

$$M_n = \widehat{M}_n + \Delta M_n \,, \tag{U4b}$$

where

$$\lim_{n\to\infty} \|\Delta M_n\| = 0 \text{ a.s.} \,, \tag{U4c}$$

$$u^T \widehat{M}_n H(x) u \geq \underline{a} \|u\|^2 \tag{U4d}$$

for each $n \in \mathbb{N}$, $x \in D$ and $u \in \mathbb{R}^\nu$. Here, $\underline{a}$ is a positive constant.

g) There exists a probability space $(\Omega, \mathcal{A}, \mathcal{P})$ and $\sigma$-algebras $\mathcal{A}_1 \subset \ldots \subset \mathcal{A}_n \subset \mathcal{A}$ such that the variables $X_n, \rho_n, M_n, \widehat{M}_n$ und $Z_{n-1}$ for $n = 1, 2, \ldots$ are $\mathcal{A}_n$-measurable.

h) For the estimation error $Z_n$ in (6.2a,b) we suppose, cf. (5.35a,b), that

$$\sqrt{n}\|E_n Z_n\| \longrightarrow 0 \text{ a.s.}, \tag{U5a}$$

$$E_n\|Z_n - E_n Z_n\|^2 \leq \sigma_{0,n}^2 + \sigma_1^2 \|X_n - x^*\|^2 \tag{U5b}$$

for all $n = 1, 2, \ldots$ . Here $E_n(\cdot) := E(\cdot \mid \mathcal{A}_n)$ denotes the conditional expectation with respect to $\mathcal{A}_n$, $\sigma_1$ is a constant and $\sigma_{0,n}$ denotes a $\mathcal{A}_n$-measurable random variable with

$$\sup_n E\, \sigma_{0,n}^2 \leq \sigma_0^2 < \infty \tag{U5c}$$

and a further constant $\sigma_0$ .

i) There exists a disjoint partition $\mathbb{N} = \mathbb{N}^{(1)} \dot{\cup} \ldots \dot{\cup} \mathbb{N}^{(\kappa)}$ such that for $k = 1, \ldots, \kappa$ the limits

$$q_k := \lim_{n\to\infty} \frac{|\{1, \ldots, n\} \cap \mathbb{N}^{(k)}|}{n} > 0 \tag{U6}$$

exist and are positive, see Section 5.5.1.

In the later sections 6.5, 6.6, 6.7 and 6.8 additional assumptions are needed, generating further convergence properties or implying some of the above assumptions. Assumptions of that type are marked by one of the letters V,W,X und Y, and hold always for the current section.

### 6.3.1 Interpretation of the Assumptions

According to condition c), the mapping $G : D \longrightarrow \mathbb{R}^\nu$ is continuously differentiable und has the Jacobian $H(x)$. Because of a) und (U1a-c) there is a number $\beta < \infty$ with

$$\left\| H(x) \right\| \leq \beta \quad \text{for all} \quad x \in D. \tag{6.3}$$

Using the mean value theorem, we have then

$$\left\| G(x) - G(y) \right\| \leq \beta \cdot \|x - y\| \tag{6.4}$$

for all $x, y \in D$. Hence, $G$ is Lipschitz continuous with the Lipschitz constant $\beta$.

Since the projection operator $p_D(y)$ and the map $G(x)$ are continuous, because of a) there is a vector $x^* \in D$ fulfilling (U2).

Conditions (U2) and (U3) guarantee, that for each $n$ we have

$$x^* = p_D\big(x^* - R_n \cdot G(x^*)\big)\,. \tag{6.5}$$

Consequently, in the *deterministic* case, i.e. if $Z_n \equiv 0$, then algorithm (6.2a,b) stays at $x^*$ provided that $X_n = x^*$ for a certain index $n$.

According to condition f), the matrices $M_n$ and $\widehat{M_n}$ are bounded, and (U4d) and the mean value theorem imply that

$$\Big\langle x - x^*, \widehat{M_n}\big(G(x) - G(x^*)\big)\Big\rangle \geq \underline{a}\|x - x^*\|^2 \tag{6.6}$$

for all $n$ and $x \in D$, where $< \cdot, \cdot >$ denotes the scalar product in $\mathbb{R}^\nu$. Especially, for each $n \in \mathbb{N}$ we have

$$\Big\langle X_n - x^*, \widehat{M_n}\big(G(X_n) - G(x^*)\big)\Big\rangle \geq \hat{a}_n\|X_n - x^*\|^2 \text{ a.s.} \tag{6.7}$$

with certain $\mathcal{A}_n$-measurable random variables $\hat{a}_n \geq \underline{a}$ . Furthermore, (U4d) yields:

$$\widehat{M_n}, H(x) \text{ are regular for each } n \in \mathbb{N} \text{ and } x \in D. \tag{6.8}$$

E.g., condition (U4d) holds in case of scalar step sizes, provided that

$$\widehat{M_n} = \widehat{m}_n \cdot I \quad \text{with} \quad \widehat{m}_n \geq \underline{d}\,,$$
$$H(x) + H(x)^T \geq 2\alpha I$$

for all $n \in \mathbb{N}$ and $x \in D$, where $\underline{d}$ and $\alpha$ are positive numbers. These conditions hold in Section 6.7.2 und Section 6.8. The second of the above conditions implies also (U2), and means in the gradient case $G(x) = \nabla F(x)$ that the function $F(x)$ is strong convex on $D$.

According to g), the random variables $\rho_n, M_n$ and $Z_n$ may depend on the *history* $X_1, \ldots, X_n, Z_1, \ldots, Z_{n-1}, \rho_1, \ldots, \rho_n, M_1, \ldots, M_n$ of the process (6.2a,b).

Condition (U5a) concerning the estimation error $Z_n$ holds e.g. if the search direction $Y_n$ is an unbiased estimation of $G(X_n)$, hence, if $E_n Y_n = G(X_n)$. From (U5a-c) and a) we obtain that the estimation error has a bounded variance.

Since for different indices $n$ the estimation error $Z_n$ may vary considerably, according to i) the set of integers $\mathbb{N}$ will be partitioned such that the variances of $Z_n$ are equal approximatively for large indices $n$ lying in the same part $\mathbb{N}^{(k)}$. Moreover, the numbers $q^{(1)}, \ldots, q^{(\kappa)}$ represent the portion of the sets $\mathbb{N}^{(1)}, \ldots, \mathbb{N}^{(\kappa)}$ with respect to $\mathbb{N}$.

### 6.3.2 Notations and Abbreviations in this Chapter

The following notations and abbreviations are used in the next sections:

$$\begin{aligned}
\widehat{a}^{(k)} &:= \liminf_{n \in \mathbb{N}^{(k)}} \widehat{a}_n \qquad (6.9)\\
\widehat{a} &:= q_1 \widehat{a}^{(1)} + \ldots + q_\kappa \widehat{a}^{(\kappa)} \ \geq \ \underline{a}\\
\Delta_n &:= X_n - x^* = \text{ actual argument error}\\
\Delta Z_n &:= Z_n - E_n Z_n\\
\sigma_k^2 &:= \limsup_{n \in \mathbb{N}^{(k)}} E\sigma_{0,n}^2\\
\tau_{n+1} &:= \tau_n(1+\gamma_n) \text{ weight factor, where}\\
&\qquad \tau_1 := 1 \text{ and } 0 \leq \gamma_n \text{ is a nonnegative number}\\
\widehat{\Delta}_n &:= \tau_n(X_n - x^*) \text{ weighted argument error}\\
G^* &:= G(x^*)\,, \quad H^* := H(x^*)\\
G_n &:= G(X_n)\,, \quad H_n := H(X_n)\\
\lambda_{min}(A) &:= \text{ minimum eigenvalue of a symmetric matrix A}\\
\lambda_{max}(A) &:= \text{ maximum eigenvalue of a symmetric matrix A}\\
A^T &:= \text{ transpose of A}\\
\|A\| &:= \sqrt{\lambda_{max}(A^T A)} = \text{ norm of the matrix A}\\
A \leq B &:\Longleftrightarrow \lambda_{min}(B-A) \geq 0\\
\alpha^* &:= \frac{1}{2}\lambda_{min}(H^* + H^{*T})\\
\mathbb{M}_{m,n} &:= \{m+1, \ldots, n\} \cap \mathbb{M} = \ (m,n)\text{-section of a set } \mathbb{M} \subset \mathbb{N}\\
a^+ &:= \max(a, 0)\,, \quad a^- := a^+ - a\\
\bar{\Omega}_0 &:= \{\omega \in \Omega : \omega \notin \Omega_0\} = \text{ complement of the set } \Omega_0 \subset \Omega\\
I_{\Omega_0} &:= \text{characteristic function of } \Omega_0 \subset \Omega
\end{aligned}$$

$E_{\Omega_0} T := E(TI_{\Omega_0})$ = expectation of the random variable $T$ restricted to the set $\Omega_0$

*limextr* := either lim sup or lim inf

## 6.4 Preliminary Results

In order to shorten the proofs of the results given in the subsequent sections, here some preliminary results are given.

### 6.4.1 Estimation of the Quadratic Error

For the proof of the convergence of the quadratic argument error $\|X_n - x^*\|^2$ the following auxiliary statements are needed.

**Lemma 6.1.** *For all $n \in \mathbb{N}$ we have*

$$\begin{aligned} E_n\|\Delta_{n+1}\|^2 \le & \left\|\Delta_n - R_n(G_n - G^*)\right\|^2 \\ & + \|R_n\|^2 \left(\|E_n Z_n\|^2 + E_n\|Z_n - E_n Z_n\|^2\right) \\ & + 2\|R_n\| \, \|E_n Z_n\| \left\|\Delta_n - R_n(G_n - G^*)\right\| . \end{aligned} \tag{6.10}$$

*Proof.* Because of (6.5), for arbitrary $n$ we get

$$\begin{aligned} & \|\Delta_{n+1}\|^2 = \left\|p_D(X_n - R_n Y_n) - p_D(x^* - R_n G^*)\right\|^2 \\ & \le \left\|X_n - R_n(G_n + Z_n) - (x^* - R_n G^*)\right\|^2 \\ & = \left\|\left[\Delta_n - R_n(G_n - G^*) - R_n E_n Z_n\right] - R_n(Z_n - E_n Z_n)\right\|^2 \\ & \le \left(\left\|\Delta_n - R_n(G_n - G^*)\right\| + \left\|R_n E_n Z_n\right\|\right)^2 + \left\|R_n(Z_n - E_n Z_n)\right\|^2 \\ & +2\langle \Delta_n - R_n(G_n - G^*) - R_n E_n Z_n \, , \, R_n(Z_n - E_n Z_n)\rangle . \end{aligned}$$

Inequality (6.10) is obtained then by taking expectations.

The first term on the right hand side of inequality (6.10) can be estimated as follows:

**Lemma 6.2.** *For each $n$ we find*

$$\begin{aligned} & \left\|\Delta_n - R_n(G_n - G^*)\right\|^2 \\ & \le \left(1 - 2\rho_n(\hat{a}_n - \beta\|\Delta M_n\|) + \rho_n^2\|M_n\|^2\beta^2\right)\|\Delta_n\|^2. \end{aligned} \tag{6.11}$$

*Proof.* According to (U4b) we have

$$\left\|\Delta_n - \rho_n M_n(G_n - G^*)\right\|^2 \le \|\Delta_n\|^2 + \rho_n^2\|M_n\|^2\|G_n - G^*\|^2$$
$$-2\rho_n\langle\Delta_n, \widehat{M}_n(G_n - G^*)\rangle + 2\rho_n\|\Delta_n\|\|\Delta M_n\|\|G_n - G^*\| .$$

Inequality (6.11) follows now by applying (6.4) and (6.7).

For the sequence of weighted errors ($\widehat{\Delta}_n = \tau_n(X_n - x^*)$) we have now an upper bound which is used several times in the following:

**Lemma 6.3.** *For arbitrary* $\varepsilon > 0$ *and* $n \in \mathbb{N}$ *we have*

$$E_n\|\widehat{\Delta}_{n+1}\|^2 \le (1 - \rho_n t_n + t_n^{(1)})\|\widehat{\Delta}_n\|^2 + t_n^{(2)} , \tag{6.12}$$

*where*

$$t_n := (1+\gamma_n)^2\left(1 + \frac{g_n}{\varepsilon}\right)u_n - \frac{\gamma_n}{\rho_n}(2+\gamma_n) ,$$
$$t_n^{(1)} := (1+\gamma_n)^2\left(\frac{g_n}{\varepsilon} + \left(\rho_n\|M_n\|\right)^2\sigma_1^2\right) ,$$
$$t_n^{(2)} := (1+\gamma_n)^2\left(g_n(\varepsilon + g_n) + \left(\tau_n\rho_n\|M_n\|\right)^2\sigma_{0,n}^2\right),$$

*and the following definitions are used:*

$$u_n := 2\left(\widehat{a}_n - \beta\|\Delta M_n\|\right) - \rho_n\|M_n\|^2\beta^2 ,$$
$$g_n := \tau_n\rho_n\|M_n\|\|E_n Z_n\| .$$

*Proof.* Because of the inequality

$$2\tau_n\left\|\Delta_n - R_n(G_n - G^*)\right\| \le \varepsilon + \frac{\tau_n^2\left\|\Delta_n - R_n(G_n - G^*)\right\|^2}{\varepsilon} ,$$

after multiplication with $\tau_{n+1}^2$, inequality (6.10) yields

$$E_n\|\widehat{\Delta}_{n+1}\|^2$$
$$\le (1+\gamma_n)^2\Bigg[\left(1 + \frac{\tau_n\|R_n\|\|E_n Z_n\|}{\varepsilon}\right)\|\Delta_n - R_n(G_n - G^*)\|^2\tau_n^2$$
$$+\left(\tau_n\|R_n\|\right)^2\left(\|E_n Z_n\|^2 + E_n\|Z_n - E_n Z_n\|^2\right) + \varepsilon\tau_n\|R_n\|\|E_n Z_n\|\Bigg] .$$

Taking into account (6.11) and (U5b), we find

$$E_n\|\widehat{\Delta}_{n+1}\|^2 \le (1+\gamma_n)^2\Bigg[\left(1 + \frac{g_n}{\varepsilon}\right)(1 - \rho_n u_n)\|\widehat{\Delta}_n\|^2$$
$$+g_n(\varepsilon + g_n) + (\tau_n\|R_n\|)^2(\sigma_{0,n}^2 + \sigma_1^2\|\Delta_n\|^2)\Bigg]$$
$$= (1+\gamma_n)^2\Bigg[\left(1 - \rho_n\left(1 + \frac{g_n}{\varepsilon}\right)u_n + \frac{g_n}{\varepsilon} + \|R_n\|^2\sigma_1^2\right)\|\widehat{\Delta}_n\|^2$$
$$+g_n(\varepsilon + g_n) + \left(\tau_n\|R_n\|\right)^2\sigma_{0,n}^2\Bigg] .$$

This proves now (6.12).

### 6.4.2 Consideration of the Weighted Error Sequence

Suppose that

$$\frac{\rho_n}{s_n} \longrightarrow 1 \quad \text{a.s.} , \tag{6.13}$$

$$M_n - D_n \longrightarrow 0 \quad \text{a.s.} \tag{6.14}$$

for given deterministic numbers $s_n$ and matrices $D_n$. The weighted error $\widehat{\Delta}_n := \tau_n(X_n - x^*)$ can be represented then as follows:

**Lemma 6.4.** *For each integer n:*

$$\begin{aligned}\widehat{\Delta}_{n+1} &= (I - s_n B_n)\widehat{\Delta}_n + \tau_n s_n D_n(E_n Z_n - Z_n) \\ &\quad + s_n \Delta B_n \widehat{\Delta}_n + \tau_n s_n \Delta D_n(E_n Z_n - Z_n) \\ &\quad + \tau_{n+1}\big(-R_n(E_n Z_n + G^*) + \Delta P_n\big) ,\end{aligned} \tag{6.15}$$

where the variables in Lemma 6.4 are defined as follows:

$$B_n := D_n H^* - \frac{\gamma_n}{s_n} I , \tag{6.16}$$

$$\Delta B_n := B_n - \frac{1}{s_n}\big((1+\gamma_n)R_n\widehat{H}_n - \gamma_n I\big) , \tag{6.17}$$

$$\Delta D_n := \frac{1}{s_n}(1+\gamma_n)R_n - D_n , \tag{6.18}$$

$$\Delta P_n := p_D(X_n - R_n Y_n) - (X_n - R_n Y_n) , \tag{6.19}$$

$$\widehat{H}_n := \int_0^1 H\big(x^* + t(X_n - x^*)\big)\, dt . \tag{6.20}$$

*Proof.* Because of assumption (U1a-c), and using the Taylor formula we get

$$Y_n = G_n + Z_n = G^* + \widehat{H}_n \Delta_n + Z_n .$$

Hence, (6.2a,b) yields

$$\begin{aligned}\Delta_{n+1} &= \Delta_n - R_n Y_n + \Delta P_n \\ &= (I - R_n\widehat{H}_n)\Delta_n + R_n(E_n Z_n - Z_n) - R_n(E_n Z_n + G^*) + \Delta P_n .\end{aligned}$$

Multiplying then with $\tau_{n+1}$, equation (6.15) follows, provided that the following relations are taken into account:

$$\begin{aligned}&(1+\gamma_n)(I - R_n\widehat{H}_n) = I - \big((1+\gamma_n)R_n\widehat{H}_n - \gamma_n I\big) = I - s_n(B_n - \Delta B_n) , \\ &(1+\gamma_n)R_n = s_n(D_n + \Delta D_n).\end{aligned}$$

For the further consideration of (6.15) some additional auxiliary lemmas are needed.

**Lemma 6.5.** *For each $n \in \mathbb{N}$ we have*

$$\widehat{\Delta}_n = \Delta_n^{(0)} + \Delta_n^{(1)} + \Delta_n^{(2)} \tag{6.21}$$

*with $\mathcal{A}_n$-measurable random variables $\Delta_n^{(i)}, i = 0, 1, 2$, where*

$$\Delta_{n+1}^{(0)} = (I - s_n B_n)\Delta_n^{(0)} + (\tau_n s_n) D_n (E_n Z_n - Z_n) , \tag{6.22}$$

$$\begin{aligned}\|\Delta_{n+1}^{(1)}\| &\leq \|I - s_n B_n\| \|\Delta_n^{(1)}\| + s_n \|\Delta B_n\| \|\widehat{\Delta}_n\| \\ &+ \tau_{n+1}\Big(\rho_n \|M_n\| (\|E_n Z_n\| + \|G^*\|) + \|\Delta P_n\|\Big) ,\end{aligned} \tag{6.23}$$

$$\begin{aligned}E_n\|\Delta_{n+1}^{(2)}\|^2 &\leq \|I - s_n B_n\|^2 \|\Delta_n^{(2)}\|^2 \\ &+ (\tau_n s_n)^2 \|\Delta D_n\|^2 (\sigma_{0,n}^2 + \sigma_1^2 \|\Delta_n\|^2) .\end{aligned} \tag{6.24}$$

*Proof.* For $i = 0, 1, 2$, select arbitrary $\mathcal{A}_1$-measurable random variables $\Delta_1^{(i)}$ such that

$$\widehat{\Delta}_1 = \Delta_1^{(0)} + \Delta_1^{(1)} + \Delta_1^{(2)} .$$

For $n \geq 1$, define $\Delta_{n+1}^{(0)}$ then by (6.22) and put

$$\begin{aligned}\Delta_{n+1}^{(1)} &:= (I - s_n B_n)\Delta_n^{(1)} + s_n \Delta B_n \widehat{\Delta}_n + \tau_{n+1}\big(-R_n(E_n Z_n + G^*) \\ &\quad + \Delta P_n\big) , &\text{(i)} \\ \Delta_{n+1}^{(2)} &:= (I - s_n B_n)\Delta_n^{(2)} + \tau_n s_n \Delta D_n (E_n Z_n - Z_n) . &\text{(ii)}\end{aligned}$$

Relation (6.21) holds then because of Lemma 6.4, and (6.23) follows directly from (i) by taking the norm. Moreover, (ii) yields

$$\begin{aligned}\|\Delta_{n+1}^{(2)}\|^2 &\leq \|I - s_n B_n\|^2 \|\Delta_n^{(2)}\|^2 + (\tau_n s_n)^2 \|\Delta D_n\|^2 \|E_n Z_n - Z_n\|^2 \\ &\quad + 2\tau_n s_n \big\langle (I - s_n B_n)\Delta_n^{(2)}, \Delta D_n (E_n Z_n - Z_n)\big\rangle .\end{aligned}$$

The random variables $\Delta_n^{(2)}$ and $\Delta D_n$ are $\mathcal{A}_n$-measurable. Thus, taking expectations in the above inequality, by means of (U5b) we get then inequality (6.24).

Suppose now that the matrices $D_n$ from (6.14) fulfill the following relations:

$$D_n H^* + (D_n H^*)^T \geq 2 a_n I , \tag{6.25}$$

$$\|D_n H^*\| \leq b_n , \tag{6.26}$$

with certain numbers $a_n, b_n \in \mathbb{R}$ . The matrices $B_n$ from (6.16) fulfill then the following inequalities:

**Lemma 6.6.** *For each $n \in \mathbb{N}$:*

$$B_n + B_n^T \geq 2\left(a_n - \frac{\gamma_n}{s_n}\right) I =: 2\tilde{a}_n I ,$$

$$\|B_n\|^2 \leq \left(\frac{\gamma_n}{s_n}\right)^2 - 2\frac{\gamma_n}{s_n} a_n + b_n^2 =: \tilde{b}_n^2 ,$$

$$\|I - s_n B_n\|^2 \leq 1 - 2s_n \left(\tilde{a}_n - \frac{1}{2} s_n \tilde{b}_n^2\right) =: 1 - 2s_n u_n ,$$

$$\|I - s_n B_n\| \leq 1 - s_n u_n .$$

*Proof.* The assertions follow by simple calculations from the relations (6.16), (6.25) und (6.26), where:

$$\sqrt{1-2t} \leq 1 - t \quad \text{for} \quad t \leq \frac{1}{2} .$$

Some useful properties of the random variables $\Delta B_n$, $\Delta D_n$ und $\Delta P_n$ arising in (6.17), (6.18) und (6.19) are contained in the next lemma:

**Lemma 6.7.** *Assuming that*

$$\gamma_n \longrightarrow 0 \text{ and } X_n \longrightarrow x^* \in \overset{\circ}{D} \text{ a.s.},$$

*we have*

*a)* $\Delta B_n$, $\Delta D_n \longrightarrow 0$ *a.s.* ,
*b)* $G^* = 0$ ,
*c) For each $\omega \in \Omega$, with exception of a null-set, there is an index $n_0(\omega)$ such that*

$$\Delta P_n(\omega) = 0 \quad \textit{for every} \quad n \geq n_0(\omega) .$$

*Proof.* Because of $R_n = \rho_n M_n$ , (6.13), (6.14), (6.18) and (U4a) we have

$$\Delta D_n = (1 + \gamma_n)\frac{\rho_n}{s_n} M_n - D_n \longrightarrow 0 \quad \text{a.s.} .$$

Furthermore, $X_n \to x^*$ a.s. and (U1a-c) yield

$$\widehat{H}_n := \int_0^1 H\big(x^* + t(X_n - X^*)\big) dt \longrightarrow H^* \quad \text{a.s.} .$$

Hence, because of (6.13), (6.14), (6.16) and (6.17) we find

$$\Delta B_n = D_n H^* - (1 + \gamma_n)\frac{\rho_n}{s_n} M_n \widehat{H}_n \longrightarrow 0 \quad \text{a.s.} .$$

Since $x^* \in \overset{\circ}{D}$, with (U2) we get $G^* = 0$ .

Consider now an element $\omega \in \Omega$ with $X_n(\omega) \longrightarrow x^*$. Because of $x^* \in \mathring{D}$ there is a number $n_0(\omega)$ such that $X_{n+1}(\omega) \in \mathring{D}$ for all $n \geq n_0(\omega)$ . Hence, for $n \geq n_0(\omega)$ we find $p_D(X_n(\omega) - R_n(\omega)Y_n(\omega)) =: X_{n+1}(\omega) \in \mathring{D}$, and therefore

$$p_D(X_n(\omega) - R_n(\omega)Y_n(\omega)) = X_n(\omega) - R_n(\omega)Y_n(\omega) ,$$

hence, $\Delta P_n(\omega) = 0$.

### 6.4.3 Further Preliminary Results

Sufficient conditions for $X_{n+1} - X_n \to 0$ a.s. are given in the following:

**Lemma 6.8.** *a)* $\|X_{n+1} - X_n\| \leq \rho_n \|M_n\| \|Y_n\|$ *for* $n = 1, 2, \ldots$ .

*b)* $\|Y_n\|^2 \leq 4 \left(\|G^*\|^2 + \|G_n - G^*\|^2 + \|Z_n - E_n Z_n\|^2 + \|E_n Z_n\|^2\right)$ *for* $n = 1, 2, \ldots$ .

*c) For each deterministic sequence of positive numbers* $(s_n)$ *such that* $\sum\limits_{n=1}^{\infty} s_n^2 < \infty$ *we have* $\sum\limits_{n=1}^{\infty} s_n^2 \|Y_n\|^2 < \infty$ *a.s..*

*d) Let* $(s_n)$ *denote a sequence according to c), and assume that* $(\rho_n)$ *fulfills* $\limsup\limits_{n\to\infty} \frac{\rho_n}{s_n} \leq 1$ *a.s.. Then* $\sum\limits_{n=1}^{\infty} \|X_{n+1} - X_n\|^2 < \infty$ *a.s., and* $(X_{n+1} - X_n)$ *converges to zero a.s..*

*Proof.* For each $n \in \mathbb{N}$, according to (6.2a,b) we find

$$\begin{aligned} \|X_{n+1} - X_n\| &= \|p_D(X_n - R_n Y_n) - p_D(X_n)\| \\ &\leq \|R_n\| \|Y_n\| = \rho_n \|M_n\| \|Y_n\| , \end{aligned} \tag{i}$$

$$\begin{aligned} \|Y_n\|^2 &= 16 \left\| \frac{1}{4} [G^* + (G_n - G^*) + (Z_n - E_n Z_n) + E_n Z_n] \right\|^2 \\ &\leq 4\Big(\|G^*\|^2 + \|G_n - G^*\|^2 + \|Z_n - E_n Z_n\|^2 + \|E_n Z_n\|^2\Big) . \end{aligned} \tag{ii}$$

Due to assumption a) in Section 6.3, inequality (6.4) and (U5a-c), for each $n \in \mathbb{N}$ we find

$$\begin{aligned} &\|G_n - G^*\|^2 \leq \beta^2 \|X_n - x^*\|^2 \leq \beta^2 \delta_0^2 , \\ &\|E_n Z_n\|^2 \longrightarrow 0 \quad \text{a.s.,} \\ &E\|Z_n - E_n Z_n\|^2 = E\Big(E_n \|Z_n - E_n Z_n\|^2\Big) \\ &\leq E\Big(\sigma_{0,n}^2 + \sigma_1^2 \|X_n - x^*\|^2\Big) \leq \sigma_0^2 + \sigma_1^2 \delta_0^2 . \end{aligned}$$

Hence, for a sequence $(s_n)$ according c) we get

$$\sum_n s_n^2 \Big(\|G^*\|^2 + \|G_n - G^*\|^2 + \|E_n Z_n\|^2\Big) < \infty \quad \text{a.s.} \tag{iii}$$

and $\sum_n s_n^2 E\|Z_n - E_n Z_n\|^2 < \infty$ . This yields then

$$\sum_n s_n^2 \|Z_n - E_n Z_n\|^2 < \infty \quad \text{a.s..} \tag{iv}$$

Now relations (ii), (iii) und (iv) imply

$$\sum_n s_n^2 \|Y_n\|^2 < \infty \quad \text{a.s..} \tag{v}$$

Because of (U4a-d) the sequence $\left(\|M_n\|\right)$ is bounded a.s.. Hence, under the assumptions in d), relation (v) yields

$$\sum_n \rho_n^2 \|M_n\|^2 \|Y_n\|^2 < \infty \quad \text{a.s..}$$

Thus, relation (i) implies now the assertion in d).

Conditions for the selection of sequence $(\widehat{a}_n)$ in inequality (6.7) are given in the next lemma:

**Lemma 6.9.** *Suppose that*

*i)* $X_n \longrightarrow x^*$ *a.s.* ,

*ii)* $\lim_{n \in \mathbb{N}^{(k)}} \widehat{M}_n = D^{(k)}$ *a.s. with deterministic matrices* $D^{(k)}$ *for* $k = 1, \ldots, \kappa$ .

*Then* $(\widehat{a}_n)$ *can be selected such that*

$$\widehat{a}^{(k)} := \liminf_{n \in \mathbb{N}^{(k)}} \widehat{a}_n \geq \frac{1}{2} \lambda_{min} \left( D^{(k)} H(x^*) + \left( D^{(k)} H(x^*) \right)^T \right) \quad a.s.,\ k = 1, \ldots, \kappa.$$

*Proof.* Consider an arbitrary $k \in \{1, \ldots, \kappa\}$, put

$$H^* := H(x^*) \quad \text{and} \quad b := \frac{1}{2} \lambda_{min} \left( D^{(k)} H^* + (D^{(k)} H^*)^T \right),$$

and for $n \in \mathbb{N}^{(k)}$ define

$$\widehat{a}_n := \begin{cases} \frac{\langle \Delta_n, \widehat{M}_n (G(X_n) - G(x^*)) \rangle}{\|\Delta_n\|^2} & , \text{ for } \Delta_n \neq 0 \\ b & , \text{ for } \Delta_n = 0 . \end{cases}$$

According to assumption (U1a), for $\Delta_n \neq 0$ we find then

$$\widehat{a}_n = \frac{\Delta_n^T D^{(k)} H^* \Delta_n}{\|\Delta_n\|^2} + \frac{\Delta_n^T \left( (\widehat{M}_n - D^{(k)}) H^* \Delta_n + \widehat{M}_n \delta(x^*; \Delta_n) \right)}{\|\Delta_n\|^2} .$$

Applying condition (U1b), we get

$$\widehat{a}_n \geq b - \left( \|\widehat{M}_n - D^{(k)}\| \, \|H^*\| + \|\widehat{M}_n\| L_0 \|\Delta_n\|^{\mu_0} \right) .$$

Finally, due to the assumptions i), ii) and (U4a-d) we find

$$\liminf_{n \in \mathbb{N}^{(k)}} \widehat{a}_n \geq b \quad \text{a.s. .}$$

## 6.5 General Convergence Results

In this section the convergence properties of the sequence $(X_n)$ generated by the algorithm (6.2a,b) are considered in case of "arbitrary" matrix step sizes of order $\frac{1}{n}$ . Especially, theorems are given for the convergence with probability one (a.s.), the rate of convergence, the convergence in the mean and the convergence in distribution. These convergence results are applied then to the algorithms having step sizes as given in Sections 6.7 and 6.8.

As stated in Section 6.3, here the assumptions (U1a-c) up to (U6) hold, and all further assumptions will be marked by the letter V.

Here, we use the step size $R_n = \rho_n M_n$ with a scalar factor $\rho_n$ such that

$$\lim_{n\to\infty} n\,\rho_n = 1 \quad \text{a.s.}\ . \tag{V1}$$

According to condition (U4a), sequence $(M_n)$ is bounded a.s.. Hence, with probability 1, the step size $R_n$ is of order $\frac{1}{n}$ .

### 6.5.1 Convergence with Probability One

A theorem about the convergence with probability 1 and the convergence rate of the sequence $(X_n)$ is given in the following. Related results can be found in the literature, see e.g. [97] and [98]. The rate of convergence depends mainly on the positive random variable $\widehat{a}$, defined in (6.9). Because of (6.7), this random variable can be interpreted as a mean lower bound of the quotients

$$\frac{\langle X_n - x^*, \widehat{M}_n \cdot (G(X_n) - G(x^*))\rangle}{\|X_n - x^*\|^2}.$$

Due to the assumptions (U5a-c), for each number $\lambda \in \left[0, \frac{1}{2}\right)$ we have the relations

$$\sum_n \frac{1}{n^2}\sigma_1^2 < \infty\,, \tag{6.27}$$

$$\sum_n \frac{1}{n^{2(1-\lambda)}}\sigma_{0,n}^2 < \infty \quad \text{a.s.}, \tag{6.28}$$

$$\sum_n \frac{1}{n^{1-\lambda}}\|E_n Z_n\| < \infty \quad \text{a.s.}. \tag{6.29}$$

**Theorem 6.1.** *If $\lambda$ is selected such that $0 \le \lambda < min \left\{\frac{1}{2}, \widehat{a}\right\}$ a.s., then*

$$\lim_{n\to\infty} n^{\lambda}(X_n - x^*) = 0 \quad a.s.\ .$$

*Proof.* Because of (V1), with $\tau_n := n^\lambda$ we find

$$\lim_{n\to\infty} \tau_n \rho_n n^{1-\lambda} = 1 \quad \text{a.s. and} \quad \tau_{n+1} = (1+\gamma_n)\tau_n \quad \text{with} \quad n\gamma_n \longrightarrow \lambda\,.$$

Hence, due to (V1), (6.27), (6.28), (6.29) and (U4a) we get

$$\sum_n \left(\rho_n \|M_n\|\right)^2 \sigma_1^2 < \infty \quad \text{a.s.,}$$

$$\sum_n \left(\tau_n \rho_n \|M_n\|\right)^2 \sigma_{0,n}^2 < \infty \quad \text{a.s.,}$$

$$\sum_n \left(\tau_n \rho_n \|M_n\|\right) \|E_n Z_n\| < \infty \quad \text{a.s..}$$

For the random variables $t_n$, $t_n^{(1)}$ and $t_n^{(2)}$ occuring in Lemma 6.3 we have therefore

$$\sum_n t_n^{(i)} < \infty \quad \text{a.s. for } i = 1, 2\,, \tag{i}$$

$$t_n - 2(\widehat{a}_n - \lambda) \longrightarrow 0 \quad \text{a.s..}$$

This yields $t^{(k)} := \liminf_{n\in\mathbb{N}^{(k)}} t_n = 2(\widehat{a}^{(k)} - \lambda)$ f.s. for $k = 1, \ldots, \kappa$ . According to the assumptions in this theorem and because of (6.9) we obtain

$$t := q_1 t^{(1)} + \ldots + q_\kappa t^{(\kappa)} = 2(\widehat{a} - \lambda) > 0 \quad \text{a.s..} \tag{ii}$$

Because of (V1), (i) und (ii), the assertion follows now from Lemma 6.3 and Lemma B.3 in the Appendix.

Replacing in Theorem 6.1 the bound $\widehat{a}$ by the number $\underline{a}$ from (U4d), then a weaker result is obtained:

**Corollary 6.1.** *If* $0 \le \lambda < \min\left\{\frac{1}{2}, \underline{a}\right\}$, *then* $\lim_{n\to\infty} n^\lambda (X_n - x^*) = 0$ *a.s.* .

*Proof.* According to (6.9) we have $\widehat{a} \ge \underline{a}$. The assertion follows now from Theorem 6.1.

Since $\underline{a} > 0$ , in Corollary 6.1 we may select also $\lambda = 0$. Hence, we always have

$$X_n \longrightarrow x^* \quad \text{a.s..} \tag{6.30}$$

### 6.5.2 Convergence in the Mean

We consider now the convergence rate of the mean square error $E\|X_n - x^*\|^2$. For the estimation of this error the following lemma is needed:

**Lemma 6.10.** *If $n \in \mathbb{N}$, then*

$$\begin{aligned} &E_n(n+1)\|\Delta_{n+1}\|^2 \\ &\leq \left(1 - \frac{1}{n}s_n^{(1)}\right) n\|\Delta_n\|^2 + \frac{1}{n}\left((n\rho_n\|M_n\|)^2\sigma_{0,n}^2 + s_n^{(2)}\right) \end{aligned} \tag{6.31}$$

*with $A_n$-measurable random variables $s_n^{(1)}$ and $s_n^{(2)}$, where*

$$s_n^{(1)} - (2\widehat{a}_n - 1) \longrightarrow 0 \quad a.s., \tag{6.32}$$

$$s_n^{(2)} \longrightarrow 0 \quad a.s.. \tag{6.33}$$

*Proof.* Putting $\tau_n := \sqrt{n}$ , we find

$$\tau_{n+1} = (1+\gamma_n)\tau_n \quad \text{with } n\gamma_n \longrightarrow \frac{1}{2}\,.$$

According to Lemma 6.3 we select

$$s_n^{(1)} := n(\rho_n t_n - t_n^{(1)}) \quad , \quad s_n^{(2)} := nt_n^{(2)} - \left(n\rho_n\|M_n\|\right)^2\sigma_{0,n}^2.$$

Then (U4a-d), (U5a-c) and (V1) yield

$$\begin{aligned} ng_n = \sqrt{n}(n\rho_n)\|M_n\|\|E_nZ_n\| &\longrightarrow 0 \quad \text{a.s.,} \\ n\left(\rho_n\|M_n\|\right)^2\sigma_1^2 &\longrightarrow 0 \quad \text{a.s..} \end{aligned}$$

Consequently,

$$\begin{aligned} t_n - (2\widehat{a}_n - 1) &\longrightarrow 0 \quad \text{a.s.,} \\ nt_n^{(1)} &\longrightarrow 0 \quad \text{a.s.,} \\ nt_n^{(2)} - \left(n\rho_n\|M_n\|\right)^2\sigma_{0,n}^2 &\longrightarrow 0 \quad \text{a.s..} \end{aligned}$$

Thus, due to (V1) also (6.32) and (6.33) hold.

The following theorem shows that the mean square error $E\|X_n - x^*\|^2$ is essentially of order $\frac{1}{n}$, and an upper bound is given for the limit of the sequence $\left(n\ E\|X_n - x^*\|^2\right)$. This bound can be found in the literature in the special case of scalar step sizes and having the trivial partition $\mathbb{N} = \mathbb{N}^{(1)}$ (cf. condition i) from Section 6.3), see [24], [149, page 28] and [138]. Also in this result, an important role is played by the random variable $\widehat{a}$ defined in (6.9).

**Theorem 6.2.** *Let $\widehat{a} > \frac{1}{2}$ a.s. . Then for each $\varepsilon_0 > 0$ there is a set $\Omega_0 \in \mathcal{A}$ such that*

$$a)\quad \mathcal{P}(\bar{\Omega}_0) < \varepsilon_0 \,,$$

$$b)\quad \limsup_{n\in\mathbf{N}} E_{\Omega_0}\, n\|\Delta_n\|^2 \le \frac{\sum_{k=1}^{\kappa} q_k(\widehat{b}^{(k)}\sigma_k)^2}{2\widehat{a}-1} \,. \tag{6.34}$$

*Here, $\widehat{b}^{(1)},\ldots,\widehat{b}^{(\kappa)}$ and $\sigma_1,\ldots,\sigma_\kappa$ are constants such that*

$$\limsup_{n\in\mathbf{N}^{(k)}} \|\widehat{M}_n\| \le \widehat{b}^{(k)} \quad a.s. \text{ and } \quad \sigma_k^2 = \limsup_{n\in\mathbf{N}^{(k)}} E\,\sigma_{0,n}^2$$

*for each $k \in \{1,\ldots,\kappa\}$.*

*Proof.* Because of Lemma 6.10 and the above assumptions, the assertions in this theorem follow directly from Lemma B.4 in the Appendix, provided there we set:

$$T_n := n\|\Delta_n\|^2 \,,\ A_n := s_n^{(1)} \,,\ B_n := \left(n\rho_n\|M_n\|\right)^2 \,,$$
$$v_n := \sigma_{0,n}^2 \,,\ C_n := s_n^{(2)} \,.$$

The following corollary is needed later on:

**Corollary 6.2.** *If $\widehat{a} > \frac{1}{2}$ a.s., then for each $\varepsilon_0 > 0$ there is a set $\Omega_0 \in \mathcal{A}$ such that*

$$a)\quad \mathcal{P}(\bar{\Omega}_0) < \varepsilon_0 \,,$$

$$b)\quad \limsup_{n\in\mathbf{N}} E_{\Omega_0} n\|\Delta_n\|^2 \le \frac{(\widehat{b}\sigma)^2}{2\widehat{a}-1} \,,$$

*provided the numbers $\widehat{b}$ und $\sigma$ are selected such that*

$$\limsup_{n\in\mathbf{N}} \|\widehat{M}_n\| \le \widehat{b} \quad a.s. \text{ and } \quad \sigma^2 = \limsup_{n\in\mathbf{N}} E\sigma_{0,n}^2 \,.$$

A convergence result in case of scalar step sizes is contained in this corollary:

**Corollary 6.3.** *Select numbers $d^{(1)},\ldots,d^{(\kappa)}$ such that*

$$\lim_{n\in\mathbf{N}^{(k)}} \widehat{M}_n = d^{(k)} I \quad f.s. \text{ for } k = 1,\ldots,\kappa \,, \tag{6.35}$$

$$2d\alpha^* > 1 \,, \tag{6.36}$$

*where $d := q_1 d^{(1)} + \ldots + q_\kappa d^{(\kappa)}$ and $\alpha^* := \frac{1}{2}\lambda_{min}(H^* + H^{*T})$ . Then for each $\varepsilon_0 > 0$ there is a set $\Omega_0 \in \mathcal{A}$ such that*

$$a) \quad P(\bar{\Omega}_0) < \varepsilon_0 ,$$

$$b) \quad \limsup_{n \in \mathbb{N}} E_{\Omega_0} n \|\Delta_n\|^2 \leq \frac{\sum_{k=1}^{\kappa} q_k (d^{(k)} \sigma_k)^2}{2d\alpha^* - 1}, \tag{6.37}$$

*where again* $\sigma_k^2 = \limsup_{n \in \mathbb{N}^{(k)}} E\sigma_{0,n}^2$ *for each* $k \in \{1, \dots, \kappa\}$ .

*Proof.* Because of Corollary 6.1 and (6.35) we may apply Lemma 6.9. Hence, we may assume that the sequence $(\widehat{a}_n)$ in (6.7) fulfills

$$\widehat{a}^{(k)} := \liminf_{n \in \mathbb{N}^{(k)}} \widehat{a}_n \geq d^{(k)} \alpha^* \quad \text{for} \quad k = 1, \dots, \kappa \quad \text{a.s.} .$$

Due to (6.9) and (6.36) we get therefore the inequality $\widehat{a} \geq d\alpha^* > \frac{1}{2}$ a.s. . The assertion of the above Corollary 6.3 follows now from Theorem 6.2.

For given positive numbers $\alpha^*$, $q_1, \dots, q_\kappa$, $\sigma_1, \dots, \sigma_\kappa$ , the right hand side of (6.37) is minimal for

$$d^{(k)} = \left[\sigma_k^2 \alpha^* \left(\frac{q_1}{\sigma_1^2} + \dots + \frac{q_\kappa}{\sigma_\kappa^2}\right)\right]^{-1} \quad \text{for} \quad k = 1, \dots, \kappa . \tag{6.38}$$

Using these limits, Corollary 6.3 yields

$$d = \frac{1}{\alpha^*} , \tag{6.39}$$

$$\limsup_{n \in \mathbb{N}} E_{\Omega_0} n \|\Delta_n\|^2 \leq \left[\alpha^{*2} \left(\frac{q_1}{\sigma_1^2} + \dots + \frac{q_\kappa}{\sigma_\kappa^2}\right)\right]^{-1} . \tag{6.40}$$

For scalar step sizes this yields now some guidelines for an asymptotic optimal selection of the factors $M_n$ in the step sizes $R_n = \rho_n M_n$.

In the special case

$$M_n \longrightarrow d \cdot I \quad \text{a.s.} ,$$

the most favorable selection reads $d = \alpha^{*-1}$ which yields then the estimate

$$\limsup_{n} E_{\Omega_0} n \|\Delta_n\|^2 \leq \frac{\sum_{k=1}^{\kappa} q_k \sigma_k^2}{\alpha^{*2}} . \tag{6.41}$$

The right hand side of (6.40), (6.41), resp., contains, divided by $\alpha^{*2}$, the $(q_1, \dots, q_\kappa)$-weighted harmonic resp. arithmetic mean of the variance limits $\sigma_1^2, \dots, \sigma_\kappa^2$ of the estimation error sequence $(Z_n)$. Hence, the bound in (6.40) is smaller than that in (6.41).

### 6.5.3 Convergence in Distribution

In this section it is shown that the weighted error $\sqrt{n}(X_n - x^*)$ converges in distribution if further assumptions on the matrices $M_n$ in step size $R_n = \rho_n M_n$ and on the estimation error $Z_n$ are made.

Hence, choose the following weighting factors:

$$\tau_n = \sqrt{n} \, . \tag{6.42}$$

Now $\tau_{n+1} = (1 + \gamma_n)\tau_n$ holds again, where

$$\gamma_n = \frac{\lambda_n}{n} \quad , \quad \lambda_n \longrightarrow \frac{1}{2} \, . \tag{6.43}$$

Furthermore, according to Corollary 6.1 it holds without further assumptions

$$X_n \longrightarrow x^* \quad \text{a.s.} \, . \tag{6.44}$$

Now we demand that matrices $M_n$ fulfill certain limit conditions and, due to condition (U3), that $x^*$ is contained in the interior of the feasible set: Hence,

a) For all $k \in \{1, \ldots, \kappa\}$ there exists a deterministic matrix $D^{(k)}$ having

$$\lim_{n \in \mathbb{N}^{(k)}} M_n = D^{(k)} \text{ a.s.} \, . \tag{V2}$$

For $n \in \mathbb{N}^{(k)}$ let $D_n := D^{(k)}$.

b) There exist $a^{(1)}, \ldots, a^{(\kappa)} \in \mathbb{R}$ such that

$$a := q_1 a^{(1)} + \ldots + q_\kappa a^{(\kappa)} > \frac{1}{2} \, , \tag{V3a}$$

$$D^{(k)} H^* + (D^{(k)} H^*)^T \geq 2a^{(k)} I \, . \tag{V3b}$$

c) For any solution $x^*$ of (P), cf. (6.1), it holds

$$x^* \in \mathring{D}. \tag{V4}$$

Now some preliminary considerations are made which are useful for showing the already mentioned central limit theorem.

According to (6.44) and assumptions (V2) and (V3a,b) as well as Lemma 6.9, for bound $\hat{a}$ in (6.9) we may assume that

$$\hat{a} \geq a > \frac{1}{2} \quad \text{a.s.} \, . \tag{6.45}$$

Moreover, due to (V1), in Section 6.4.2

$$s_n := \frac{1}{n} \quad \text{for} \quad n = 1, 2, \ldots \tag{6.46}$$

can be chosen, and (6.43) yields

$$\frac{\gamma_n}{s_n} = \lambda_n \longrightarrow \frac{1}{2} \,. \tag{6.47}$$

Now, according to these relations and Lemma 6.6, for the matrices

$$B_n := D_n H^* - \frac{\gamma_n}{s_n} I = D_n H^* - \lambda_n I \tag{6.48}$$

arising in (6.16), the inequalities

$$\|I - s_n B_n\|^2 \leq 1 - \frac{2u_n}{n} \,, \tag{6.49}$$

$$\|I - s_n B_n\| \leq 1 - \frac{u_n}{n} \,, \tag{6.50}$$

hold, where

$$u_n := (a_n - \lambda_n) - \frac{1}{2n}(\lambda_n^2 - 2\lambda_n a_n + b_n^2) \,, \tag{6.51}$$

and for $n \in \mathbb{N}^{(k)}$ we put $a_n := a^{(k)}$ and $b_n := b^{(k)} := \|D^{(k)} H^*\|$. Hence, because of (6.47), for any $k = 1, \ldots, \kappa$ the limit

$$\lim_{n \in \mathbb{N}^{(k)}} u_n = u^{(k)} := a^{(k)} - \frac{1}{2} \tag{6.52}$$

exists. Moreover, using (6.45), we get

$$\bar{u} := q_1 u^{(1)} + \ldots + q_\kappa u^{(\kappa)} = a - \frac{1}{2} > 0 \,. \tag{6.53}$$

### Stochastic Equivalence

By further examination of sequence $\big(\widehat{\Delta}_n = \sqrt{n}(X_n - x^*)\big)$ we find that it can be replaced by a simpler sequence. This is guaranteed by the following result:

**Theorem 6.3.** $\big(\sqrt{n}(X_n - x^*)\big)$ *is stochastically equivalent to sequence*

$$\Delta_{n+1}^{(0)} = (I - \frac{1}{n} B_n)\Delta_n^{(0)} + \frac{1}{\sqrt{n}} D_n (E_n Z_n - Z_n), n = 1, 2, \ldots,$$

*where* $\Delta_1^{(0)}$ *denotes an arbitrary* $\mathcal{A}_1$*-measurable random variable.*

Two stochastic sequences are called *stochastically equivalent*, if their difference converges stochastically to zero.

The reduction onto the basic parts of the sequence $(\sqrt{n}(X_n - x^*))$ in Theorem 6.3 is already mentioned in a similar form in [35] (Equation 2.2.10, $\alpha = 1$). But there the matrix $B_n \equiv \Lambda$ is independent of index $n$.

Because of Lemma 6.5 it is sufficient for the proof of the theorem to show that sequences $(\Delta_n^{(1)})$ and $(\Delta_n^{(2)})$ in (6.21) converge stochastically to 0.

**Lemma 6.11.** $(\Delta_n^{(1)})$ *converges stochastically to* 0 .

*Proof.* According to the upper remarks and Lemma 6.5 it holds

$$T_{n+1} \le (1 - \frac{u_n}{n})T_n + \frac{1}{n}(\|\Delta B_n\| \|\widehat{\Delta}_n\| + C_n) \tag{i}$$

with the nonnegative random variables

$$T_n := \|\Delta_n^{(1)}\| ,$$
$$C_n := n\sqrt{n+1}\Big(\rho_n \|M_n\| (\|E_n Z_n\| + \|G^*\|) + \|\Delta P_n\|\Big) .$$

Furthermore, according to (V1), (V4), (6.44), (U4a-d), (U5a-c) and Lemma 6.7 it holds

$$\|\Delta B_n\| \, , \, C_n \longrightarrow 0 \quad \text{a.s.} \, . \tag{ii}$$

Let $\varepsilon, \delta > 0$ be arbitrary positive numbers. According to (ii) and Lemma B.5 in the Appendix there exists a set $\Omega_1 \in \mathcal{A}$ having

$$\mathcal{P}(\bar{\Omega}_1) < \frac{\delta}{4} \, , \tag{iii}$$
$$\|\Delta B_n\| \, , \, C_n \longrightarrow 0 \quad \text{uniformly on } \Omega_1 \, . \tag{iv}$$

Due to (6.45) and Corollary 6.2 there exists a constant $K < \infty$ and a set $\Omega_2 \in \mathcal{A}$ having

$$\mathcal{P}(\bar{\Omega}_2) < \frac{\delta}{4} \, , \tag{v}$$
$$\sup_n E_{\Omega_2} \|\widehat{\Delta}_n\| \le K \, . \tag{vi}$$

For $\varepsilon_0 := \delta\varepsilon\bar{u}[4(K+1)]^{-1} > 0$ , with $\bar{u}$ from (6.53), according to (iv) there exists an index $n_0$ such that for any $n \ge n_0$ on $\Omega_0 := \Omega_1 \cap \Omega_2$ we have

$$\|\Delta B_n\| \, , \, C_n \le \varepsilon_0 \, .$$

Hence, with (i) and (iv) for $n \ge n_0$ it follows

$$E_{\Omega_0} T_{n+1} \le \left(1 - \frac{u_n}{n}\right) E_{\Omega_0} T_n + \frac{\varepsilon_0 (K+1)}{n} \, ,$$

and therefore because of (6.52), (6.53) and Lemma A.7, Appendix,

$$\limsup_n E_{\Omega_0} T_n \le \frac{\varepsilon_0(K+1)}{\bar{u}} = \frac{\delta\varepsilon}{4} \, .$$

Thus, there exists an index $n_1 \ge n_0$ having

$$E_{\Omega_0} T_n \le \frac{\delta\varepsilon}{2} \quad \text{for} \quad n \ge n_1 \, . \tag{vii}$$

Finally, from (iii), (v), the Markov inequality and (vii) for indices $n \geq n_1$ holds

$$\begin{aligned}\mathcal{P}(T_n > \varepsilon) &\leq \mathcal{P}(\bar{\Omega}_0) + \mathcal{P}(\{T_n > \varepsilon\} \cap \Omega_0) \\ &\leq \mathcal{P}(\bar{\Omega}_0) + \frac{E_{\Omega_0} T_n}{\varepsilon} < \frac{\delta}{2} + \frac{\delta}{2} = \delta\,.\end{aligned}$$

Now the corresponding conclusion for the sequence $(\Delta_n^{(2)})$ is shown.

**Lemma 6.12.** $(\Delta_n^{(2)})$ *converges stochastically to* $0$ .

*Proof.* According to above remarks and Lemma 6.5 we have

$$E_n T_{n+1} \leq \left(1 - \frac{2u_n}{n}\right) T_n + \frac{1}{n}(\|\Delta D_n\|^2 \sigma_{0,n}^2 + C_n)\,,$$

where $T_n := \|\Delta_n^{(2)}\|^2$ and $C_n := \|\Delta D_n\|^2 \sigma_1^2 \|\Delta_n\|^2$ . Due to Lemma 6.7 it holds

$$\|\Delta D_n\|\,,\ C_n \longrightarrow 0 \quad \text{f.s.}\,.$$

Let $\varepsilon, \delta > 0$ be arbitrary fixed. Here, choosing $B^{(k)} \equiv C^{(k)} \equiv 0$, Lemma B.4, Appendix, can be applied. Hence, there exists a set $\Omega_0 \in \mathcal{A}$ where

$$\mathcal{P}(\bar{\Omega}_0) < \frac{\delta}{2} \quad \text{und} \quad \limsup_n E_{\Omega_0}\, T_n = 0\,. \tag{i}$$

Hence, there exists an index $n_0$ with

$$E_{\Omega_0}\, T_n \leq \frac{\varepsilon\delta}{2} \quad \text{for} \quad n \geq n_0\,. \tag{ii}$$

Like in the previous proof from (i) and (ii) follows now

$$\mathcal{P}(T_n > \varepsilon) < \delta \quad \text{for all} \quad n \geq n_0\,.$$

For sequence $(\Delta_n^{(0)})$ in Theorem 6.3 the following representation can be given

$$\Delta_{n+1}^{(0)} = B_{0,n}\Delta_1^{(0)} - \sum_{m=1}^{n} \frac{1}{\sqrt{m}} B_{m,n} D_m \Delta Z_m \tag{6.54}$$

for $n = 1, 2, \ldots$ . Here,

$$B_{n,n} := I \quad , \quad B_{m,n} := \left(I - \frac{1}{n}B_n\right) \cdot \ldots \cdot \left(I - \frac{1}{m+1}B_{m+1}\right) . \tag{6.55}$$

is defined for all indices $m < n$. Finally, according to (6.49) and (6.50) for matrices $B_{m,n}$ it follows

$$\|B_{m,n}\| \leq \prod_{j=m+1}^{n} \left(1 - \frac{u_j}{j}\right) =: b_{m,n}\,, \tag{6.56}$$

$$\|B_{m,n}\|^2 \leq \prod_{j=m+1}^{n} \left(1 - \frac{2u_j}{j}\right) =: \bar{b}_{m,n}\,. \tag{6.57}$$

### Asymptotic Distribution

Assume for the estimation error $Z_n$ the following conditions:

a) There are symmetric deterministic matrices $W^{(1)}, \ldots, W^{(\kappa)}$ such that for any $k = 1, \ldots, \kappa$ it holds either

$$W^{(k)} = \lim_{m \in \mathbb{N}^{(k)}} E(\Delta Z_m \Delta Z_m^T), \tag{V5a}$$

$$\lim_{m \to \infty} E\left\| E_m(\Delta Z_m \Delta Z_m^T) - E(\Delta Z_m \Delta Z_m^T) \right\| = 0 \tag{V5b}$$

or

$$W^{(k)} = \lim_{m \in \mathbb{N}^{(k)}} E_m(\Delta Z_m \Delta Z_m^T) \quad \text{a.s.}\,. \tag{V6}$$

b) It holds either

$$\lim_{m \to \infty} E\left(\|\Delta Z_m\|^2 I_{\{\|\Delta Z_m\|^2 > tm\}}\right) = 0 \quad \text{for all } t > 0 \tag{V7}$$

or

$$\lim_{m \to \infty} E_m\left(\|\Delta Z_m\|^2 I_{\{\|\Delta Z_m\|^2 > tm\}}\right) = 0 \quad \text{a.s.} \quad \text{for all } t > 0\,. \tag{V8}$$

The convergence in distribution of $\left(\sqrt{n}(X_n - x^*)\right)$ is guaranteed then by the following result:

**Theorem 6.4.** *The sequence $\left(\sqrt{n}(X_n - x^*)\right)$ is asymptotically $N(0, V)$-distributed. The covariance matrix $V$ is the unique solution of the Lyapunov-matrix equation*

$$\left(DH^* - \frac{1}{2}I\right)V + V\left(DH^* - \frac{1}{2}I\right)^T = C, \tag{6.58}$$

*where*

$$D := q_1 D^{(1)} + \ldots + q_\kappa D^{(\kappa)}\,, \tag{6.59}$$

$$C := q_1 D^{(1)} W^{(1)} D^{(1)T} + \ldots + q_\kappa D^{(\kappa)} W^{(\kappa)} D^{(\kappa)T} \tag{6.60}$$

*and matrices $D^{(1)}, \ldots, D^{(\kappa)}$ from condition (V2).*

Central limit theorems of the above type can be found in many papers on stochastic approximation (cf. e.g. [1,12,35,62,97,98,119,147,148] and [118]). But in these convergence theorems it is always assumed that matrices $M_n$ or the covariance matrices of the estimation error $Z_n$ converge to a single limit $D$, $W$, respectively. This special case is enclosed in Theorem 6.4, if assumptions (V2), (V3a,b) and (V5a,b)/(V6) for $\kappa = 1$ are fulfilled (cf. Corollary 6.4).

Before proving Theorem 6.4, some lemmas are shown first. For the deterministic matrices

$$A_{m,n} := \frac{1}{\sqrt{m}} B_{m,n} D_m \quad \text{for} \quad m \le n\,, \tag{6.61}$$

arising in the important equation (6.54) the following result holds:

**Lemma 6.13.** *a) For $m = 1, 2, \ldots$ it holds* $\lim_{n\to\infty} \|A_{m,n}\| = 0$ .

*b)* $\limsup_{n\to\infty} \sum_{m=1}^{n} \|A_{m,n}\|^2 < \infty$ .

*c) It exists $n_0 \in \mathbb{N}$ with* $\|A_{m,n}\| \leq \frac{2L}{\sqrt{m}}$ *for* $n_0 \leq m \leq n$ , *where* $L := \sup_n \|D_n\| < \infty$.

*d) For indices $m \in \mathbb{N}^{(k)}$, let $W_m := W^{(k)}$ . Then*

$$\lim_{n\to\infty} \sum_{m=1}^{n} A_{m,n} W_m A_{m,n}^T = V$$

*with the matrix $V$ according to (6.58).*

*Proof.* Because of (6.56) and (6.57) for $m \leq n$ it holds:

$$\|A_{m,n}\| \leq \frac{L}{\sqrt{m}} b_{m,n} , \tag{i}$$

$$\|A_{m,n}\|^2 \leq \frac{L^2}{m} \bar{b}_{m,n} . \tag{ii}$$

Hence, according to (6.52) and (6.53) Lemma A.6, Appendix, can be applied. Therefore

$$\lim_{n\to\infty} b_{m,n} = 0 \quad \text{for} \quad m = 0, 1, 2, \ldots , \tag{iii}$$

$$\lim_{n\to\infty} \sum_{m=1}^{n} \bar{b}_{m,n} \frac{1}{m} = \frac{1}{2\bar{u}} < \infty, \tag{iv}$$

and because of Lemma A.5 a), Appendix, for any $0 < \delta < \bar{u}$ we have

$$b_{m,n} \ll \phi_{m,n}(\bar{u} - \delta) \sim \left(\frac{m}{n}\right)^{\bar{u}-\delta} \ll 1 . \tag{v}$$

From (i) to (v) propositions a), b) und c) are obtained easily. Symmetric matrices $V_1 := 0$ and

$$V_{n+1} := \sum_{m=1}^{n} A_{m,n} W_m A_{m,n}^T \quad \text{for} \quad n = 1, 2, \ldots \tag{6.62}$$

by definition fulfill equation

$$V_{n+1} = \left(I - \frac{1}{n} B_n\right) V_n \left(I - \frac{1}{n} B_n\right)^T + \frac{1}{n} C_n, n \geq 1, \tag{vi}$$

where $C_n := D_n W_n D_n^T$. Hence, because of (6.47) and (6.48) for all $k$ the following limits exist:

$$B^{(k)} := \lim_{n \in \mathbb{N}^{(k)}} B_n = D^{(k)} H^* - \frac{1}{2} I .$$

Furthermore, according to (V3b) it holds

$$B^{(k)} + B^{(k)T} \geq 2\left(a^{(k)} - \frac{1}{2}\right) I = 2u^{(k)} I .$$

Therefore and because of (6.53), Lemma A.9, Appendix, can be applied to equation (vi). Hence, $V_n \longrightarrow V$, where $V$ is unique solution of

$$BV + VB^T = C .$$

Here, $B := q_1 B^{(1)} + \ldots + q_\kappa B^{(\kappa)}$, and $C$ is given by (6.60). Henceforth, $B = DH^* - \frac{1}{2}I$, where $D$ is obtained from (6.59), and part d) is proven.

**Lemma 6.14.** *a) If (V5a,b) or (V6) holds, then*

$$\sum_{m=1}^{n} A_{m,n} E_m(\Delta Z_m \Delta Z_m^T) A_{m,n}^T \stackrel{stoch}{\longrightarrow} V \tag{KV}$$

*with matrix $V$ from (6.58)*
*b) If (V7) or (V8) is fulfilled, then for each $\varepsilon > 0$*

$$\sum_{m=1}^{n} \|A_{m,n}\|^2 E_m\left(\|\Delta Z_m\|^2 I_{\{\|A_{m,n}\Delta Z_m\| > \varepsilon\}}\right) \stackrel{stoch}{\longrightarrow} 0 \tag{KL}$$

*holds.*

*Proof.* Let $\widehat{V}_0 := 0$, and for $n = 1, 2, \ldots$

$$\widehat{V}_{n+1} := \sum_{m=1}^{n} A_{m,n} E_m(\Delta Z_m \Delta Z_m^T) A_{m,n}^T .$$

Having matrix $V_{n+1}$ according to (6.62), for each $n$

$$\|\widehat{V}_{n+1} - V\| \leq \|\widehat{V}_{n+1} - V_{n+1}\| + \|V_{n+1} - V\| , \tag{i}$$

$$E\|\widehat{V}_{n+1} - V\| \leq E\|\widehat{V}_{n+1} - E\widehat{V}_{n+1}\| + \|E\widehat{V}_{n+1} - V_{n+1}\| + \|V_{n+1} - V\| \tag{ii}$$

holds. Furthermore,

$$\|\widehat{V}_{n+1} - V_{n+1}\| \leq \sum_{m=1}^{n} \|A_{m,n}\|^2 \|E_m(\Delta Z_m \Delta Z_m^T) - W_m\| , \tag{iii}$$

$$E\|\widehat{V}_{n+1} - E\widehat{V}_{n+1}\| \leq \sum_{m=1}^{n} \|A_{m,n}\|^2 E\left\|E_m(\Delta Z_m \Delta Z_m^T) - E(\Delta Z_m \Delta Z_m^T)\right\| \tag{iv}$$

and

$$\|E\widehat{V}_{n+1} - V_{n+1}\| \leq \sum_{m=1}^{n} \|A_{m,n}\|^2 \big\|E(\Delta Z_m \Delta Z_m^T) - W_m\big\|. \tag{v}$$

For any real stochastic sequence $(b_n)$ with $b_n \longrightarrow 0$ a.s., Lemma 6.13 a), b) and Lemma A.2, Appendix, provide

$$\sum_{m=1}^{n} \|A_{m,n}\|^2 b_m \longrightarrow 0 \quad \text{a.s.} \,. \tag{vi}$$

Besides, according to Lemma 6.13 d) it is

$$\|V_{n+1} - V\| \longrightarrow 0\,. \tag{vii}$$

Hence, under condition (V5a,b) we get because of (ii), (iv), (v), (vi) and (vii)

$$E\|\widehat{V}_{n+1} - V\| \longrightarrow 0\,.$$

In addition, with (V6), because of (i), (iii), (vi) and (vii) we have

$$\|\widehat{V}_{n+1} - V\| \longrightarrow 0 \quad \text{a.s.} \,.$$

Therefore, (KV) is fulfilled in both cases.

Select an arbitrary $\varepsilon > 0$. Due to Lemma 6.13 c),

$$\|\Delta Z_m\|^2 I_{\{\|A_{m,n}\Delta Z_m\| > \varepsilon\}} \leq \|\Delta Z_m\|^2 I_{\left\{\frac{2L}{\sqrt{m}}\|\Delta Z_m\| > \varepsilon\right\}}$$

holds for all indices $n_0 \leq m \leq n$. Hence, for $n > n_0$, where $t := \left(\frac{\varepsilon}{2L}\right)^2 > 0$

$$\begin{aligned} U_n &:= \sum_{m=1}^{n} \|A_{m,n}\|^2 E_m(\|\Delta Z_m\|^2 I_{\{\|A_{m,n} Z_m\| > \varepsilon\}}) \\ &\leq U_{n_0} + \sum_{m=n_0+1}^{n} \|A_{m,n}\|^2 \cdot E_m\big(\|\Delta Z_m\|^2 I_{\{\|\Delta Z_m\|^2 > tm\}}\big)\,. \end{aligned} \tag{viii}$$

Therefore, from (vi) and (viii) follows $EU_n \longrightarrow 0$ at assumption (V7) and $U_n \longrightarrow 0$ a.s. under assumption (V8). Hence, in any case (KL) is fulfilled.

Now we are able to show Theorem 6.4.

*Proof (of Theorem 6.4).* Let $(\Delta_n^{(0)})$ be defined according to (6.54) with $\Delta_1^{(0)} := 0$. Thus,

$$\Delta_{n+1}^{(0)} = -\sum_{m=1}^{n} A_{m,n}\Delta Z_m \quad \text{for} \quad n = 1, 2, \ldots\,.$$

Due to Lemma 6.14 all conditions for the limit theorem Lemma B.7 in the Appendix are fulfilled for this sequence. Hence, $(\Delta_n^{(0)})$ is asymptotically $N(0,V)$-distributed. According to Theorem 6.3 this holds also true for sequence $\left(\sqrt{n}(X_n - x^*)\right)$ .

In the special case that the matrix sequence $(M_n)$ converges a.s. to a unique limit matrix $D$, we get the following corollary:

**Corollary 6.4.** *Let*

$$M_n \longrightarrow \Theta \quad a.s. \ , \tag{6.63}$$

*where $\Theta$ denotes a deterministic matrix having*

$$\Theta H^* + (\Theta H^*)^T > I \ . \tag{6.64}$$

*Then, $\left(\sqrt{n}(X_n - x^*)\right)$ is asymptotically $N(0,V)$-distributed. The covariance-matrix $V$ is unique solution of*

$$\left(\Theta H^* - \frac{1}{2}I\right) V + V \left(\Theta H^* - \frac{1}{2}I\right)^T = \Theta W \Theta^T \ , \tag{6.65}$$

*if*

$$W := q_1 W^{(1)} + \ldots + q_\kappa W^{(\kappa)} \ . \tag{6.66}$$

*Proof.* In the present case, $\Theta^{(k)} = \Theta$ and $a^{(k)} = a$ for $k = 1, \ldots, \kappa$ , where $a$ fulfills

$$\Theta H^* + (\Theta H^*)^T \geq 2aI \ .$$

Due to (6.64) choose $a > \frac{1}{2}$. Hence, conditions (V2) and (V3) are fulfilled for matrix sequence $(M_n)$. The assertion is now obtained from Theorem 6.4

Special limit theorems are obtained for different limit-matrices $D$ in Corollary 6.4:

a) For $\Theta = H^{*\,-1}$ condition (6.64) is always fulfilled. Moreover, due to (6.65) matrix $V$ is given by

$$V = H^{*-1} W (H^{*-1})^T \ .$$

This special case was already discussed for $\kappa = 1$ in [147], [97], [12] and [98] (Corollary 9.1).

b) Let $\Theta = d \cdot I$ , where $d$ is chosen to fulfill $d \cdot (H^* + H^{*^T}) > I$ . Hence, according to (6.65) for $V$ the following formula is obtained:

$$d \cdot (H^* V + V H^{*^T}) - V = d^2 \, W \ .$$

This equation for the covariance matrix $V$ was discussed, in case $\kappa = 1$, in [119] (Theorem 5) und in [35].

Matrix $W$ in (6.66) can be interpreted as "mean limit covariance matrix" of the estimation error sequence $(Z_n)$.

## 6.6 Realisation of Search Directions $Y_n$

The approach used in this section for constructing search directions $Y_n$ of method (6.2a,b), allows adaptive estimation of important unknown items like Jacobi matrix $H(x^*)$ and limit covariance matrix of error $Z_n$. These estimators will be used in Section 6.7 and 6.8 for the construction of adaptive stepsizes.

J.H. Venter (cf. [147]) takes in the one–dimensional case ($\nu = 1$) at stage $n$ the search direction

$$Y_n := \frac{1}{2}(Y_n^{(1)} + Y_n^{(2)}) .$$

Here, $Y_n^{(1)}$ and $Y_n^{(2)}$ are given stochastic approximations of $G(X_n + c_n)$ and $G(X_n - c_n)$, and $(c_n)$ is a suitable deterministic sequence converging to zero. The advantage of this approach is that using the stochastic difference quotient

$$\widetilde{h}_n := \frac{1}{2c_n}(Y_n^{(1)} - Y_n^{(2)}),$$

a known approximation for the derivative of $G(X_n)$ is obtained. This difference quotient plays a fundamental role in the adaptive step size of Venter.

In [12,97] and [98] this approach for the determination of search directions was generalized to the multi–dimensional case. There,

$$Y_n := \frac{1}{2\nu}\sum_{j=1}^{2\nu} Y_n^{(j)}$$

is assumed as search direction, where vectors $Y_n^{(i)}$ and $Y_n^{(\nu+i)}$ are given stochastic approximations of vectors $G(X_n + c_n e_i)$ and $G(X_n - c_n e_i)$, and $e_1, \ldots, e_\nu$ denote the canonic unit vectors of $\mathbb{R}^\nu$. Then,

$$\widetilde{H}_n := \frac{1}{2c_n}(Y_n^{(1)} - Y_n^{(\nu+1)}, \ldots, Y_n^{(\nu)} - Y_n^{(2\nu)})$$

is a known estimator of the Jacobian $H_n := H(X_n)$ of $G$ at stage $n$.

The main disadvantage of this approach for $Y_n$ is that at each iteration $n$ exactly $2\nu$ realizations $Y_n^{(1)}, \ldots, Y_n^{(2\nu)}$ have to be determined. This can be avoided corresponding to the Venter approach if at stage $n$ only *two* measurements $Y_n^{(1)}$ and $Y_n^{(2)}$ are determined. Hence, only one column vector

$$\widetilde{h}_n := \frac{1}{2c_n}(Y_n^{(1)} - Y_n^{(2)})$$

of the matrix $H(X_n)$ is estimated. Cycling through all columns, after $\nu$ iterations a new estimator for the whole matrix $H(X_n)$ is obtained.

This possibility is contained in the following approach by means of choosing $l_n \equiv 1$ and $\varepsilon_n \equiv 0$:

For indices $n = 1, 2, \ldots$, let $l_n \in \mathbb{N}_0$ and $\varepsilon_n \in \{0, 1\}$ be integers with

$$\varepsilon_n + l_n > 0\,. \tag{W1a}$$

Furthermore, use at each iteration stage $n \in \mathbb{N}$

$$Y_n := \frac{1}{2l_n + \varepsilon_n} \sum_{j=1-\varepsilon_n}^{2l_n} Y_n^{(j)} \tag{W1b}$$

as stochastic approximation of $G_n := G(X_n)$. Here, $Y_n^{(j)}$ denotes a known stochastic estimator of $G$ at $X_n^{(j)}$, that is

$$Y_n^{(j)} = G_n^{(j)} + Z_n^{(j)}, \tag{6.67}$$

where $G_n^{(j)} := G(X_n^{(j)})$, and $Z_n^{(j)}$ denotes the estimation error for $j = 0, \ldots, 2l_n$. The arguments $X_n^{(0)}, \ldots, X_n^{(2l_n)}$ are given by

$$X_n^{(0)} := X_n\,,\ X_n^{(i)} := X_n + c_n d_n^{(i)}\,,\ X_n^{(l_n+i)} := X_n - c_n d_n^{(i)} \tag{W1c}$$

for $i = 1, \ldots, l_n$, where $(c_n)$ is a positive sequence converging to zero. As direction vectors $d_n^{(i)}$ one after another the unit vectors $e_1, e_2, \ldots, e_\nu, e_1, \ldots$ of $\mathbb{R}^\nu$ are chosen. Formally, this leads to

$$d_n^{(i)} = e_{l(n,i)} \quad \text{for} \quad i = 1, \ldots, l_n\,, \tag{W1d}$$

where indices $l(n, i)$ are calculated as follows:

$$\begin{aligned} &l(1, 0) := 0\,,\\ &l(n, i) := l(n, i-1) \bmod \nu + 1 \quad \text{for} \quad i = 1, \ldots, l_n\,,\\ &l(n+1, 0) := l(n, l_n)\,. \end{aligned} \tag{W1e}$$

Using the above choice of $Y_n$, we get

$$Y_n = G_n + Z_n \tag{6.68}$$

with estimation error

$$Z_n := \frac{1}{2l_n + \varepsilon_n} \sum_{j=1-\varepsilon_n}^{2l_n} (G_n^{(j)} + Z_n^{(j)}) - G_n\,. \tag{6.69}$$

In order to guarantee that this total error $Z_n$ fulfills conditions (U5a-c) in Section 6.3, let the error terms $Z_n^{(0)}, \ldots, Z_n^{(2l_n)}$ denote $\mathcal{A}_{n+1}$-measurable and stochastically independent random variables with given $\mathcal{A}_n$. Furthermore, for $j = 0, \ldots, 2l_n$ assume that

$$\|E_n Z_n^{(j)}\| \le \frac{L_1}{n^{\mu_1}} \quad \text{with} \quad \mu_1 > \frac{1}{2}\,, \tag{W2a}$$

$$E_n\|Z_n^{(j)} - E_n Z_n^{(j)}\|^2 \le s_{0,n}^2 + s_1^2\|X_n^{(j)} - x^*\|^2 \tag{W2b}$$

with positive constants $L_1$, $\mu_1$ and $s_1$ and $\mathcal{A}_n$-measurable random variables $s_{0,n}^2 \geq 0$, for which with an additional constant $s_0$

$$\sup_n E s_{0,n}^2 \leq s_0^2 \tag{W2c}$$

is fulfilled. Hence, due to (W2b) and (W1c)

$$E_n \|Z_n^{(j)} - E_n Z_n^{(j)}\|^2 \leq t_{0,n}^2 + t_1^2 \|X_n - x^*\|^2 , \tag{6.70}$$

$$\sup_n E t_{0,n}^2 \leq s_0^2 + 2 s_1^2 \sup_n c_n^2 =: t_0^2 , \tag{6.71}$$

if we denote

$$t_{0,n}^2 := s_{0,n}^2 + 2 c_n^2 s_1^2 , \tag{6.72}$$

$$t_1^2 := 2 s_1^2 . \tag{6.73}$$

The error $Z_n$ has further properties:

**Lemma 6.15.** *For each $n \in \mathbb{N}$, $Z_n$ is $\mathcal{A}_{n+1}$-measurable, and the following properties hold:*

*a)* $E_n Z_n = \dfrac{1}{2l_n + \varepsilon_n} \displaystyle\sum_{j=1-\varepsilon_n}^{2l_n} (G_n^{(j)} + E_n Z_n^{(j)}) - G_n ,$

*b)* $E_n \Delta Z_n \Delta Z_n^T = \dfrac{1}{(2l_n + \varepsilon_n)^2} \displaystyle\sum_{j=1-\varepsilon_n}^{2l_n} E_n \Delta Z_n^{(j)} \Delta Z_n^{(j)T} ,$

*c)* $E_n \|\Delta Z_n\|^2 = \dfrac{1}{(2l_n + \varepsilon_n)^2} \displaystyle\sum_{j=1-\varepsilon_n}^{2l_n} E_n \|\Delta Z_n^{(j)}\|^2$

$$\leq \frac{1}{2l_n + \varepsilon_n} \left( t_{0,n}^2 + t_1^2 \|X_n - x^*\|^2 \right) ,$$

*d)* $\|E_n Z_n\| \leq L_0 c_n^{1+\mu_0} + \dfrac{L_1}{n^{\mu_1}}$ *with $L_0$ and $\mu_0$ from (U1b).*

Here, the following abbreviations for $n = 1, 2, \ldots$ and $j = 0, \ldots, 2l_n$ are used in Lemma 6.15:

$$\Delta Z_n := Z_n - E_n Z_n , \; \Delta Z_n^{(j)} := Z_n^{(j)} - E_n Z_n^{(j)} . \tag{6.74}$$

*Proof.* Formula a) is a direct consequence of (6.69). Hence, it holds

$$\Delta Z_n := \frac{1}{2l_n + \varepsilon_n} \sum_{j=1-\varepsilon_n}^{2l_n} \Delta Z_n^{(j)},$$

and therefore due to the conditional independence of $Z_n^{(0)}, \ldots, Z_n^{(2l_n)}$ we get

$$E_n \Delta Z_n \Delta Z_n^T = \frac{1}{(2l_n + \varepsilon_n)^2} \sum_{j=1-\varepsilon_n}^{2l_n} E_n \Delta Z_n^{(j)} \Delta Z_n^{(j)T} .$$

Taking the trace of the matrix, we get

$$\begin{aligned} E_n \|\Delta Z_n\|^2 &= E_n \text{ trace } (\Delta Z_n \Delta Z_n^T) = \text{ trace } (E_n \Delta Z_n \Delta Z_n^T) \\ &= \frac{1}{(2l_n + \varepsilon_n)^2} \sum_{j=1-\varepsilon_n}^{2l_n} E_n \|\Delta Z_n^{(j)}\|^2 , \end{aligned}$$

and using (6.70) assertion c) follows. Due to condition (W2a) we have

$$\left\| \frac{1}{2l_n + \varepsilon_n} \sum_{j=1-\varepsilon_n}^{2l_n} E_n Z_n^{(j)} \right\| \le \frac{L_1}{n^{\mu_1}} ,$$

and therfore by applying a) we find

$$\begin{aligned} \|E_n Z_n\| &\le \left\| \frac{2}{2l_n + \varepsilon_n} \sum_{i=1}^{l_n} \left( \frac{1}{2}(G_n^{(i)} + G_n^{(l_n+i)}) - G_n \right) \right\| + \frac{L_1}{n^{\mu_1}} \\ &\le \frac{2}{2l_n + \varepsilon_n} \sum_{i=1}^{l_n} \left\| \frac{1}{2}(G_n^{(i)} + G_n^{(l_n+i)}) - G_n \right\| + \frac{L_1}{n^{\mu_1}} . \end{aligned}$$

Because of conditions (U1a-c) and (W1a,b), for each $i = 1, \dots, l_n$

$$\begin{aligned} &\left\| \frac{1}{2}(G_n^{(i)} + G_n^{(l_n+i)}) - G_n \right\| = \left\| \frac{1}{2} \left( \delta(X_n; c_n d_n^{(i)}) + \delta(X_n; -c_n d_n^{(i)}) \right) \right\| \\ &\le \frac{1}{2} L_0 \left( \|c_n d_n^{(i)}\|^{1+\mu_0} + \|c_n d_n^{(i)}\|^{1+\mu_0} \right) = L_0 c_n^{1+\mu_0} \end{aligned}$$

holds. Using the previously obtained inequality, assertion d) follows.

Due to Lemma 6.15 c), (6.70) and (6.71), conditions (U5b) and (U5c) are fulfilled by means of

$$\sigma_{0,n}^2 := \frac{1}{2l_n + \varepsilon_n} t_{0,n}^2 , \ \sigma_0 := t_0 , \ \sigma_1 := t_1 . \tag{6.75}$$

For given positive numbers $c_0$ and $\mu$, choose for any $n = 1, 2, \dots$

$$c_n = \frac{c_0}{n^\mu} , \quad \text{where} \quad \mu > \frac{1}{2(1 + \mu_0)} \tag{W3a}$$

with $\mu_0$ from condition (U1a-c). Due to Lemma 6.15 d) and (W2a) we get

$$\sqrt{n} \|E_n Z_n\| \longrightarrow 0 \quad \text{a.s. .} \tag{6.76}$$

Therefore the estimation error sequence $(Z_n)$ fulfills all conditions (U5a,b).

For convergence of the covariance matrices of the estimation error sequence $(Z_n)$ (cf. Lemma 6.18), we need the existence of

$$l^{(k)} := \lim_{n \in \mathbf{N}^{(k)}} l_n < \infty , \tag{W4a}$$

$$\varepsilon^{(k)} := \lim_{n \in \mathbf{N}^{(k)}} \varepsilon_n \in \{0,1\} \tag{W4b}$$

for sequences $(l_n)$ and $(\varepsilon_n)$ for each $k = 1, \dots, \kappa$. In particular this leads to

$$\widehat{l} := \sup_{n \in \mathbf{N}} l_n < \infty , \tag{6.77}$$

$$\bar{l} := \lim_{n \to \infty} \frac{l_1 + \ldots + l_n}{n} = q_1 l^{(1)} + \ldots + q_\kappa l^{(\kappa)} . \tag{6.78}$$

By means of approach (W1a,b) for search directions $Y_n$ it is easy to get convergent estimators for $G(x^*)$ , $H(x^*)$ as well as the limit covariance matrix of $(Z_n)$.

### 6.6.1 Estimation of $G^*$

Vectors

$$G^*_{n+1} := \frac{1}{n}(Y_1 + \ldots + Y_n) \quad \text{for} \quad n = 1, 2, \ldots \tag{6.79}$$

can be interpreted as $\mathcal{A}_{n+1}$-measurable estimators of $G^* := G(x^*)$. This is shown in the next result:

**Theorem 6.5.** *If $X_n \longrightarrow x^*$ a.s., then $G^*_n \longrightarrow G^*$ a.s. .*

*Proof.* For $n = 1, 2, \ldots$ we have

$$G^*_{n+1} = \left(1 - \frac{1}{n}\right) G^*_n + \frac{1}{n} Y_n . \tag{i}$$

Due to (U5a), our assumption and (6.4) it holds

$$\|E_n Y_n - G_n\| = \|E_n Z_n\| \longrightarrow 0 , \; G_n \longrightarrow G^* \quad \text{a.s.} ,$$

and therefore we get

$$E_n Y_n \longrightarrow G^* \quad \text{a.s.} . \tag{ii}$$

Moreover, due to (U5b), (U5c) and condition a) in Section 6.3 for each $n$

$$E\|Y_n - E_n Y_n\|^2 = E\|Z_n - E_n Z_n\|^2 \le \sigma_0^2 + (\sigma_1 \delta_0)^2$$

holds and therefore

$$\sum_n \frac{1}{n^2} E_n \|Y_n - E_n Y_n\|^2 < \infty \text{ a.s.} . \tag{iii}$$

Using (i), (ii) and (iii), the proposition follows from Lemma B.6 in the Appendix.

### 6.6.2 Update of the Jacobian

For numbers $l_n$ in definition (W1a-e) we demand now

$$l_n > 0 \quad \text{for} \quad n = 1, 2, \dots . \tag{W5}$$

Furthermore, for exponent $\mu$ in (W3a)

$$\mu < \frac{1}{2} \tag{W3b}$$

is assumed. Due to the following lemma, the "stochastic central difference quotients"

$$h_n^{(i)} := \frac{1}{2c_n}(Y_n^{(i)} - Y_n^{(l_n+i)}) \quad \text{for} \quad i = 1, \dots, l_n, \tag{6.80}$$

known at iteration stage $n$, are estimators of $H_n \cdot d_n^{(i)}$, where $H_n := H(X_n)$.

**Lemma 6.16.** *For each $n \in \mathbb{N}$ and $i \in \{1, \dots, l_n\}$, $h_n^{(i)}$ is $\mathcal{A}_{n+1}$-measurable, and it holds:*

*a)* $E_n h_n^{(i)} = \frac{1}{2c_n}(G_n^{(i)} - G_n^{(l_n+i)}) + \frac{1}{2c_n}(E_n Z_n^{(i)} - E_n Z_n^{(l_n+i)})$ ,

*b)* $E_n \|h_n^{(i)} - E_n h_n^{(i)}\|^2 \le \frac{1}{2c_n^2}\left(t_{0,n}^2 + t_1^2\|X_n - x^*\|^2\right)$ ,

*c)* $\|E_n h_n^{(i)} - H_n d_n^{(i)}\| \longrightarrow 0$ .

*Proof.* Part a) holds because of $E_n Y_n^{(j)} = G_n^{(j)} + E_n Z_n^{(j)}$ for $j = 1, \dots, 2l_n$. Furthermore, with $\Delta Z_n^{(j)} := Z_n^{(j)} - E_n Z_n^{(j)}$ we have

$$E_n\|h_n^{(i)} - E_n h_n^{(i)}\|^2 = E_n\|\frac{1}{2c_n}(\Delta Z_n^{(i)} - \Delta Z_n^{(l_n+i)})\|^2$$
$$= \frac{1}{4c_n^2}\left(E_n\|\Delta Z_n^{(i)}\|^2 + E_n\|\Delta Z_n^{(l_n+i)}\|^2\right) \le \frac{1}{2c_n^2}\left(t_{0,n}^2 + t_1^2\|X_n - x^*\|^2\right) .$$

Here, the independence of the variables $Z_n^{(j)}$ was taken into account, and (6.70) was used. Now, because of (U1a-c)

$$0G(X_n \pm c_n d_n^{(i)}) = G(X_n) \pm c_n H_n d_n^{(i)} + \delta(X_n; \pm c_n d_n^{(i)}) ,$$

and therefore due to a), (W2a) (U1b) and (W3a) we find

$$\|E_n h_n^{(i)} - H_n d_n^{(i)}\| = \left\| \frac{1}{2c_n}\left(\delta(X_n; c_n d_n^{(i)}) - \delta(X_n; -c_n d_n^{(i)})\right) \right\| + \frac{L_1}{c_n n^{\mu_1}}$$
$$\le \frac{L_0}{2c_n}\left(\|c_n d_n^{(i)}\|^{1+\mu_0} + \| - c_n d_n^{(i)}\|^{1+\mu_0}\right) + \frac{L_1}{c_n n^{\mu_1}} = L_0 c_n^{\mu_0} + \frac{L_1}{c_0 n^{\mu_1 - \mu}} .$$

Finally, since $c_n \longrightarrow 0$ , $\mu_0 > 0$ and $\mu_1 > \frac{1}{2} > \mu$, proposition c) follows.

Since for directions $d_n^{(i)}$ we use cyclically all unit vectors $e_1, \dots, e_\nu$, the Jacobian $H^* := DG(x^*)$ can be estimated easily by means of vectors $h_n^{(i)}$.

For this recursive estimate we start with a $\mathcal{A}_1$-measurable $\nu \times \nu$-Matrix $H_1^{(0)}$. In an iteration step $n \geq 1$ we calculate for $i = 1, \dots, l_n$ $\mathcal{A}_{n+1}$-measurable matrices

$$H_n^{(i)} := H_n^{(i-1)} + \frac{1}{n}(h_n^{(i)} - H_n^{(i-1)} d_n^{(i)})\, d_n^{(i)T} \tag{6.81a}$$

and put

$$H_{n+1}^{(0)} := H_n^{(l_n)}. \tag{6.81b}$$

According to (W1d), (6.81a,b) leads to

$$h_{n,l}^{(i)} := \begin{cases} \left(1 - \frac{1}{n}\right) h_{n,l}^{(i-1)} + \frac{1}{n} h_n^{(i)} & , \text{for } \; l = l(n,i) \\ h_{n,l}^{(i-1)} & , \text{for } \; l \neq l(n,i) \,, \end{cases}$$

if $H_n^{(i)} = (h_{n,1}^{(i)}, \dots, h_{n,\nu}^{(i)})$.

Hence, at transition of $H_n^{(i-1)}$ to $H_n^{(i)}$ only the $l(n,i)$-th column vector is changed (updated) by averaging with vector $h_n^{(i)}$ of (6.80).

Due to the following theorem, matrices $H_n^{(0)}$ are known approximations of $H^*$, if $X_n$ converges to $x^*$.

Besides $H_n^{(i)}$ this theorem uses further matrices: Let $\bar{H}_1^{(0)}$ denote an arbitrary matrix, and for $n = 1, 2, \dots$ and $i = 1, \dots, l_n$ let

$$\bar{H}_n^{(i)} := \bar{H}_n^{(i-1)} + \frac{1}{n}(H_n - \bar{H}_n^{(i-1)}) d_n^{(i)} d_n^{(i)T} \,, \tag{6.82a}$$

$$\bar{H}_{n+1}^{(0)} := \bar{H}_n^{(l_n)} \,. \tag{6.82b}$$

**Theorem 6.6.** *a)* $(\bar{H}_n^{(i)})$ *is bounded, and it holds* $H_n^{(0)} - \bar{H}_n^{(0)} \longrightarrow 0$ *a.s.* .
*b) If "*$X_n \longrightarrow x^*$ *a.s.", then* $H_n^{(0)} \longrightarrow H^*$ *a.s.* .

*Proof.* To arbitrary fixed $l \in \{1, \dots, \nu\}$ let $(n_1, i_1), (n_2, i_2), \dots$ denote all ordered index pairs $(n, i)$, for which $d_n^{(i)} = e_l$ is chosen. Hence, $1 \leq n_1 \leq n_2 \leq \dots$, and for each $k$ it holds $1 \leq i_k \leq l_{n_k}$ as well as $i_k < i_{k+1}$ if $n_k = n_{k+1}$ . Moreover, for $n \geq 1$ and $i \leq l_n$ we have

$$d_n^{(i)} = e_l \Longleftrightarrow (n,i) \in \Big\{(n_1, i_1), (n_2, i_2), \dots\Big\} .$$

Due to (6.81a,b) and (6.82a,b), for $k = 1, 2, \dots$ we get

$$V_{k+1} = (1 - \alpha_k) V_k + \alpha_k U_k \,, \tag{i}$$

$$\bar{V}_{k+1} = (1 - \alpha_k) \bar{V}_k + \alpha_k \bar{U}_k \,, \tag{ii}$$

if

$$\alpha_k := \frac{1}{n_k}, \quad U_k := h_{n_k}^{(i_k)}, \quad V_k := H_{n_{k-1}}^{(i_{k-1})} e_l,$$
$$\bar{U}_k := H_{n_k} e_l, \quad \bar{V}_k := \bar{H}_{n_{k-1}}^{(i_{k-1})} e_l.$$

Because of (W1e), at latest after each $\nu$–th iteration stage $n$, direction vector $e_l$ is chosen. Therefore it holds $n_{k+1} \le n_k + \nu$ and $n_k \le k\nu$ for each index $k$. Consequently, we get

$$\sum_k \alpha_k = \infty. \tag{iii}$$

Because of (W1e), and (6.77), at each iteration stage $n$, direction vector $e_l$ is chosen at most $l_0 := [\frac{\hat{l}-1}{\nu}] + 1$ times and therefore $k \le l_0 n_k$ for $k = 1, 2, \ldots$. Hence, using condition (W3a,b) we get

$$\sum_k \frac{\alpha_k^2}{c_{n_k}^2} = \frac{1}{c_0^2} \sum_k \frac{1}{n_k^{2-2\mu}} \le \frac{l_0^{2-2\mu}}{c_0^2} \sum_k \frac{1}{k^{2-2\mu}} < \infty. \tag{iv}$$

Now, for $k \ge 1$ let denote $\mathcal{B}_k$ the $\sigma$–algebra generated by $\mathcal{A}_{n_k}$ and $U_1, \ldots, U_{k-1}$. Obviously now holds:

$$\mathcal{B}_1 \subset \mathcal{B}_2 \subset \ldots \subset \mathcal{B}_k \subset \mathcal{A}, \tag{v}$$
$$\Delta U_k := U_k - \bar{U}_k \quad \text{is} \quad \mathcal{B}_{k+1}\text{-measurable}, \tag{vi}$$
$$E(\Delta U_k \mid \mathcal{B}_k) = E_{n_k}(\Delta U_k) = E_{n_k} h_{n_k}^{(i_k)} - H_{n_k} e_l, \tag{vii}$$
$$E\Big(\big\|\Delta U_k - E(\Delta U_k \mid \mathcal{B}_k)\big\|^2 \mid \mathcal{B}_k\Big) = E_{n_k}\Big(\|h_{n_k}^{(i_k)} - E_{n_k} h_{n_k}^{(i_k)}\|^2\Big). \tag{viii}$$

From (iv), (viii) and Lemma 6.16 b) we have

$$\sum_k \alpha_k^2 E\Big(\|\Delta U_k - E(\Delta U_k \mid \mathcal{B}_k)\|^2 \mid \mathcal{B}_k\Big)$$
$$\le \sum_k \frac{\alpha_k^2}{2c_{n_k}^2} \Big(t_{0,n_k} + t_1^2 \|X_{n_k} - x^*\|^2\Big) < \infty \quad \text{a.s.}. \tag{ix}$$

From (vii) and Lemma 6.16 c) we obtain

$$E(\Delta U_k \mid \mathcal{B}_k) \longrightarrow 0 \quad \text{a.s.}. \tag{x}$$

Equations (i) and (ii) yield

$$V_{k+1} - \bar{V}_{k+1} = (1 - \alpha_k)(V_k - \bar{V}_k) + \alpha_k \Delta U_k.$$

Due to (iii), (v), (vi), (ix) and (x), Lemma B.6 in the Appendix can be applied. Hence, it holds

$$V_{k+1} - \bar{V}_{k+1} = (H_{n_k}^{(i_k)} - \bar{H}_{n_k}^{(i_k)}) e_l \longrightarrow 0 \quad \text{a.s.}.$$

Since $l \in \{1, \ldots, \nu\}$ was chosen arbitrarily, we have shown proposition a).

In case of "$X_n \longrightarrow x^*$ a.s." follows according to (U1c) also $H_n \longrightarrow H^*$ a.s. . Therfore, because of (ii), (iii) and Lemma B.6 (Appendix)

$$\bar{V}_{k+1} = \bar{H}_{n_k}^{(i_k)} e_l \longrightarrow H^* e_l \quad \text{a.s. .}$$

Hence, it holds $\bar{H}_n^{(0)} \longrightarrow H^*$ a.s. and therefore because of a) also $H_n^{(0)} \longrightarrow H^*$ a.s..

Matrices $H_{n+1}^{(0)}$ from (6.81a,b) can also be approximated by means of special convex combinations of Jacobians $H_1, \ldots, H_n$ This is shown next.

**Theorem 6.7.** *Having "$X_{n+1} - X_n \longrightarrow 0$ a.s", we get*

$$H_n^{(0)} - \widehat{H}_n^{(0)} \longrightarrow 0 \quad a.s. \ .$$

*Here, $\mathcal{A}_n$-measurable matrices $\widehat{H}_n^{(i)}$ are chosen as follows: Choose $\widehat{H}_1^{(0)} := H_1^{(0)}$, and for $n = 1, 2, \ldots$ as well as $i = 1, \ldots, l_n$ let*

$$\widehat{H}_n^{(i)} := \widehat{H}_n^{(i-1)} + \frac{1}{n\nu}(H_n - \widehat{H}_n^{(i-1)}) \ , \tag{6.83a}$$

$$\widehat{H}_{n+1}^{(0)} := \widehat{H}_n^{(l_n)} \ . \tag{6.83b}$$

*Proof.* According to Theorem 6.6 a) only the following property has to be shown:

$$\bar{H}_n^{(0)} - \widehat{H}_n^{(0)} \longrightarrow 0 \quad \text{a.s. .} \tag{$*$}$$

Simple computation yields for $n = 1, 2, \ldots$ and $1 \le i \le l_n$

$$\bar{H}_n^{(i)} - \widehat{H}_n^{(i)} = \left(1 - \frac{1}{n\nu}\right)(\bar{H}_n^{(i-1)} - \widehat{H}_n^{(i-1)}) + \frac{1}{n\nu}(H_n - \bar{H}_n^{(i-1)})(\nu d_n^{(i)} d_n^{(i)T} - I) \ . \tag{i}$$

Defining now

$$L(n,i) := l_1 + \ldots + l_{n-1} + i \ , \ \alpha_{L(n,i)} := \frac{L(n,i)}{n\nu} \ ,$$
$$U_{L(n,i)} := (H_n - \bar{H}_n^{(i-1)})(\nu d_n^{(i)} d_n^{(i)T} - I) \ , \ V_{L(n,i)} := \bar{H}_n^{(i)} - \widehat{H}_n^{(i)} \ ,$$

(i) can be rewritten as:

$$V_L = \left(1 - \frac{\alpha_L}{L}\right) V_{L-1} + \frac{\alpha_L}{L} U_L \tag{ii}$$

for $L = 1, 2, \ldots$ . Due to (6.3) and Theorem 6.6 a) we have

$$(U_L) \text{ bounded a.s. ,} \tag{iii}$$

and according to (6.78) we get

$$\lim_{L\to\infty} \alpha_L = \frac{\bar{l}}{\nu} =: \alpha > 0 \,. \tag{iv}$$

Hence, because of (ii), (iii), (iv) and Lemma A.4, Appendix, it is

$$(V_L) \text{ bounded and } V_L - V_{L-1} \longrightarrow 0 \quad \text{a.s.} \,. \tag{v}$$

Because of (ii), for $k = 0, 1, 2, \dots$ it holds

$$V_{(k+1)\nu} = p_{0,k} V_{k\nu} + \sum_{m=1}^{\nu} p_{m,k} \frac{\alpha_{k\nu+m}}{k\nu+m} U_{k\nu+m} = \left(1 - \frac{\alpha}{k}\right) V_{k\nu} + \frac{\alpha}{k} \bar{U}_k \,, \tag{vi}$$

if we define

$$p_{m,k} := \begin{cases} \prod_{j=m+1}^{\nu} \left(1 - \frac{\alpha_{k\nu+j}}{k\nu+j}\right) & , \text{ for } m = 0, \dots, \nu - 1 \\ 1 & , \text{ for } m = \nu \quad , \end{cases}$$

$$\bar{U}_k := \frac{1}{\nu} \sum_{m=1}^{\nu} p_{m,k} \frac{\alpha_{k\nu+m}}{\alpha} \cdot \frac{k\nu}{k\nu+m} U_{k\nu+m} + \left( (p_{0,k} - 1) \frac{k}{\alpha} + 1 \right) V_{k\nu} \,.$$

Now we show

$$\bar{U}_k \longrightarrow 0 \quad \text{a.s.} \,. \tag{**}$$

Because of (vi) and Lemma A.4, Appendix, from this follows $V_{k\nu} \longrightarrow 0$ a.s.. Hence, if (**) holds, then due to (v), also (*) holds true.

First of all, from the definition of $\bar{U}_k$, for each $k = 0, 1, 2, \dots$ the following approximation is obtained:

$$\begin{aligned} \|\bar{U}_k\| \le \frac{1}{\nu} &\left[ \left\| \sum_{m=1}^{\nu} U_{k\nu+m} \right\| + \sum_{m=1}^{\nu} \|U_{k\nu+m}\| \cdot \left| p_{m,k} \frac{\alpha_{k\nu+m}}{\alpha} \cdot \frac{k\nu}{k\nu+m} - 1 \right| \right] \\ &+ \left| (p_{0,k} - 1) \frac{k}{\alpha} + 1 \right| \cdot \|V_{k\nu}\| \,. \end{aligned} \tag{vii}$$

For matrices

$$T_{L(n,i)} := H_n - \bar{H}_n^{(i-1)} ,$$

due to "$X_{n+1} - X_n \longrightarrow 0$ a.s. ", inequality (6.3) and Theorem 6.6 a), we have

$$T_{L+1} - T_L \longrightarrow 0 \quad \text{a.s.} \,. \tag{viii}$$

Since for an index $L = k\nu + m$ direction $d_n^{(i)} = e_m$ is chosen, for each $k = 0, 1, 2, \dots$ holds

$$\begin{aligned} \left\| \sum_{m=1}^{\nu} U_{k\nu+m} \right\| &= \left\| \sum_{m=1}^{\nu} \Big( T_{k\nu} + (T_{k\nu+m} - T_{k\nu}) \Big) \cdot (\nu e_m e_m^T - I) \right\| \\ &= \left\| \sum_{m=1}^{\nu} (T_{k\nu+m} - T_{k\nu}) \cdot (\nu e_m e_m^T - I) \right\| \le \nu \sum_{m=1}^{\nu} \|T_{k\nu+m} - T_{k\nu}\| \,. \end{aligned} \tag{ix}$$

Now, let $\varepsilon \in (0, \alpha)$ denote a given arbitrary number. According to (iv) and (viii) there exists an index $k_0 \in \mathbb{N}$ such that

$$\alpha - \varepsilon \leq \alpha_{k\nu+m} \leq \alpha + \varepsilon\,, \tag{x}$$

$$\sum_{m=1}^{\nu} \|T_{k\nu+m} - T_{k\nu}\| \leq \varepsilon\,, \tag{xi}$$

$$\left| p_{m,k} \frac{\alpha_{k\nu+m}}{\alpha} \cdot \frac{k\nu}{k\nu+m} - 1 \right| \leq \varepsilon \tag{xii}$$

for all $k \geq k_0$ and $m = 1, \ldots, \nu$. From (x), for $k \geq k_0$ we find

$$\left(1 - \frac{\alpha+\varepsilon}{k\nu}\right)^{\nu} \leq p_{0,k} \leq \left(1 - \frac{\alpha-\varepsilon}{(k+1)\nu}\right)^{\nu}.$$

Since for each $t \in [0,1]$ the inequality $0 \leq (1-t)^{\nu} - (1-\nu t) \leq \frac{\nu(\nu-1)}{2} t^2$ holds, for indices $k \geq k_0$ we get

$$-\frac{\alpha+\varepsilon}{k} \leq p_{0,k} - 1 \leq \frac{\alpha-\varepsilon}{k+1}(-1 + \frac{\nu-1}{2} \cdot \frac{\alpha-\varepsilon}{(k+1)\nu})$$

or

$$-\frac{\alpha+\varepsilon}{\alpha} + 1 \leq (p_{0,k} - 1)\frac{k}{\alpha} + 1 \leq \frac{k}{\alpha(k+1)}\left(\alpha\left(\frac{k+1}{k} - \frac{\alpha-\varepsilon}{\alpha}\right) + \frac{(\alpha-\varepsilon)^2}{2(k+1)}\right),$$

respectively. Hence, there exists an index $k_1 \geq k_0$ with

$$\left|(p_{0,k} - 1)\frac{k}{\alpha} + 1\right| \leq 2\frac{\varepsilon}{\alpha} \quad \text{for } k \geq k_1\,. \tag{xiii}$$

Using now in (vii) the approximations (ix), (xi), (xii) and (xiii), we finally get for all $k \geq k_1$

$$\begin{aligned} \|\bar{U}_k\| &\leq \frac{1}{\nu}\left[\nu\varepsilon + \varepsilon \sum_{m=1}^{\nu} \|U_{k\nu+m}\|\right] + \frac{2\varepsilon}{\alpha}\|V_{k\nu}\| \\ &\leq \varepsilon \cdot \left[1 + \sup_L \|U_L\| + \frac{2}{\alpha} \sup_L \|V_L\|\right] \quad \text{a.s.}\,. \end{aligned}$$

Since $\varepsilon > 0$ was chosen arbitrarily, relation (**) follows now by means of (iii) and (v).

### 6.6.3 Estimation of Error Variances

Special approach (W1a-e) for search directions $Y_n$ also allows estimation of limit covariance matrices $W^{(1)}, \ldots, W^{(\kappa)}$ of error sequence $(Z_n)$ in procedure (6.2a,b). Again the following is assumed:

$$l_n > 0 \quad \text{für} \quad n = 1, 2, \ldots \,. \tag{W5}$$

Since at stage $n$ the vectors $Y_n^{(1)}, \ldots, Y_n^{(2l_n)}$ are known, also

$$\widehat{Z}_n := \frac{1}{2l_n} \sum_{i=1}^{l_n} (Y_n^{(i)} - Y_n^{(l_n+i)}) \tag{6.84}$$

can be calculated. These random vectors behave similarily to the estimation error $Z_n$. This can be seen by comparing Lemma 6.15 with the following result:

**Lemma 6.17.** *For each $n \in \mathbb{N}$ holds*

*a)* $E_n\widehat{Z}_n = \frac{1}{2l_n} \sum_{i=1}^{l_n} (G_n^{(i)} - G_n^{(l_n+i)} + E_n Z_n^{(i)} - E_n Z_n^{(l_n+i)})\,,$

*b)* $E_n(\widehat{Z}_n - E_n\widehat{Z}_n)(\widehat{Z}_n - E_n\widehat{Z}_n)^T = \frac{1}{(2l_n)^2} \sum_{j=1}^{2l_n} E_n(\Delta Z_n^{(j)} \Delta Z_n^{(j)T})\,,$

*c)* $E_n\|\widehat{Z}_n - E_n\widehat{Z}_n\|^2 = \frac{1}{(2l_n)^2} \sum_{j=1}^{2l_n} E_n\|\Delta Z_n^{(j)}\|^2\,,$

*d)* $\|E_n\widehat{Z}_n\| \longrightarrow 0\,.$

In the following we put $\Delta\widehat{Z}_n := \widehat{Z}_n - E_n\widehat{Z}_n$.

*Proof.* From relation $Y_n^{(j)} = G_n^{(j)} + E_n Z_n^{(j)} + \Delta Z_n^{(j)}$ we obtain representation in a) and the following formula

$$\Delta\widehat{Z}_n = \frac{1}{2l_n} \sum_{i=1}^{l_n} (\Delta Z_n^{(i)} - \Delta Z_n^{(l_n+i)}).$$

Hence, due to the conditional independence of $Z_n^{(1)}, \ldots, Z_n^{(2l_n)}$ we get

$$\begin{aligned}
E_n\Delta\widehat{Z}_n\Delta\widehat{Z}_n^T &= \frac{1}{(2l_n)^2} \sum_{i=1}^{l_n} E_n(\Delta Z_n^{(i)} - \Delta Z_n^{(l_n+i)})(\Delta Z_n^{(i)} - \Delta Z_n^{(l_n+i)})^T \\
&= \frac{1}{(2l_n)^2} \sum_{i=1}^{l_n} (E_n\Delta Z_n^{(i)} \Delta Z_n^{(i)T} + E_n\Delta Z_n^{(l_n+i)} \Delta Z_n^{(l_n+i)T})\,.
\end{aligned}$$

This is proposition b), and by taking the trace we also get c). According to (U1a-c) and (6.3), for $i = 1, \ldots, l_n$ follows

$$\begin{aligned}
\|G_n^{(i)} - G_n^{(l_n+i)}\| &= \left\| 2c_n H_n d_n^{(i)} + \delta(X_n; c_n d_n^{(i)}) - \delta(X_n; -c_n d_n^{(i)}) \right\| \\
&\le 2c_n\beta + 2L_0 c_n^{1+\mu_0}\,.
\end{aligned}$$

Because of (W2a) we have $\left\|\frac{1}{2l_n}\sum_{i=1}^{l_n}(E_nZ_n^{(i)} - E_nZ_n^{(l_n+i)})\right\| \le \frac{L_1}{n^{\mu_1}}$ and therefore by means of a)

$$\|E_n\widehat{Z}_n\| \le c_n(\beta + L_0 c_n^{\mu_0}) + \frac{L_1}{n^{\mu_1}} .$$

The righthand side of this inequality converges to 0, since due to (W2a) and (W3a,b) it holds $c_n \longrightarrow 0$ and $\mu_1 > 0$ .

Now, for the existence of limit covariance matrices $W^{(1)}, \dots, W^{(\kappa)}$ of $(Z_n)$ required by (V6) we assume:

Under assumption "$X_n \longrightarrow x^*$ a.s. " suppose

$$\lim_{n\in\mathbf{N}^{(k)}} E_n \Delta Z_n^{(j_n)} \Delta Z_n^{(j_n)T} = V^{(k)} \quad \text{a.s.} \tag{W6}$$

exist for each $k = 1, \dots, \kappa$ and for each sequence $(j_n)$ with $j_n \in \{0, \dots, 2l_n\}$, $n = 1, 2, \dots$. Here, $V^{(1)}, \dots, V^{(k)}$ are deterministic matrices.

Due to the following lemma, sequence $(Z_n)$ fulfills condition (V6) from Section 6.5.3 with

$$W^{(k)} := \frac{1}{2l^{(k)} + \varepsilon^{(k)}} V^{(k)} \quad \text{for} \quad k = 1, \dots, \kappa . \tag{6.85}$$

**Lemma 6.18.** *If "$X_n \longrightarrow x^*$ f.s.", then for each $k = 1, \dots, \kappa$*

$$a) \lim_{n\in\mathbf{N}^{(k)}} E_n \Delta Z_n \Delta Z_n^T = \frac{1}{2l^{(k)} + \varepsilon^{(k)}} V^{(k)} \quad a.s. ,$$

$$b) \lim_{n\in\mathbf{N}^{(k)}} E_n \widehat{Z}_n \widehat{Z}_n^T = \frac{1}{2l^{(k)}} V^{(k)} \quad a.s. .$$

*Proof.* Due to conditions (W4a,b) and (W6), with Lemma 6.15 b) and Lemma 6.17 b) we have

$$\lim_{n\in\mathbf{N}^{(k)}} E_n \Delta Z_n \Delta Z_n^T = \frac{1}{2l^{(k)} + \varepsilon^{(k)}} V^{(k)} \quad \text{a.s.} ,$$

$$\lim_{n\in\mathbf{N}^{(k)}} E_n \Delta \widehat{Z}_n \Delta \widehat{Z}_n^T = \frac{1}{2l_n^{(k)}} V^{(k)} \quad \text{a.s.}.$$

By means of

$$E_n \Delta\widehat{Z}_n \Delta\widehat{Z}_n^T = E_n \widehat{Z}_n \widehat{Z}_n^T - (E_n\widehat{Z}_n)(E_n\widehat{Z}_n)^T ,$$
$$\|(E_n\widehat{Z}_n)(E_n\widehat{Z}_n)^T\| = \|E_n\widehat{Z}_n\|^2$$

from the second equation and Lemma 6.17 d) we obtain proposition b).

Because of Lemma 6.18 the following *averaging method* for the calculation of matrices $W^{(1)}, \dots, W^{(\kappa)}$ is proposed:

At step $n \in \mathbb{N}^{(k)}$ with unique $k \in \{1, \dots, \kappa\}$, let matrices $W_n^{(1)}, \dots, W_n^{(\kappa)}$ already be calculated. Furthermore, let

$$W_{n+1}^{(k)} := (1 - \frac{1}{n})W_n^{(k)} + \frac{1}{n} \cdot \frac{2l_n}{2l_n + \varepsilon_n} \widehat{Z}_n \widehat{Z}_n^T \tag{6.86a}$$

$$W_{n+1}^{(i)} := W_n^{(i)} \quad \text{for} \quad i \neq k \, , \, i \in \{1, \dots, \kappa\} \, . \tag{6.86b}$$

Here, at the beginning let matrices $W_1^{(1)}, \dots, W_1^{(\kappa)}$ be chosen arbitrarily. Hence, if at stage $n$ index $n$ lies in $\mathbb{N}^{(k)}$ with a $k \in \{1, \dots, \kappa\}$, only matrix $W_n^{(k)}$ of $W_n^{(1)}, \dots, W_n^{(\kappa)}$ is updated.

For the convergence of matrix sequence $(W_n^{(i)})_n$ the following assumption on the fourth moments of sequence $(Z_n^{(j)})$ is needed:

There exists $\mu_2 \in [0, 1)$ and a random variable $L_2 < \infty$ a.s. such that

$$E_n \|\Delta Z_n^{(j)}\|^4 \leq L_2 \, n^{\mu_2} \text{ a.s.} \tag{W7}$$

for all $n \in \mathbb{N}$ and $j = 0, \dots, 2l_n$ .

Hence, initially we get

**Lemma 6.19.** $\sum\limits_n \frac{1}{n^2} E_n \|\widehat{Z}_n \widehat{Z}_n^T - E_n \widehat{Z}_n \widehat{Z}_n^T\|^2 < \infty$ *a.s.* .

*Proof.* For each $n \in \mathbb{N}$ with matrix $C_n := \widehat{Z}_n \widehat{Z}_n^T$ we have

$$\begin{aligned}
E_n \|C_n - E_n C_n\|^2 &\leq E_n \, tr \left( (C_n - E_n C_n)^T (C_n - E_n C_n) \right) \\
&= tr \left( E_n (C_n - E_n C_n)^T (C_n - E_n C_n) \right) \\
&= tr \left( E_n C_n^T C_n - (E_n C_n)^T (E_n C_n) \right) \\
&\leq tr \, (E_n C_n^T C_n) = E_n \, tr \, C_n^T C_n = E_n \, tr \left( \|\widehat{Z}_n\|^2 C_n \right) \\
&= E_n \|\widehat{Z}_n\|^4 = E_n \|\Delta \widehat{Z}_n + E_n \widehat{Z}_n\|^4 \\
&= E_n \left\| \frac{1}{2l_n} \sum_{i=1}^{l_n} (\Delta Z_n^{(i)} - \Delta Z_n^{(l_n+i)}) + E_n \widehat{Z}_n \right\|^4 \\
&\leq E_n \left( \frac{1}{2l_n} \sum_{j=1}^{2l_n} \|\Delta Z_n^{(j)}\| + \|E_n \widehat{Z}_n\| \right)^4 \\
&\leq 8 \left( \frac{1}{2l_n} \sum_{j=1}^{2l_n} E_n \|\Delta Z_n^{(j)}\|^4 + \|E_n \widehat{Z}_n\|^4 \right) \, .
\end{aligned}$$

where "*tr*" denotes the trace of a matrix.

Hence, with (W7) and Lemma 6.17 d) the proposition follows.

Finally, we show that matrices $W_n^{(1)}, \dots, W_n^{(\kappa)}$ are approximations of limit covariance matrices $W^{(1)}, \dots, W^{(\kappa)}$:

**Theorem 6.8.** *If "$X_n \longrightarrow x^*$ a.s.", then for any $k = 1, \dots, \kappa$*

$$\lim_{n \in \mathbb{N}^{(k)}} W_n^{(k)} = \frac{1}{2l^{(k)} + \varepsilon^{(k)}} V^{(k)} = \lim_{n \in \mathbb{N}^{(k)}} E_n \, \Delta Z_n \Delta Z_n^T \quad a.s. \,. \tag{6.87}$$

*Proof.* For given fixed $k \in \{1, \dots, \kappa\}$ let $\mathbb{N}^{(k)} = \{n_1, n_2, \dots\}$ with indices $n_1 < n_2 < \dots$. With abbreviations

$$C_i := \frac{2l_{n_i}}{2l_{n_i} + \varepsilon_{n_i}} \widehat{Z}_{n_i} \widehat{Z}_{n_i}^T \,, \quad T_i := W_{n_i}^{(k)} \,, \quad \mathcal{B}_i := \mathcal{A}_{n_i},$$

by definition we have

$$T_{i+1} = \left(1 - \frac{1}{n_i}\right) T_i + \frac{1}{n_i} C_i \quad \text{for} \quad i = 1, 2, \dots \,.$$

Due to Lemma 6.18 and (W4a,b) it holds

$$\lim_{i \to \infty} E(C_i \mid \mathcal{B}_i) = \frac{1}{2l^{(k)} + \varepsilon^{(k)}} V^{(k)} = \lim_{n \in \mathbb{N}^{(\kappa)}} E_n \, \Delta Z_n \Delta Z_n^T \quad \text{a.s.} \,. \tag{i}$$

Because of (U6) we have $\dfrac{i}{n_i} \longrightarrow q_k > 0$ for $i \longrightarrow \infty$ and therfore

$$\sum_i \frac{1}{n_i} = \infty \,. \tag{ii}$$

Lemma 6.19 yields

$$\sum_i \frac{1}{n_i^2} E\left( \left\| C_i - E(C_i \mid \mathcal{B}_i) \right\|^2 \mid \mathcal{B}_i \right) < \infty \quad \text{a.s.} \,. \tag{iii}$$

Due to (i), (ii) and (iii) Lemma B.6, Appendix, can be applied to sequence $(T_n)$. Hence, we have shown the proposition.

In a similar way approximations $w_n^{(1)}, \dots, w_n^{(\kappa)}$ of the mean quadratic limit errors of sequence $(Z_n)$ can be given:

At step $n \in \mathbb{N}^{(k)}$ with unique $k \in \{1, \dots, \kappa\}$, let

$$w_{n+1}^{(k)} := \left(1 - \frac{1}{n}\right) w_n^{(k)} + \frac{1}{n} \cdot \frac{2l_n}{2l_n + \varepsilon_n} \|\widehat{Z}_n\|^2 \tag{6.88a}$$

$$w_{n+1}^{(i)} := w_n^{(i)} \quad \text{for} \quad i \neq k \,,\, i \in \{1, \dots, \kappa\} \,. \tag{6.88b}$$

Here, initially let arbitrary $w_1^{(1)}, \ldots, w_1^{(\kappa)}$ be given. Then, the following convergence corollary can be shown:

**Corollary 6.5.** *If "$X_n \longrightarrow x^*$ a.s. ", then for any $k = 1, \ldots, \kappa$*

$$\lim_{n \in \mathbb{N}^{(k)}} w_n^{(k)} = \frac{1}{2l^{(k)} + \varepsilon^{(k)}} tr(V^{(k)}) = \lim_{n \in \mathbb{N}^{(k)}} E_n \|\Delta Z_n\|^2 \quad a.s. \, . \tag{6.89}$$

*Proof.* If $w_1^{(k)} = tr(W_1^{(k)})$ for $k = 1, \ldots, \kappa$ , then for each $n \in \mathbb{N}$ and $k = 1, \ldots, \kappa$

$$w_n^{(k)} = tr(W_n^{(k)})$$

with matrix $W_n^{(k)}$ from (6.86a,b). Hence, the proposition follows from Theorem 6.8 by means of taking the trace.

## 6.7 Realization of Adaptive Step Sizes

Using the estimators for the unknown Jacobian $H^* := H(x^*)$ and the limit covariance matrices $W^{(1)}, \ldots, W^{(\kappa)}$ of the sequence of estimation errors $(Z_n)$, developed in Section 6.6, it is possible to select the step size $R_n$ in algorithm (6.2a,b) adaptively such that certain convergence conditions are fulfilled.

For the scalar $\rho_n$ in the step size $R_n = \rho_n M_n$ we suppose, as in Section 6.5,

$$\lim_{n \to \infty} n \, \rho_n = 1 \quad \text{a.s..} \tag{X1}$$

Moreover, let the step directions $(Y_n)$ be defined by relations (W1a,b) in Section 6.6.

Starting with a matrix $H_0$, and putting $H_1^{(0)} := H_0$, $\mathcal{A}_n$-measurable matrices $H_n^{(0)}$ can be determined recursively by (6.81a,b).

Depending on $H_n^{(0)}$, the factor $M_n$ in the step size $R_n$ is selected here such that conditions (U4a-d) from Section 6.3 are fulfilled.

Independent of the convergence behavior of $(X_n)$, for large index $n$ the matrices $H_n^{(0)}$ do not deviate considerably from $\widehat{H}_n^{(0)}$ given by (6.83a,b). This is shown next:

**Lemma 6.20.** $\Delta H_n := H_n^{(0)} - \widehat{H}_n^{(0)} \longrightarrow 0$ *a.s.* .

*Proof.* Because of (X1), the assumption of Lemma 6.8 d) holds. Hence, $(X_{n+1} - X_n)$ converges to zero a.s.. Using Theorem 6.7, the assertion follows.

According to the definition we have $\widehat{H}_n^{(0)} \in \mathcal{H}$, where $\mathcal{H}$ denotes the set of all matrices $H \in \mathbb{M}(\nu \times \nu, \mathbb{R})$ such that

$$H = \sum_{i \in I} \lambda_i H^{(i)} \;, I \text{ is finite}, \quad \sum_{i \in I} \lambda_i = 1 \;, \lambda_i > 0 \text{ and}$$
$$H^{(i)} \in \{H(x) : x \in D\} \cup \{H_0\} \text{ for } i \in I \,.$$

For the starting matrix $H_0$ we assume $\|H_0\| \leq \beta$ with a number $\beta$ according to (6.3). Then, for all $H \in \mathcal{H}$ we get

$$\|H\| \leq \beta \,. \tag{6.90}$$

### 6.7.1 Optimal Matrix Step Sizes

If in (6.2a,b) a true matrix step size is used, then due to condition (U3) from Section 6.3 we suppose

$$x^* \in \overset{\circ}{D} \,. \tag{X2}$$

In case that the covariance matrix of the estimation error $Z_n$ converges towards a unique limit matrix $W$ in the sense of (V5a,b) or (V6), see Section 6.5, according to [106] for

$$M_n := (H^*)^{-1} \quad \text{with} \quad H^* := H(x^*)$$

we obtain the minimum asymptotic covariance matrix of the process $(\sqrt{n}(X_n - x^*))$. Since $H^*$ is unknown in general, this approach is not possible directly.

However, according to Theorem 6.6, the matrix $H_n^{(0)}$ is a known $\mathcal{A}_n$-measurable approximate to $H^*$ at stage $n$, provided that $(X_n)$ converges to $x^*$. Hence, similar to [97], [12] and [98], for $n = 1, 2, \ldots$ we select

$$M_n := \begin{cases} (H_n^{(0)})^{-1} & , \text{ if } H_n^{(0)} \in \mathcal{H}_0 \\ H_0^{-1} & , \text{ else.} \end{cases} \tag{6.91}$$

Here, $\mathcal{H}_0$ is a set of matrices $H \in \mathbb{M}(\nu \times \nu, \mathbb{R})$ with appropriate properties. Concerning $H(x)$, $H_0$ and $\mathcal{H}_0$ we suppose:

a) There is a positive number $\alpha$ such that

$$H(x)H(y)^T + H(y)H(x)^T \geq 2\alpha^2 I \,, \tag{X3a}$$
$$H(x)H_0^T + H_0 H(x)^T \geq 2\alpha^2 I \,, \tag{X3b}$$
$$H_0 H_0^T \geq \alpha^2 I \tag{X3c}$$

for all $x, y \in D$ .

b) There is a positive number $\alpha_0$ with

$$\mathcal{H}_0 \subset \mathcal{H}_\oplus := \left\{ H \in \mathbb{M}(\nu \times \nu, \mathbb{R}) : HH^T \geq \alpha_0^2 I \right\} . \tag{X4}$$

Assumption (X3a-c) is also used in [98] and in [97]. As is shown later, see Lemma 6.22, conditions (X3a-c) and (X4) guarantee that sequence $(M_n)$ given by (6.91) fulfills condition (U4a-d) in Section 6.3.

If $G(x)$ is a gradient function, hence, $G(x) = \nabla F(x)$ with a function $F : D_0 \longrightarrow \mathbb{R}$, then (X3a-c) does not guarantee the convexity of $F$. Conversely, the Hessian $H(x)$ of a strongly convex function $F(x)$ does not fulfill condition (X3a-c) in general. Counter examples can be found in [104].

Some properties of the matrix sets $\mathcal{H}_\oplus$ and $\mathcal{H}$ are described in the following:

**Lemma 6.21.**
*a) For a matrix $H \in \mathbb{M}(\nu \times \nu, \mathbb{R})$ the following properties are equivalent:*

$$i)\ H \in \mathcal{H}_\oplus\,,$$
$$ii)\ H \text{ is regular and } \|H^{-1}\| \le \frac{1}{\alpha_0}\,.$$

*b) For $H \in \mathcal{H}$ and $x \in D$ we have*

$$\begin{aligned}
&i)\ H(x)H^T + HH(x)^T \ge 2\alpha^2 I\,,\\
&ii)\ HH^T \ge \alpha^2 I\,,\\
&iii)\ H \text{ is regular and } \|H^{-1}\| \le \frac{1}{\alpha}\,,\\
&iv)\ H^T H \le \beta^2 I\,,\\
&v)\ (H^{-1})(H^{-1})^T \ge \frac{1}{\beta^2} I\,,\\
&vi)\ H^{-1}H(x) + \left(H^{-1}H(x)\right)^T \ge 2\frac{\alpha^2}{\beta^2} I\,.
\end{aligned}$$

*Proof.* If $H$ is regular, then

$$\begin{aligned}
&(HH^T)^{-1} = (H^{-1})^T H^{-1}\,,\\
&\|H^{-1}\|^2 = \lambda_{\max}\left((H^{-1})^T(H^{-1})\right)\,,\\
&HH^T \ge \alpha_0^2 I \Longleftrightarrow (HH^T)^{-1} \le \frac{1}{\alpha_0^2} I\,.
\end{aligned}$$

This yields the equivalence of i) and ii) in a).

Let now $H = \sum_{i\in I} \lambda_i H^{(i)} \in \mathcal{H}$ and $x \in D$ be given arbitrarily. Because of (X3a-c) we have

$$H(x)H^T + HH(x)^T = \sum_{i\in I} \lambda_i \left(H(x)H^{(i)T} + H^{(i)}H(x)^T\right) \ge 2\alpha^2 I$$

and

$$HH^T = \sum_{i,k \in I} \lambda_i \lambda_k H^{(i)} H^{(k)T} \geq \alpha^2 I \; .$$

Hence, $H$ is regular, and $(H^{-1})^T(H^{-1}) = (HH^T)^{-1} \leq \alpha^{-2} I$. This yields $\|H^{-1}\|^2 = \lambda_{\max}\Big((H^{-1})^T(H^{-1})\Big) \leq \alpha^{-2}$. Due to (6.90) we find $H^T H \leq \lambda_{\max}(H^T H) I = \|H\|^2 I \leq \beta^2 I$ and therefore $\beta^{-2} I \leq (H^T H)^{-1} = (H^{-1})(H^{-1})^T$. Because of

$$H^{-1}H(x) + \Big(H^{-1}H(x)\Big)^T = (H^{-1})\Big(H(x)H^T + HH(x)^T\Big)(H^{-1})^T \, ,$$

assumptions i) and v) of b) finally yield

$$H^{-1}H(x) + \Big(H^{-1}H(x)\Big)^T \geq 2\alpha^2 (H^{-1})(H^{-1})^T \geq 2\frac{\alpha^2}{\beta^2} I \; .$$

Using Lemma 6.21, we obtain, as announced, the following result:

**Lemma 6.22.**

*a)* $\sup_{n \in \mathbb{N}} \|M_n\| \leq \max\{\alpha_0^{-1} \, , \, \alpha^{-1}\}$ .

*b) For each $n \in \mathbb{N}$ there are $\mathcal{A}_n$-measurable matrices $\widehat{M}_n$ and $\Delta M_n$ such that*

*i)* $M_n = \widehat{M}_n + \Delta M_n$ ,

*ii)* $\lim_{n \to \infty} \|\Delta M_n\| = 0 \quad a.s.$ ,

*iii)* $u^T \, \widehat{M}_n H(x) \, u \geq \frac{\alpha^2}{\beta^2} \|u\|^2$ *for* $x \in D$ *and* $u \in \mathbb{R}^\nu$ .

*Proof.* Since $\mathcal{H}_0 \subset \mathcal{H}_\oplus$ and $H_0 \in \mathcal{H}$, Lemma 6.21 a) ii) and b) iii) for $n = 1, 2, \ldots$ yields

$$\|M_n\| \leq \begin{cases} {\alpha_0}^{-1} & , \quad \text{if } H_n^{(0)} \in \mathcal{H}_0 \\ \alpha^{-1} & , \quad \text{else} \end{cases} \, ,$$

hence the first assertion. Furthermore, according to Lemma 6.20 it is

$$\Delta H_n := H_n^{(0)} - \widehat{H}_n^{(0)} \longrightarrow 0 \quad \text{a.s.} \, , \tag{i}$$

with matrix $\widehat{H}_n^{(0)} \in \mathcal{H}$ given by (6.83a,b). Because of $\widehat{H}_n^{(0)} \in \mathcal{H}$ for each $n \in \mathbb{N}$, Lemma 6.21 b) iii) yields

$$\widehat{H}_n^{(0)} \text{ is regular and } \left\| (\widehat{H}_n^{(0)})^{-1} \right\| \leq \frac{1}{\alpha} \; . \tag{ii}$$

Define now

$$\widehat{M}_n := \begin{cases} (\widehat{H}_n^{(0)})^{-1} & , \text{ if } H_n^{(0)} \in \mathcal{H}_0 \\ H_0^{-1} & , \text{ else,} \end{cases} \tag{6.92}$$

and put $\Delta M_n := M_n - \widehat{M}_n$. Because of (i) and (ii), the assumptions of Lemma C.2, Appendix, hold for $A_n := \widehat{H}_n^{(0)}$ and $B_n := H_n^{(0)}$. Hence,

$$\Delta M_n \longrightarrow 0 \quad \text{a.s. .}$$

Furthermore, according to Lemma 6.21 b) vi), and because of $\widehat{H}_n^{(0)}$ , $H_0 \in \mathcal{H}$ we get

$$u^T \widehat{M}_n H(x)\, u \geq \frac{\alpha^2}{\beta^2}\|u\|^2$$

for all $n \in \mathbb{N}$ , $x \in D$ and $u \in \mathbb{R}^\nu$ .

Due to Lemma 6.22, sequence $(M_n)$ fulfills condition (U4a-d) from Section 6.3. Consequently, we have now this result:

**Theorem 6.9.**

*a)* $X_n \longrightarrow x^*$ *a.s.* ,

*b)* $H_n^{(0)}$ , $\widehat{H}_n^{(0)} \longrightarrow H^* := H(x^*)$ *a.s.* .

*Proof.* The first assertion follows immediately from Corollary 6.1. The second assertion follows then from a), Theorem 6.6 b) and Theorem 6.7.

Supposing still

$$H^* \in \overset{\circ}{\mathcal{H}}_0 := \text{ interior of } \mathcal{H}_0 , \tag{X5}$$

we get the main result of this section:

**Theorem 6.10.**

*a)* $M_n \longrightarrow (H^*)^{-1}$ *a.s.* .

*b)* $n^\lambda(X_n - x^*) \longrightarrow 0$ *a.s. for each* $\lambda \in \left[0, \frac{1}{2}\right)$.

*c) For each* $\epsilon_0 > 0$ *there is a set* $\Omega_0 \in \mathcal{A}$ *such that*

*i)* $\mathcal{P}(\bar{\Omega}_0) < \epsilon_0$ ,

*ii)* $\limsup_{n\in\mathbb{N}} E_{\Omega_0}\, n\|X_n - x^*\|^2 \leq \left\|(H^*)^{-1}\right\|^2 \sum_{k=1}^{\kappa} q_k \sigma_k^2$ ,

*where* $\sigma_k^2 := \limsup_{n\in\mathbb{N}^{(k)}} E\, \sigma_{0,n}^2$ *for* $k = 1, \ldots, n$ .

d) *If the covariance matrices of* $(Z_n)$ *fulfill (V6) and (V8) in Section 6.5.3, then* $\left(\sqrt{n}(X_n - x^*)\right)$ *has an asymptotic normal distribution with the asymptotic covariance matrix given by*

$$V = (H^*)^{-1}\Big(\sum_{k=1}^{\kappa} q_k W^{(k)}\Big)\left((H^*)^{-1}\right)^T.$$

*Proof.* According to Theorem 6.9 b) and assumption (X5) there is a (stochastic) index $n_0 \in \mathbb{N}$ such that $H_n^{(0)} \in \mathcal{H}_0$ for all $n \geq n_0$ a.s.. Hence, according to the definition, for $n \geq n_0$ we get

$$M_n = (H_n^{(0)})^{-1} \quad \text{and} \quad \widehat{M}_n = (\widehat{H}_n^{(0)})^{-1} \quad \text{a.s.}\,.$$

Theorem 6.9 b) yields then

$$M_n\,,\ \widehat{M}_n \longrightarrow \Theta := (H^*)^{-1} \quad \text{a.s.}\,. \tag{i}$$

Because of Lemma 6.9 and (i) we suppose that

$$\hat{a} := \liminf_{n\in\mathbb{N}} \hat{a}_n \geq a = 1 \tag{ii}$$

for sequence $(\hat{a}_n)$ arising in (6.7). Assertion b) follows by applying Theorem 6.1. Moreover, assertion c) follows from (i), (ii) and Theorem 6.2. Because of (i), assertion d) is obtained directly from Corollary 6.4.

### Practical computation of $(H_n^{(0)})^{-1}$

In formula (6.91) defining matrices $M_n$ we have to compute $(H_n^{(0)})^{-1}$. According to (6.81a,b), matrix $H_{n+1}^{(0)} := H_n^{(l_n)}$ follows from $H_n^{(0)}$ by an $l_n$-fold rank-1-update. Also the inverse of $H_n^{(0)}$ can be obtain by rank-1-updates.

**Lemma 6.23.** *Let* $n \in \mathbb{N}$ *and* $i \in \{1, \ldots, l_n\}$ *with a regular* $H_n^{(i-1)}$. *Then*

a) $\det(H_n^{(i)}) = \det(H_n^{(i-1)})\dfrac{\delta_n^{(i)}}{n}$ *with* $\delta_n^{(i)} := n - 1 + d_n^{(i)T}(H_n^{(i-1)})^{-1}h_n^{(i)}$ .

b) $H_n^{(i)}$ *is regular if and only if* $\delta_n^{(i)} \neq 0$. *In this case*
$$(H_n^{(i)})^{-1} = (H_n^{(i-1)})^{-1} + \frac{1}{\delta_n^{(i)}}\left(d_n^{(i)} - (H_n^{(i-1)})^{-1}h_n^{(i)}\right)d_n^{(i)T}(H_n^{(i-1)})^{-1}\,.$$

*Proof.* The assertion follows directly from Lemma C.3, Appendix, if there we put

$$d := d_n^{(i)}\,,\quad e := \frac{1}{n}(h_n^{(i)} - H_n^{(i-1)}d_n^{(i)})$$

and take into account that $\|d_n^{(i)}\|^2 = 1$ .

Of course, at stage $n$ matrix $(H_{n+1}^{(0)})^{-1}$ follows from $(H_n^{(0)})^{-1}$ at minimum expenses for $l_n = 1$. Indeed, according to (6.81a,b) we have

$$H_{n+1}^{(0)} = H_n^{(0)} + \frac{1}{n}(h_n^{(1)} - H_n^{(0)}d_n^{(1)})d_n^{(1)T}.$$

## Choice of set $\mathcal{H}_0$

We still have to clarify how to select the set $\mathcal{H}_0$ occuring in (6.91) such that conditions (X4) and (X5) are fulfilled. Theoretically, $\mathcal{H}_0 := \mathcal{H}_\oplus$ could be taken. However, this is not possible since the minimum eigenvalue of a matrix $HH^T$ cannot be computed exactly in general.

Hence, for the choice of $\mathcal{H}_0$, two practicable cases are presented:

*Example 6.1.* Let denote $\|\cdot\|_0$ any simple matrix norm with $\|I\|_0 = 1$. Furthermore, define

$$\mathcal{H}_0 := \left\{H \in \mathbb{M}(\nu \times \nu, \mathbb{R}) : \det(H) \neq 0 \quad \text{and} \quad \|H^{-1}\|_0 \leq b_0\right\} ,$$

where $b_0$ is a positive number such that

$$\left\|(H^*)^{-1}\right\|_0 < b_0 \, . \tag{6.93}$$

For this set $\mathcal{H}_0$ we have the following result:

**Lemma 6.24.**

*a)* $\mathcal{H}_0 \subset \mathcal{H}_\oplus$ *with* $\alpha_0 := \dfrac{1}{\kappa_0 b_0}$ *and* $\kappa_0$ *such that* $\|A\| \leq \kappa_0 \|A\|_0$ *for each matrix* $A \in \mathbb{M}(\nu \times \nu, \mathbb{R})$ .

*b)* $H^* \in \mathring{\mathcal{H}}_0$ .

*Proof.* Each matrix $H \in \mathcal{H}_0$ is regular, and according to the above assumption we have

$$\|H^{-1}\| \leq \kappa_0 \|H^{-1}\|_0 \leq \kappa_0 b_0 =: \frac{1}{\alpha_0} \, .$$

Thus, Lemma 6.21 a) yields $H \in \mathcal{H}_\oplus$ . Defining $a_0 := \left\|(H^*)^{-1}\right\|_0$ und $\epsilon_0 := \frac{1}{b_0}(1 - \frac{a_0}{b_0}) > 0$, formula (6.93) yields

$$0 < \frac{a_0}{b_0} \leq 1 - a_0\epsilon_0 \quad \text{and} \quad \epsilon_0 < \frac{1}{a_0} \, .$$

If $\|H - H^*\|_0 \leq \epsilon_0$, then according to Lemma C.1, Appendix, we know that $H$ is regular and

$$\|H^{-1}\|_0 \leq \frac{a_0}{1 - a_0\epsilon_0} \leq b_0 \, .$$

Hence, $\left\{H\colon \|H - H^*\|_0 \leq \epsilon_0\right\} \subset \mathcal{H}_0$ , or $H^* \in \mathring{\mathcal{H}}_0$.

In the above, the following norms $\|\cdot\|_0$ can be selected :

a) $\|A\|_0 := \max_{i,k\le\nu} | a_{i,k} |$ with $\kappa_0 := \nu$ ,

b) $\|A\|_0 := \sqrt{\frac{1}{\nu} tr(A^T A)} = \sqrt{\frac{1}{\nu}\sum_{i,k\le\nu} a_{i,k}^2}$ , where $\kappa_0 := \sqrt{\nu}$ .

By means of Lemma 6.23 the determinant and the inverse of $H_n^{(0)}$ can be computed easily. Hence, the condition "$H_n^{(0)} \in \mathcal{H}_0$" arising in (6.91) can be verified easily.

*Example 6.2.* For a given positive number $\beta_0$ with

$$\beta_0 < \frac{(\det(H^*))^2}{(tr(H^* H^{*T}))^{\nu-1}}, \tag{6.94}$$

define $\mathcal{H}_0 := \left\{ H \in \mathbb{M}(\nu\times\nu, \mathbb{R}) : \frac{(\det(H))^2}{(tr(HH^T))^{\nu-1}} \ge \beta_0 \right\}$. According to Lemma 6.23 we easily can decide whether $H_n^{(0)} \in \mathcal{H}_0$ is fulfilled. Also in this case $\mathcal{H}_0$ fulfills conditions (X4) and (X5):

**Lemma 6.25.** *If $\nu > 1$, then*

*a)* $\mathcal{H}_0 \subset \mathcal{H}_\oplus$ *with* $\alpha_0 := \sqrt{\beta_0(\nu-1)^{\nu-1}} > 0$ ,

*b)* $H^* \in \overset{\circ}{\mathcal{H}_0}$ .

*Proof.* If for given matrix $H \in \mathcal{H}_0$ Lemma C.4, Appendix, is applied to $A := HH^T$ , then

$$HH^T \ge \det(HH^T)\left(\frac{\nu-1}{tr(HH^T)}\right)^{\nu-1} I \ge \beta_0(\nu-1)^{\nu-1} I =: \alpha_0^2 I ,$$

hence, $H \in \mathcal{H}_\oplus$ . Since $\det(\cdot)$ and $tr(\cdot)$ are continuous functions on $\mathbb{M}(\nu\times\nu, \mathbb{R})$, because of (6.94) there is a positive number $\epsilon_0$ such that

$$\frac{(\det(H))^2}{(tr(HH^T))^{\nu-1}} \ge \beta_0$$

for each matrix $H$ with $\|H - H^*\| \le \epsilon_0$. Consequently, also in the present case we have $H^* \in \overset{\circ}{\mathcal{H}_0}$ .

### 6.7.2 Adaptive Scalar Step Size

We consider now algorithm (6.2a,b) with a scalar step size $R_n = r_n I = \rho_n M_n$ where $r_n \in \mathbb{R}_+$ , and $\rho_n$ fulfills (X1). Moreover, we suppose that $M_n$ is defined by

$$M_n = \frac{\pi_n}{\alpha_n^*} I \quad \text{for} \quad n = 1, 2, \ldots . \tag{6.95}$$

Here, $\pi_n$ and $\alpha_n^*$ are positive $\mathcal{A}_n$-measurable random variables such that $\pi_n$ controls the variance of the estimation error $Z_n$, and $\alpha_n^*$ takes into account the influence of the Jacobian $H(x)$. In the following $\pi_n$ and $\alpha_n$are selected such that the optimal limit condition (5.83) is fulfilled.

In this section we first assume that the Jacobian $H(x)$ and the random numbers $\pi_n$ and $\alpha_n^*$ fulfill the following conditions:

a) Suppose that there is a positive number $\alpha$ such that

$$H(x) + H(x)^T \geq 2\alpha I \quad \text{for} \quad x \in D . \tag{X6}$$

b) There are positive numbers $\underline{d}$ and $\bar{d}$ with

$$\underline{d} \leq \frac{\pi_n}{\alpha_n^*} \leq \bar{d} \quad \text{a.s.} \quad \text{for all } n = 1, 2, \ldots . \tag{X7}$$

Corresponding to Lemma 6.22, here we have this result:

**Lemma 6.26.**

*i)* $\sup_n \|M_n\| \leq \bar{d}$,

*ii)* $u^T \, M_n H(x) \, u \geq \underline{d}\alpha \|u\|^2$ *for each* $x \in D$ , $u \in \mathbb{R}^\nu$ *and* $n \in \mathbb{N}$ .

*Proof.* According to the definition of $M_n$, the assertions i), ii) follow immediately from the above assumptions a), b).

Put now $\widehat{M_n} :\equiv M_n$ and $\Delta M_n :\equiv 0$ . According to Lemma 6.26, also in the present case condition (U4a-d) from Section 5.3 is fulfilled. Thus, corresponding to the last section, cf. Theorem 6.9, we obtain the following result:

**Theorem 6.11.**

*a)* $X_n \longrightarrow x^*$ *a.s.* ,

*b)* $H_n^{(0)} \longrightarrow H^*$ *a.s.* .

Further convergence results are obtained under additional assumptions on the random parameters $\pi_n$ and $\alpha_n^*$:

c) There is a positive number $\widehat{\alpha}$ with

$$\lim_{n \to \infty} \alpha_n^* = \widehat{\alpha} \quad \text{a.s.} , \tag{X8}$$

$$\widehat{\alpha} < 2\alpha^* := \lambda_{min}(H^* + H^{*T}) . \tag{X9}$$

d) There are positive numbers $\pi^{(1)}, \ldots, \pi^{(\kappa)}$ such that

$$\lim_{n \in \mathbb{N}^{(k)}} \pi_n = \pi^{(k)} \text{ a.s., } k = 1, \ldots, \kappa , \tag{X10a}$$

$$q_1 \pi^{(1)} + \ldots + q_\kappa \pi^{(\kappa)} = 1 . \tag{X10b}$$

Because of c) and d), for $k = 1, \ldots, \kappa$ we have

$$\lim_{n \in \mathbb{N}^{(k)}} \frac{\pi_n}{\alpha_n^*} = \frac{\pi^{(k)}}{\widehat{\alpha}} =: d^{(k)} \quad \text{a.s.} , \tag{6.96a}$$

$$d := q_1 d^{(1)} + \ldots + q_\kappa d^{(\kappa)} = \frac{1}{\widehat{\alpha}} , \tag{6.96b}$$

$$d\alpha^* > \frac{1}{2} , \tag{6.96c}$$

especially

$$\lim_{n \in \mathbb{N}^{(k)}} M_n = d^{(k)} I \quad \text{a.s.} . \tag{6.97}$$

Results on the rate of convergence and the convergence in the quadratic mean of $(X_n)$ are obtained now:

**Theorem 6.12.**
*a) If $\lambda \in [0, \frac{1}{2})$, then $n^\lambda (X_n - x^*) \longrightarrow 0$ a.s. .*
*b) For $\epsilon_0 > 0$ there is a set $\Omega_0 \in \mathcal{A}$ such that*

*i)* $P(\bar{\Omega}_0) < \epsilon_0$ ,

*ii)* $$\limsup_{n \in \mathbb{N}} E_{\Omega_0}\, n \|X_n - x^*\|^2 \le \frac{\sum_{k=1}^{\kappa} q_k (\pi^{(k)} \sigma_k)^2}{\widehat{\alpha}(2\alpha^* - \widehat{\alpha})} , \tag{6.98}$$

*where $\sigma_1, \ldots, \sigma_\kappa$ are selected as in Theorem 6.10 c).*

*Proof.* Because of (6.97) and Lemma 6.9 we may assume that

$$\widehat{a}^{(k)} \ge d^{(k)} \alpha^*, \quad k = 1, \ldots, \kappa.$$

Hence, due to (6.96b,c) we have $\widehat{a} \ge d\alpha^* > \frac{1}{2}$ . The first assertion follows then from Theorem 6.1. Because of (6.96a) and (6.97), assertion b) is obtained from Corollary 6.3.

As already mentioned in Section 6.5.2, the right hand side of (6.98) takes a minimum for

$$\widehat{\alpha} = \alpha^* , \tag{6.99a}$$

$$\pi^{(k)} = \left(\sigma_k^2 \left(\frac{q_1}{\sigma_1^2} + \ldots + \frac{q_\kappa}{\sigma_\kappa^2}\right)\right)^{-1}, \quad k = 1, \ldots, \kappa . \tag{6.99b}$$

Corresponding to Theorem 6.10 d), also in the present case we have a central limit theorem:

**Theorem 6.13.** *In addition to the above assumptions, suppose*

*i)* $x^* \in \mathring{D}$ ,

*ii) the covariance matrices of $(Z_n)$ fulfill conditions (V6) and (V8) in Section 6.5.3.*

*Then $(\sqrt{n}(X_n - x^*))$ is asymptotically $N(0,V)$-distributed, where the matrix $V$ is determined by the equation*

$$\frac{1}{\widehat{\alpha}}(H^*V + VH^{*T}) - V = \frac{1}{\widehat{\alpha}^2}\sum_{k=1}^{\kappa} q_k(\pi^{(k)})^2 W^{(k)}. \tag{6.100}$$

*Proof.* For $k = 1, \ldots, \kappa$ we have $d^{(k)}(H^* + H^*)^T \geq 2(d^{(k)}\alpha^*)I$, and because of conditions (6.96b,c) we obtain

$$a := q_1(d^{(1)}\alpha^*) + \ldots + q_\kappa(d^{(\kappa)}\alpha^*) = d\alpha^* > \frac{1}{2}\,.$$

Hence, according to (6.96a) and (6.97), conditions (V2) and (V3a,b) in Section 6.5 hold. The assertion follows now from Theorem 6.4.

### Choice of sequence $(\pi_n)$

Of course, the standard definition

$$\pi_n :\equiv 1, \quad n = 1, 2, \ldots, \tag{6.101}$$

fulfills condition (X10), but does not yield the minimum asymptotic bound of $(E_{\Omega_0}\, n\|X_n - x^*\|^2)$.

Using the assumptions of Section 6.6.3 concerning the estimation error $(Z_n)$, the $\mathcal{A}_n$-measurable random variables $w_n^{(1)}, \ldots, w_n^{(\kappa)}$, given by (6.88a,b), are approximations of $\sigma_1^2, \ldots, \sigma_\kappa^2$. Indeed, according to Corollary 6.5 we have

$$\lim_{n\in\mathbb{N}^{(k)}} w_n^{(k)} = \lim_{n\in\mathbb{N}^{(k)}} E_n\|\Delta Z_n\|^2 \text{ a.s.}, \quad k = 1, \ldots, \kappa\,. \tag{6.102}$$

For given positive numbers $\underline{w}$ and $\bar{w}$ with

$$\underline{w} \leq \min_{k\leq\kappa} \sigma_k^2 \leq \max_{k\leq\kappa} \sigma_k^2 \leq \bar{w}, \tag{X11}$$

for $n \in \mathbb{N}$ and $k = 1, \ldots, \kappa$ we define

$$\widehat{w}_n^{(k)} := \begin{cases} w_n^{(k)} & , \quad \text{for} \quad w_n^{(k)} \in [\underline{w}, \bar{w}] \\ \underline{w} & , \quad \text{for} \quad w_n^{(k)} < \underline{w} \\ \bar{w} & , \quad \text{for} \quad w_n^{(k)} > \bar{w} \end{cases}\,. \tag{6.103}$$

Then, for indices $n \in \mathbb{N}^{(k)}$ with $k \in \{1, \ldots, \kappa\}$ we consider

$$\pi_n := \left(\widehat{w}_n^{(k)}\left(\frac{q_1}{\widehat{w}_n^{(1)}} + \ldots + \frac{q_\kappa}{\widehat{w}_n^{(\kappa)}}\right)\right)^{-1}. \tag{6.104}$$

Because of (6.102) and (X11), for each $n \in \mathbb{N}$ and $k = 1, \ldots, \kappa$ we find

$$\frac{\underline{w}}{\bar{w}} \leq \pi_n \leq \frac{\bar{w}}{\underline{w}}, \tag{6.105a}$$

$$\lim_{n \in \mathbb{N}^{(k)}} \pi_n = \left( \sigma_k^2 \left( \frac{q_1}{\sigma_1^2} + \ldots + \frac{q_\kappa}{\sigma_\kappa^2} \right) \right)^{-1} \quad \text{a.s.} \,. \tag{6.105b}$$

Hence, condition (X10a,b) holds, and because of the optimality condition (6.99b), this choice of $(\pi_n)$ yields the minimum asymptotic bound for $E_{\Omega_0}\, n\|X_n - x^*\|^2$.

### Choice of sequence $(\alpha_n^*)$

Obviously, at stage $n$ the factor $\alpha_n^*$ in the step size (6.95) may be chosen as a function of $H_n^{(0)}$. Hence, for $n = 1, 2, \ldots$ we put

$$\alpha_n^* := \begin{cases} \vartheta_n(H_n^{(0)}) &, \ \text{if } H_n^{(0)} \in \mathcal{H}_1 \text{ and } \vartheta_n(H_n^{(0)}) \in [\underline{\vartheta}\,, \bar{\vartheta}] \\ \vartheta_0 &, \ \text{else} \end{cases} \tag{6.106}$$

Here, $\mathcal{H}_1$ is an appropriate subset of $\mathbb{M}(\nu \times \nu, \mathbb{R})$, $\vartheta_n(H)$ is a real value function on $\mathcal{H}_1$, and $\vartheta_0$, $\underline{\vartheta}$, $\bar{\vartheta}$ denote positive constants with $\vartheta_0 \in [\underline{\vartheta}\,, \bar{\vartheta}]$.

If sequence $(\pi_n)$ is chosen according to (6.104), then (X7) holds with

$$\underline{d} := \frac{\underline{w}}{\bar{w} \cdot \bar{\vartheta}} \quad \text{and} \quad \bar{d} := \frac{\bar{w}}{\underline{w} \cdot \underline{\vartheta}}\,. \tag{6.107}$$

Consequently, without any further assumptions, the convergence properties according to Theorem 6.11 hold. Now we still require the following:

e) Suppose

$$H^* \in \overset{\circ}{\mathcal{H}_1} \ = \text{interior of } \mathcal{H}_1\,. \tag{X12}$$

f) In case "$H_n^{(0)} \longrightarrow H^*$ a.s." we assume that there is number $\widehat{\alpha}$ such that

$$\underline{\vartheta} < \widehat{\alpha} < \bar{\vartheta}\,, \tag{X13}$$

$$\vartheta_n(H_n^{(0)}) \longrightarrow \widehat{\alpha} \quad \text{a.s.} \,. \tag{X14}$$

In the important case (see the following examples)

$$\vartheta_n(H) \equiv \vartheta(H) \quad \text{is continuous on} \quad \mathcal{H}_1,$$

condition (X14) holds with $\widehat{\alpha} = \vartheta(H^*)$. In the following we show that (X12), (X13) and (X14) imply condition (X8):

**Lemma 6.27.** $\lim\limits_{n \to \infty} \alpha_n^* = \widehat{\alpha}$ *a.s.* .

*Proof.* According to Theorem 6.11b)

$$H_n^{(0)} \longrightarrow H^* \quad \text{a.s.} \,. \tag{i}$$

Thus, because of (X12), (X13) and (X14) there is an index $n_0$ such that

$$H_n^{(0)} \in \mathcal{H}_1 \quad \text{and} \quad \vartheta_n(H_n^{(0)}) \in [\underline{\vartheta}\,, \bar{\vartheta}]$$

for all $n \geq n_0$ a.s.. By definition for $n \geq n_0$ we have

$$\alpha_n^* = \vartheta_n(H_n^{(0)}) \quad \text{a.s.} \,.$$

Hence, because of (i) and (X14) we get $\lim\limits_{n\to\infty} \alpha_n^* = \widehat{\alpha}$ a.s..

## Examples for $\vartheta_n(H)$

We consider now several examples for the function $\vartheta_n(H)$ to be needed in (6.106) fulfilling conditions (X14) and (X9). In these cases the convergence properties of Theorem 6.12 and 6.13 hold.

According to (X6) we have

$$H^* \in \overset{\circ}{\mathcal{H}}_+ \,, \tag{6.108}$$

provided that $\mathcal{H}_+ := \{H \in \mathbb{M}(\nu \times \nu, \mathbb{R}) : H + H^T > 0\}$. In the following examples we take $\mathcal{H}_1 := \mathcal{H}_+$. Hence, condition (X12) holds.

*Example 6.3.* The function

$$\vartheta^{(1)}(H) := \frac{1}{2}\lambda_{\min}(H + H^T)$$

is continuous on $\mathcal{H}_+$ and $\vartheta^{(1)}(H^*) = \alpha^*$. Therefore, $\vartheta_n(H) :\equiv \vartheta^{(1)}(H)$ is an appropriate choice still guaranteeing the optimality condition (6.99a).

*Example 6.4.* $\vartheta^{(2)}(H) := (tr((H+H^T)^{-1}))^{-1}$ is also continuous on $\mathcal{H}_+$, where

$$\vartheta^{(2)}(H) < \frac{1}{\lambda_{\max}((H + H^T)^{-1})} = \lambda_{\min}(H + H^T)$$

for each matrix $H \in \mathcal{H}_+$, especially $\vartheta^{(2)}(H^*) < 2\alpha^*$. Hence, also $\vartheta_n(H) :\equiv \vartheta^{(2)}(H)$ is a possible approach.

*Example 6.5.* According to C.4, Appendix, on $\mathcal{H}_+$ the inequality $\vartheta^{(3)}(H) < \lambda_{\min}(H + H^T)$ holds with the continuous function

$$\vartheta^{(3)}(H) := \det(H + H^T) \cdot \left(\frac{\nu - 1}{2tr(H)}\right)^{\nu-1} .$$

Thus, also $\vartheta_n(H) :\equiv \vartheta^{(3)}(H)$ is possible.

*Example 6.6.* For given $\mathcal{A}_n$-measurable random vectors $u_n$ from $\mathbb{R}^\nu$ such that $\|u_n\| = 1, n = 1, 2, \ldots$, for $H \in \mathcal{H}_+$ we define

$$\vartheta_n^{(4)}(H) := \left(2u_n^T(H + H^T)^{-1}u_n\right)^{-1} .$$

If $u_n$ is an approximate eigenvector to the maximum eigenvalue $(2\alpha^*)^{-1}$ of the matrix $(H^* + H^{*T})^{-1}$, then $\vartheta_n^{(4)}(H^*)$ is an approximation of $\alpha^*$ .

Since $H^* \in \overset{\circ}{\mathcal{H}}_+$, according to Theorem 6.11 there is a stochastic index $n_0$ such that

$$H_n^{(0)} \in \mathcal{H}_+, \quad n \geq n_0 \quad \text{a.s.} , \tag{6.109}$$

and

$$(H_n^{(0)} + H_n^{(0)T})^{-1} \longrightarrow (H^* + H^{*T})^{-1} \quad \text{a.s.} . \tag{6.110}$$

Regarding to the v. Mises-algorithm for the computation of an eigenvector related to the maximum eigenvalue of $(H^* + H^{*T})^{-1}$, the sequence of vectors $(u_n)$ is defined recursively as follows (see Appendix C.2):

a) Select a vector $0 \neq v_0 \in \mathbb{R}^\nu$ and put $u_1 := \|v_0\|^{-1}v_0$.
b) For a given $\mathcal{A}_n$-measurable vector $u_n$ at stage $n$, define

$$v_n := \begin{cases} (H_n^{(0)} + H_n^{(0)T})^{-1}u_n \text{ , if } H_n^{(0)} \in \mathcal{H}_+ \\ v_0 \text{ , else} \end{cases} , \; u_{n+1} := \frac{v_n}{\|v_n\|} .$$

From (6.109), with probability one for $n \geq n_0$ we get

$$v_n = (H_n^{(0)} + H_n^{(0)T})^{-1}u_n \, , \; \vartheta_n^{(4)}(H_n^{(0)}) = \frac{1}{2 < u_n, v_n >} \, , \; u_{n+1} = \frac{v_n}{\|v_n\|} .$$

If the matrices $A := (H^* + H^{*T})^{-1}$ and $A_n := (H_n^{(0)} + H_n^{(0)T})^{-1}$ fulfill a.s. the assumptions in Lemma C.5 b), cf. Appendix, then due to (6.110) and Lemma C.5 sequence $\vartheta_n^{(4)}(H_n^{(0)})$ converges a.s. towards

$$\frac{1}{2\lambda_{\max}\left((H^* + H^{*T})^{-1}\right)} = \alpha^*.$$

Therefore, $\vartheta_n(H) :\equiv \vartheta_n^{(4)}(H)$ fulfills (X14) with $\hat{\alpha} = \alpha^*$. Hence, also (X9) holds in this case.

## Some remarks on the computation of $\vartheta_n(H)$

If sequence $(\alpha_n^*)$ in (6.95) is chosen according to the above examples, in each iteration step condition

$$H_n^{(0)} \in \mathcal{H}_+ \Longleftrightarrow H_n^{(0)} + H_n^{(0)T} > 0$$

must be verified. This holds if and only if (see [129]) the matrix $H_n^{(0)} + H_n^{(0)T}$ has a Cholesky-decomposition. Hence, there is a matrix $S_n$ such that

$$H_n^{(0)} + H_n^{(0)T} = S_n S_n^T \,, \tag{6.111a}$$

$$S_n = \begin{pmatrix} s_{11}^{(n)} & & 0 \\ & \ddots & \\ * & & s_{\nu\nu}^{(n)} \end{pmatrix} , \tag{6.111b}$$

$$s_{11}^{(n)}, \ldots, s_{\nu\nu}^{(n)} > 0 \,. \tag{6.111c}$$

In [129] an algorithm for the decomposition (6.111a-c) is presented. If this algorithm stops without result, then $H_n^{(0)} \notin \mathcal{H}_+$.

The decomposition (6.111a-c) is also useful for the computation of $\vartheta_n(H_n^{(0)})$. This is shown in the following lemma:

**Lemma 6.28.** *For $H_n^{(0)} \in \mathcal{H}_+$, let $S_n$ be given by (6.111b). Then*

*a)* $\det(H_n^{(0)} + H_n^{(0)T}) = (s_{11}^{(n)} \cdot \ldots \cdot s_{\nu\nu}^{(n)})^2$ .

*b)* $(H_n^{(0)} + H_n^{(0)T})^{-1} = (S_n^{-1})^T S_n^{-1}$ , *where*

$$S_n^{-1} = \begin{pmatrix} t_{11}^{(n)} & & 0 \\ & \ddots & \\ * & & t_{\nu\nu}^{(n)} \end{pmatrix} .$$

*c)* $tr((H_n^{(0)} + H_n^{(0)T})^{-1}) = \sum_{i \geq j} (t_{ij}^{(n)})^2$ .

*d) For given $u_n \in \mathbf{R}^\nu$ determine $v_n := (H_n^{(0)} + H_n^{(0)T})^{-1} u_n$ as follows:*

*i) find $w_n$ with* $\quad S_n w_n = u_n$ ,

*ii) determine $v_n$ such that* $\quad S_n^T v_n = w_n$ .

### Symmetric $H^*$

If (e.g. in case of a gradient $G(x) = \nabla F(x)$):

$$H^* := H(x^*) \quad \text{is symmetric,} \tag{X15}$$

then

$$\alpha^* := \frac{1}{2}\lambda_{\min}(H^* + H^{*T}) = \lambda_{\min}(H^*) \,. \tag{6.112}$$

Hence, in the present case the function $\vartheta_n(H)$ in (6.106) can be chosen to depend only on $H$. In addition, conditions (X14) and (X9) can be fulfilled.

Due to (X6) we have

$$H^* \in \mathring{\mathcal{H}}_{++}, \tag{6.113}$$

provided that

$$\mathcal{H}_{++} := \{H \in \mathbb{M}(\nu \times \nu, \mathbb{R}) : \det(H) > 0 \text{ and } tr(H) > 0\}.$$

In the following examples we take $\mathcal{H}_1 := \mathcal{H}_{++}$, hence, condition (X12) holds.

*Example 6.7.* (see Example 6.4). The function

$$\vartheta^{(5)}(H) := \frac{2}{tr(H^{-1})}$$

is continuous on $\mathcal{H}_{++}$ and $\vartheta^{(5)}(H^*) < \frac{2}{\lambda_{\max}((H^*)^{-1})} = 2\alpha^*$. Thus, $\vartheta_n(H) :\equiv \vartheta^{(5)}(H)$ is a suitable function.

*Example 6.8.* (see Example 6.5)

$$\vartheta^{(6)}(H) := 2\det(H) \cdot (\frac{\nu - 1}{tr(H)})^{\nu-1}$$

is also continuous on $\mathcal{H}_{++}$, and Lemma C.4, Appendix, yields $\vartheta^{(6)}(H^*) < 2\alpha^*$. Hence, also $\vartheta_n(H) :\equiv \vartheta^{(6)}(H)$ is an appropriate choice.

*Example 6.9.* (see example 6.6) For $H \in \mathcal{H}_{++}$ define

$$\vartheta_n^{(7)}(H) := \frac{1}{u_n^T H^{-1} u_n},$$

where $u_1 := \|v_0\|^{-1} v_0$ and $(u_n)$ is given recursively by

$$v_n := \begin{cases} (H_n^{(0)})^{-1} u_n & , \text{ for } H_n^{(0)} \in \mathcal{H}_{++} \\ v_0 & , \text{ else} \end{cases}, \quad u_{n+1} := \frac{v_n}{\|v_n\|},$$

$n = 1, 2, \ldots$. Here, $0 \neq v_0 \in \mathbb{R}^\nu$ is an appropriate start vector. According to C.5, Appendix, in many cases the following equation holds:

$$\lim_{n\to\infty} \vartheta_n^{(7)}(H_n^{(0)}) = \frac{1}{\lim\limits_{n\to\infty} < u_n, v_n >} = \alpha^* \quad \text{a.s.}\,.$$

Thus, also $\vartheta_n(H) :\equiv \vartheta_n^{(7)}(H)$ is a suitable function.

*Remark 6.2.* Computation of $\vartheta_n(H)$:

In the above Examples 6.7, 6.8 and 6.9 the expressions

$$tr((H_n^{(0)})^{-1}) \quad , \quad \det(H_n^{(0)}) \quad , \quad (H_n^{(0)})^{-1}$$

must be determined. This can be done very efficiently by means of Lemma 6.23.

**Non optimal $\vartheta_n(H)$**

The quantities $\vartheta_n(H_n^{(0)})$ in the Examples 6.3-6.9 can be determined only with larger expenses, since these examples were chosen such that condition (X9) holds. If this is not demanded, then simpler functions $\vartheta_n(H)$ can be taken into account. However, in this case the convergence properties in Theorem 6.12 and 6.13 can not be guaranteed.

On the other hand, if all conditions hold, expect (X9), then we have this result:

**Theorem 6.14.** *It is $n^\lambda(X_n - x^*) \longrightarrow 0$ a.s. for each $\lambda$ such that*

$$0 \leq \lambda < \min\left\{\frac{1}{2}, \frac{\alpha^*}{\widehat{\alpha}}\right\} .$$

*Proof.* We may assume (see the proof of Theorem 6.12) that $\widehat{a} \geq \alpha^*\widehat{\alpha}^{-1}$. The assertion follows then from Theorem 6.1.

According to Section 6.5.2, it would be favorable if the limit $\widehat{\alpha}$ in (X14) fulfills

$$\widehat{\alpha} \approx \alpha^* .$$

This condition helps to find appropriate functions $\vartheta_n(H)$.

Among many other examples, the following one shows how such "non optimal" functions $\vartheta_n(H)$ can be selected.

*Example 6.10.* The function $\vartheta^{(8)}(H) := \min\limits_{i \leq \nu} h_{ii}$ is continuous on $\mathbb{M}(\nu \times \nu, \mathbb{R})$, where $h_{ii}$ denotes the $i$-th diagonal element of $H$. Hence, define

$$\vartheta_n(H) := \vartheta^{(8)}(H) \quad , \quad \mathcal{H}_1 := \mathbb{M}(\nu \times \nu, \mathbb{R}).$$

According to (6.106) we have

$$\alpha_n^* = \begin{cases} \vartheta^{(8)}(H_n^{(0)}) & , \quad \text{for } \vartheta^{(8)}(H_n^{(0)}) \in [\underline{\vartheta}\,, \bar{\vartheta}] \\ \vartheta_0 & , \quad \text{else} \end{cases} ,$$

and in case $\widehat{\alpha} := \min\limits_{i \leq \nu} h_{ii}^* \in (\underline{\vartheta}\,, \bar{\vartheta})$ we get

$$\lim_{n \to \infty} \alpha_n^* = \widehat{\alpha} \quad \text{a.s.} \, .$$

Here, inequality $\widehat{\alpha} \geq \alpha^*$, holds, but condition (X9) does not hold in general.

*Remark 6.3.* Reduction of the expenses in the computation of $(\alpha_n^*)$. Formula (6.106) for sequence $(\alpha_n^*)$ can be modified such that $\alpha_n^*$ is changed only at stages $n \in \mathbb{M}$. Here, $\mathbb{M}$ is an arbitrary infinite subset of $\mathbb{N}$. The advantage of this procedure is that the time consuming Cholesky-decomposition of $H_n^{(0)} + H_n^{(0)T}$ must be calculated only for "some" indices $n \in \mathbb{M}$.

Using a sequence $(\alpha_n^*)$ as described above, the above convergence properties hold true.

## 6.8 A Special Class of Adaptive Scalar Step Sizes

In contrary to the step size rule considered in Section 6.7.2, in the present section scalar step sizes of the type

$$R_n = r_n I \quad \text{with} \quad r_n \in \mathbb{R}_+ , \tag{6.114a}$$

are studied for algorithm (6.2a,b), where the $\mathcal{A}_n$-measurable random variables $r_n$ are defined recursively as follows:

For $n = 1$ we select positive $\mathcal{A}_1$-measurable random variables $r_1, w_1$, and for $n > 1$ we put

$$r_n = \widehat{r}_{n-1}\, Q_n(\widehat{r}_{n-1})\, \frac{w_{n-1}}{w_n}\, K_n. \tag{6.114b}$$

Here,

$$\widehat{r}_{n-1} := \min\{\bar{r}_n, r_{n-1}\} , \tag{6.114c}$$

$$Q_n(r) = \frac{1}{1 + r\, v_n(r)} \quad \text{if } r \in [0, \bar{r}_n], \tag{6.114d}$$

with positive $\mathcal{A}_n$-measurable random variables $\bar{r}_n, w_n, K_n$ and $v_n(r)$, $r \in [0, \bar{r}_n]$. We assume that the following a priori assumptions hold:

A1) For each $n > 1$ and $r \in [0, \bar{r}_n]$ suppose

$$\bar{r}_0 \le \inf_n \bar{r}_n \le \infty \quad \text{a.s.} , \tag{Y1}$$

$$\underline{v} \le v_n(r) \le \bar{v} \quad \text{a.s.} , \tag{Y2}$$

$$\underline{w} \le w_n \le \bar{w} \quad \text{a.s.} , \tag{Y3}$$

$$\lim_{n\to\infty} (K_2 \cdot \ldots \cdot K_n) = P^* \in (0, \infty) \quad \text{a.s.} \tag{Y4}$$

with positive numbers $\bar{r}_0$ , $\underline{v}$ , $\bar{v}$ , $\underline{w}$ , $\bar{w}$ and a random variable $P^*$.

As can be seen from the following example, the definition (6.114a-d) is motivated by the standard step size $r_n = \frac{d}{n}$.

*Example 6.11.* Obviously, $r_n = \frac{d}{n}$ can be represented by

$$r_n = \frac{r_{n-1}}{1 + r_{n-1}\frac{1}{d}} , n > 1 .$$

However, this corresponds to (6.114a-d) with $\bar{r}_n := \infty$ , $v_n(r) := \frac{1}{d}$ and $w_n = K_n := 1$.

In (6.114a-d), the factor $Q_n(r)$ guarantees the convergence of $r_n$ to 0, the variance of the estimation error $Z_n$ is taken into consideration by $w_n$, and the starting behavior of $r_n$ can be influenced by $K_n$.

Concrete examples for $Q_n(r)$, $w_n$ and $K_n$ are given later on. Next to the convergence behavior of $(r_n)$ and $(X_n)$ is studied. We suppose here that the step directions $Y_n$ in algorithm (6.2a,b) are given by (W1a-e) in Section 6.6.

### 6.8.1 Convergence Properties

The following simple lemma is a key tool for the analysis of the step size rule (6.114a-d).

**Lemma 6.29.** *Let $t_1, u, u_2, u_3, \ldots$ be given positive numbers.*

*a) The function $\varphi(t) := \frac{t}{1+tu}$ is monotoneous increasing on $[0,\infty)$ and $\varphi(t) \in [0,t)$ for each $t > 0$.*
*b) If the sequence $(t_n)$ is defined recursively by $t_n := \frac{t_{n-1}}{1+t_{n-1}u_n}, n = 2,3,\ldots,$ then $t_n = \frac{t_1}{1+t_1(u_2+\ldots+u_n)}, n = 2,3,\ldots.$*

*Proof.* The first part is clear, and the formula in b) follows by induction with respect to $n$.

We show now that the step size $r_n$ according to (6.114a-d) is of order $\frac{1}{n}$.

**Lemma 6.30.** $\operatorname*{limextr}_n n \cdot r_n \in \left[\frac{\underline{w}}{\bar{w}\,\bar{v}}, \frac{\bar{w}}{\underline{w}\,\underline{v}}\right]$ *a.s.* .

*Proof.* Putting

$$P_n := K_2 \cdot \ldots \cdot K_n \ , \ t_n := \frac{\widehat{r}_n w_n}{P_n} \ , \ u_n := \frac{P_{n-1}}{w_{n-1}} v_n(\widehat{r}_{n-1}) \ ,$$

after multiplication with $w_n P_n^{-1}$, definition (6.114a-d) yields

$$t_n \le \frac{t_{n-1}}{1+t_{n-1}u_n}, n = 2,3,\ldots \ .$$

Because of Lemma 6.29 we get $t_n \le \frac{t_1}{1+t_1(u_2+\ldots+u_n)}$. Hence, (Y2) and (Y3) yield

$$\begin{aligned} n\widehat{r}_n &= nt_n\frac{P_n}{w_n} \le nt_n\frac{P_n}{\underline{w}} \le nt_1\left(1+t_1\frac{\underline{v}}{\bar{w}}(P_1+\ldots+P_{n-1})\right)^{-1}\frac{P_n}{\underline{w}} \\ &= \left(\frac{1}{nt_1}+\frac{\underline{v}}{\bar{w}\,n}(P_1+\ldots+P_{n-1})\right)^{-1}\frac{P_n}{\underline{w}} \quad \text{a.s.} \ . \end{aligned}$$

Therefore, by (Y4) and Lemma A.8, Appendix, we have

$$\limsup_n n \cdot \widehat{r}_n \le \frac{\bar{w}}{\underline{w}\,\underline{v}} \quad \text{a.s.} \ , \tag{i}$$

and $(\widehat{r}_n)$ converges to zero a.s. . Because of (Y1) there exists a stochastic index $n_0$ with

$$r_n = \widehat{r}_n, n \ge n_0 \quad \text{a.s.} \ . \tag{ii}$$

From (i), (ii) and with (6.114a-d) we get

$$\limsup_n \, n \cdot r_n \le \frac{\bar{w}}{\underline{w}\,\underline{v}} \quad \text{a.s.}\,,$$

$$t_n = \frac{t_{n-1}}{1 + t_{n-1} u_n}, n \ge n_0 \quad \text{a.s.}\,.$$

Thus, for $n \ge n_0$, Lemma 6.29 yields

$$t_n = \frac{t_{n_0}}{1 + t_{n_0}(u_{n_0+1} + \ldots + u_n)} \quad \text{a.s.}\,. \tag{6.115}$$

Applying now (ii), (Y2) and (Y3), we get

$$\begin{aligned} nr_n = nt_n \frac{P_n}{w_n} &\ge nt_{n_0} \left(1 + t_{n_0} \frac{\bar{v}}{\underline{w}} (P_{n_0} + \ldots + P_{n-1})\right)^{-1} \frac{P_n}{\bar{w}} \\ &= \left(\frac{1}{nt_{n_0}} + \frac{\bar{v}}{\underline{w}\, n} (P_{n_0} + \ldots + P_{n-1})\right)^{-1} \frac{P_n}{\bar{w}} \quad \text{a.s.}\,. \end{aligned}$$

Finally, condition (Y4) and Lemma A.8, Appendix, yield

$$\liminf_n \, (n \cdot r_n) \ge \frac{\underline{w}}{\bar{v}\,\bar{w}} \quad \text{a.s.}.$$

Obviously, the step size $R_n$ in (6.114a) can be represented by $R_n = \rho_n M_n$, where

$$\rho_n := \frac{1}{n} \quad \text{and} \quad M_n := (nr_n)\, I\,. \tag{6.116}$$

Concerning the Jacobian $H(x)$ of $G(x)$ we demand as in Section 6.7.2 (cf. condition (X6)):

A2) There is a positive number $\alpha$ such that

$$H(x) + H(x)^T \ge 2\alpha\, I \quad \text{for all } x \in D\,. \tag{Y5}$$

The sequence $(M_n)$ fulfills condition (U4a-d) from Section 6.3:

**Lemma 6.31.** *a)* $\sup_n \|M_n\| < \infty$ *a.s.* .

*b) For each $n \in \mathbb{N}$ there are $\mathcal{A}_n$-measurable matrices $\widehat{M}_n$ and $\Delta M_n$ such that*

*i )* $M_n = \widehat{M}_n + \Delta M_n$ ,

*ii )* $\lim_{n\to\infty} \|\Delta M_n\| = 0$ *a.s.* ,

*iii)* $u^T\, \widehat{M}_n H(x)\, u \ge \alpha \frac{\underline{w}}{\bar{w}\,\bar{v}} \|u\|^2$ *for each* $x \in D$ *and* $u \in \mathbb{R}^\nu$.

*Proof.* For $n \in \mathbb{N}$ define

$$\widehat{M}_n := \max\left\{nr_n\,, \frac{\underline{w}}{\bar{w}\,\bar{v}}\right\} I \quad \text{and} \quad \Delta M_n := M_n - \widehat{M}_n\,. \tag{6.117}$$

Because of condition (Y5), the assertions follows then immediately from Lemma 6.30.

Because of Lemma 6.31, corresponding to Theorem 6.9 and 6.11, here we have the following result:

**Theorem 6.15.**

$$a)\ X_n \longrightarrow x^* \quad a.s.\ ,$$
$$b)\ H_n^{(0)} \longrightarrow H^* \quad a.s.\ .$$

In order to derive further convergence properties, one needs again additional a posteriori assumption on the random functions $v_n(r)$ and random quantities $w_n$:

A3) There is a constant $\bar{L}$ such that for all $n \in \mathbb{N}$

$$| v_n(r) - v_n(0) | \leq \bar{L}\, r \quad \text{for} \quad r \in [0, \bar{r}_n] \ \text{a.s.}\ . \tag{Y6}$$

A4) If " $X_n \longrightarrow x^*$ a.s.", then there are positive numbers $\widehat{\alpha}$ and $w^{(1)}, \ldots, w^{(\kappa)}$ with

$$\lim_{n\to\infty} v_n(0) = \widehat{\alpha} \quad \text{a.s.}\ , \tag{Y7}$$
$$\lim_{n \in \mathbf{N}^{(k)}} w_n = w^{(k)} \quad \text{a.s.}, \quad k = 1, \ldots, \kappa\ . \tag{Y8}$$

Now, a stronger form of Lemma 6.30 can be given:

**Lemma 6.32.** *For each $k \in \{1, \ldots, \kappa\}$ we have*

$$\lim_{n \in \mathbf{N}^{(k)}} n \cdot r_n = \left( \widehat{\alpha} w^{(k)} \left( \frac{q_1}{w^{(1)}} + \ldots + \frac{q_\kappa}{w^{(\kappa)}} \right) \right)^{-1} \quad a.s.\ .$$

*Proof.* According to (6.115) and (ii) from the proof of Lemma 6.30 we obtain, for $n \geq n_0$,

$$\begin{aligned} n r_n = n t_n \frac{P_n}{w_n} &= \frac{n t_{n_0}}{1 + t_{n_0}(u_{n_0+1} + \ldots + u_n)} \frac{P_n}{w_n} \\ &= \left( \frac{1}{n t_{n_0}} + \frac{1}{n}(u_{n_0+1} + \ldots + u_n) \right)^{-1} \frac{P_n}{w_n}\ . \end{aligned} \tag{i}$$

Because of Lemma 6.30, $(\widehat{r}_n)$ converges to zero a.s., and by Theorem 6.15 we have $X_n \longrightarrow x^*$ a.s.. Hence, (Y4), (Y6), (Y7) and (Y8) yield

$$\lim_{n \in \mathbf{N}^{(k)}} (P_n\, , w_n\, , u_n) = \left( P^*, w^{(k)}, \frac{P^*}{w^{(k)}} \widehat{\alpha} \right) \quad \text{a.s.}\ . \tag{ii}$$

Having (i) and (ii), with Lemma A.8, Appendix, we obtain

$$\lim_{n \in \mathbf{N}^{(k)}} n r_n = \left( P^* \widehat{\alpha} \left( \frac{q_1}{w^{(1)}} + \ldots + \frac{q_\kappa}{w^{(\kappa)}} \right) \right)^{-1} \frac{P^*}{w^{(k)}} \quad \text{a.s.}\ .$$

According to Lemma 6.32 we have now

$$\lim_{n \in \mathbb{N}^{(k)}} M_n = d^{(k)} I \text{ a.s., } k = 1, \ldots, \kappa \,, \tag{6.118}$$

$$d := q_1 d^{(1)} + \ldots + q_\kappa d^{(\kappa)} = \frac{1}{\widehat{\alpha}} \,, \tag{6.119}$$

where

$$d^{(k)} := \left( \widehat{\alpha} w^{(k)} \left( \frac{q_1}{w^{(1)}} + \ldots + \frac{q_\kappa}{w^{(\kappa)}} \right) \right)^{-1}, k = 1, \ldots, \kappa \tag{6.120}$$

Hence, the constant $\widehat{\alpha}$ has the same meaning as in Section 6.7.2, see (6.97) and (6.96b). Thus, corresponding to condition (X9), here we demand:

$$\widehat{\alpha} < 2\alpha^* := \lambda_{\min}(H^* + H^{*T}) \,. \tag{Y9}$$

Corresponding to Theorem 6.12, here we have the following result:

**Theorem 6.16.** *a) If $\lambda \in [0, \frac{1}{2})$, then $n^\lambda (X_n - x^*) \longrightarrow 0$ a.s. .*
*b) For each $\epsilon_0 > 0$ there is a set $\Omega_0 \in \mathcal{A}$ such that*

*i )* $\mathcal{P}(\bar{\Omega}_0) < \epsilon_0$ ,
*ii)* $\limsup_n E_{\Omega_0} n \|X_n - x^*\|^2$

$$\leq \left[ \widehat{\alpha}(2\alpha^* - \widehat{\alpha}) \left( \frac{q_1}{w^{(1)}} + \ldots + \frac{q_\kappa}{w^{(\kappa)}} \right)^2 \right]^{-1} \sum_{k=1}^{\kappa} q_k \left( \frac{\sigma_k}{w^{(k)}} \right)^2 . \tag{6.121}$$

*Proof.* Because of (6.119), (6.120) and (Y9), as in the proof of Theorem 6.12, for the number $\widehat{\alpha}$, $d$ according to (6.9), (6.120), resp., we have

$$\widehat{\alpha} \geq d\alpha^* > \frac{1}{2} \,.$$

Based on (6.118-6.120), the assertions follow now from Theorem 6.1 and Corollary 6.3.

According to Section 6.5.2, the right hand side of (6.121) takes a minimum at the values

$$\widehat{\alpha} = \alpha^* \,, \tag{6.122}$$

$$w^{(k)} = \sigma_k^2 := \limsup_{n \in \mathbb{N}^{(k)}} E\, \sigma_{0,n}^2, k = 1, \ldots, \kappa \,. \tag{6.123}$$

Also the limit properties in Theorem 6.13 can be transfered easily to the present situation:

**Theorem 6.17.** *Under the additional assumptions*

*i)* $x^* \in \overset{\circ}{D}$,
*ii) the covariance matrices of* $(Z_n)$ *fulfill conditions (V6), (V8) in Section 6.5.3,*

*sequence* $(\sqrt{n}(X_n - x^*))$ *has an asymptotic* $N(0, V)$*-distribution, where the matrix* $V$ *is determined by the equation*

$$\frac{1}{\bar{\hat{\alpha}}}(H^*V + VH^{*T}) - V = \left(\hat{\alpha}\left(\frac{q_1}{w^{(1)}} + \ldots + \frac{q_\kappa}{w^{(\kappa)}}\right)\right)^{-2} \sum_{k=1}^{\kappa} \frac{q_k}{(w^{(k)})^2} W^{(k)}. \tag{6.124}$$

Similar to Theorem 6.13, the above result can be shown by using Theorem 6.4, where the numbers $d^{(k)}$ are not given by (6.96a), but by (6.118).

### 6.8.2 Examples for the Function $Q_n(r)$

Several formulas are presented now for the function $Q_n(r)$ in (6.114a-d) such that conditions (Y1), (Y2) and (Y6) for the sequences $(\bar{r}_n)$ and $(v_n(r))$ hold true. In each case the function $Q_n(r)$ depends on two $\mathcal{A}_n$-measurable random parameters $\alpha_n^*$ and $\Gamma_n$. Suppose that $\alpha_n^*, \Gamma_n$ fulfill the following condition:

A5) There are positive constants $\underline{\vartheta}$, $\bar{\vartheta}$ and $\bar{\gamma}$ such that for $n > 1$

$$\underline{\vartheta} \le \alpha_n^* \le \bar{\vartheta} \quad \text{a.s.}, \tag{Y10}$$

$$|\, \Gamma_n \,| \le \bar{\gamma} \quad \text{a.s.}. \tag{Y11}$$

*Example 6.12.* Standard step size

Corresponding to Example 6.11 we take

$$Q_n(r) = \frac{1}{1 + r\alpha_n^* - r^2\Gamma_n}, \quad r \in [0, \bar{r}_n] \tag{6.125a}$$

$$\bar{r}_n := \alpha_n^* \min\left\{\frac{\bar{K}}{\Gamma_n^-}, \frac{\bar{q}}{\Gamma_n^+}\right\}. \tag{6.125b}$$

Here, $\bar{K} > 0$ and $\bar{q} \in (0, 1)$ are arbitrary constants. The following lemma shows the feasibility of (6.125a,b):

**Lemma 6.33.** *According to (6.125a,b) function* $Q_n(r)$ *has the form (6.114d), and for each* $r \in [0, \bar{r}_n]$ *we have*

*a)* $v_n(r) = \alpha_n^* - r\Gamma_n$,
*b)* $\underline{\vartheta}(1 - \bar{q}) \le v_n(r) \le \bar{\vartheta}(1 + \bar{K})$ *a.s.*,
*c)* $\bar{r}_n \ge \bar{r}_0 := \dfrac{\underline{\vartheta}}{\bar{\gamma}} \min\{\bar{K}, \bar{q}\} > 0$ *a.s.*.

*Proof.* Because of (Y10), (Y11) and (6.125b), the inequality in c) holds for each $n > 1$ and $r \in [0, \bar{r}_n]$. Furthermore,

$$\begin{aligned}\underline{\vartheta}\,(1-\bar{q}) \le \alpha_n^*(1-\bar{q}) \le \alpha_n^* - \bar{r}_n {\Gamma_n}^+ \le \alpha_n^* - r\Gamma_n \\ \le \alpha_n^* + \bar{r}_n {\Gamma_n}^- \le \alpha_n^*(1+\bar{K}) \le \bar{\vartheta}\,(1+\bar{K}) \text{ a.s.,}\end{aligned}$$

hence, also b) holds.

*Example 6.13.* "Optimal" step size

If $E_n Z_n = 0$, $n = 1, 2, \ldots$ (cf. Robbins-Monro process), then by Lemma 6.1 and assumption (U5b) for each $n \in \mathbb{N}$ we have

$$\begin{aligned}E_n\|\Delta_{n+1}\|^2 &\le \left\|\Delta_n - r_n(G_n - G^*)\right\|^2 + r_n^2 E_n\|Z_n\|^2 \\ &\le \left(1 - 2\alpha r_n + (\beta^2 + \sigma_1^2) r_n^2\right)\|\Delta_n\|^2 + r_n^2 \sigma_{0,n}^2 .\end{aligned}$$

Here, $\alpha$ and $\beta$ are positive numbers with

$$\begin{aligned}\langle X_n - x^*, G_n - G^*\rangle &\ge \alpha\|X_n - x^*\|^2 , \\ \|G_n - G^*\| &\le \beta\|X_n - x^*\| \quad \text{(see (6.4)).}\end{aligned}$$

Based on the above estimate for the quadratic error $\|\Delta_n\|^2$, in [80] an "optimal" deterministic step size $r_n$ is given which is defined recursively as follows:

$$r_n = \frac{1 - r_{n-1}\alpha}{1 - r_{n-1}^2(\beta^2 + \sigma_1^2)}\, r_{n-1} \quad \text{for } n = 2, 3, \ldots .$$

This suggests the following function $Q_n(r)$:

Let $\bar{q} \in (0,1)$, and for $n > 1$ define

$$Q_n(r) = \frac{1 - r\alpha_n^*}{1 - r^2\Gamma_n} , \; r \in [0, \bar{r}_n] \tag{6.126a}$$

$$\bar{r}_n := \bar{q} \min\left\{\frac{\alpha_n^*}{{\Gamma_n}^+}, \frac{1}{\alpha_n^*}\right\} . \tag{6.126b}$$

The feasibility of this approach is shown next:

**Lemma 6.34.** *If $Q_n(r)$ and $\bar{r}_n$ are defined by (6.126a,b), then $Q_n(r)$ has the form (6.114d), and for each $r \in [0, \bar{r}_n]$ we have*

$$a)\;\; v_n(r) = \frac{\alpha_n^* - r\Gamma_n}{1 - r\alpha_n^*} = \alpha_n^* - \frac{\Gamma_n - \alpha_n^{*2}}{1 - r\alpha_n^*}\, r ,$$

$$b)\;\; (1-\bar{q})\,\underline{\vartheta} \le v_n(r) \le \bar{\vartheta} + \frac{\bar{q}}{1-\bar{q}}\,\frac{\bar{\vartheta}^2 + \bar{\gamma}}{\underline{\vartheta}} \quad a.s. ,$$

$$c)\;\; \left|\frac{\Gamma_n - \alpha_n^{*2}}{1 - r\alpha_n^*}\right| \le \frac{\bar{\vartheta}^2 + \bar{\gamma}}{1 - \bar{q}} \quad a.s. ,$$

$$d)\;\; \bar{r}_n \ge \bar{r}_0 := \bar{q} \min\left\{\frac{\underline{\vartheta}}{\bar{\gamma}}, \frac{1}{\bar{\vartheta}}\right\} > 0 \quad a.s. .$$

*Proof.* Assertion a) follows by a simple computation.

If $\alpha_n^{*2} < \Gamma_n$, then (6.126b) yields

$$0 \leq \frac{\Gamma_n - \alpha_n^{*2}}{1 - r\alpha_n^*} \leq \frac{\Gamma_n - \alpha_n^{*2}}{1 - \bar{q}\, \alpha_n^{*2} \Gamma_n^{-1}} \leq \Gamma_n \,.$$

Hence, because of a) we get

$$\alpha_n^* \geq v_n(r) \geq \alpha_n^* - r\Gamma_n \geq \alpha_n^* - \bar{r}_n {\Gamma_n}^+ \geq \alpha_n^*(1 - \bar{q}) \,.$$

Let $\Gamma_n \leq \alpha_n^{*2}$. Then (6.126b) yields

$$0 \leq \frac{\alpha_n^{*2} - \Gamma_n}{1 - r\alpha_n^*} \leq \frac{\alpha_n^{*2} + {\Gamma_n}^-}{1 - \bar{q}} \,,$$

and therefore with a)

$$\alpha_n^* \leq v_n(r) \leq \alpha_n^* + \frac{\alpha_n^{*2} + {\Gamma_n}^-}{1 - \bar{q}} \frac{\bar{q}}{\alpha_n^*} \,.$$

From the above relations and assumptions (Y10), (Y11) we obtain now the assertions in b), c) and d).

*Example 6.14.* A further approach reads

$$Q_n(r) = 1 - r\alpha_n^* + r^2\Gamma_n \,, \; r \in [0, \bar{r}_n] \tag{6.127a}$$

$$\bar{r}_n := \min\left\{ \frac{\bar{K}\alpha_n^*}{{\Gamma_n}^-}, \frac{\bar{q}}{\alpha_n^*(1 + \bar{K})}, \frac{\bar{q}\alpha_n^*}{{\Gamma_n}^+} \right\} , \tag{6.127b}$$

where $\bar{K} > 0$ and $\bar{q} \in (0, 1)$ are arbitrary constants. In this case we have the following result:

**Lemma 6.35.** *Let $Q_n(r)$ and $\bar{r}_n$ be given by (6.127a,b). Then function $Q_n(r)$ can be represented by (6.114d), and for each $r \in [0, \bar{r}_n]$ we have*

*a)* $$v_n(r) = \frac{\alpha_n^* - r\Gamma_n}{1 - r\alpha_n^* + r^2\Gamma_n} = \alpha_n^* - \frac{\Gamma_n - \alpha_n^*(\alpha_n^* - r\Gamma_n)}{1 - r(\alpha_n^* - r\Gamma_n)} r \,,$$

*b)* $$(1 - \bar{q})\, \underline{\vartheta} \leq v_n(r) \leq \frac{\bar{\vartheta}\,(1 + \bar{K})}{1 - \bar{q}} \quad a.s. \,,$$

*c)* $$\left| \frac{\Gamma_n - \alpha_n^*(\alpha_n^* - r\Gamma_n)}{1 - r(\alpha_n^* - r\Gamma_n)} \right| \leq \frac{\bar{\gamma} + \bar{\vartheta}^2(1 + \bar{K})}{1 - \bar{q}} \quad a.s. \,,$$

*d)* $$\bar{r}_n \geq \bar{r}_0 := \min\left\{ \frac{\bar{K}\, \underline{\vartheta}}{\bar{\gamma}}, \frac{\bar{q}}{\bar{\vartheta}\,(1 + \bar{K})}, \frac{\bar{q}\, \underline{\vartheta}}{\bar{\gamma}} \right\} > 0 \quad a.s. \,.$$

*Proof.* Obviously a) holds. Using (6.127b) we find

$$\begin{aligned} u_n &:= \alpha_n^* - r\Gamma_n \geq \alpha_n^* - \bar{r}_n {\Gamma_n}^+ \geq \alpha_n^*(1-\bar{q}) \,, \\ u_n &\leq \alpha_n^* + \bar{r}_n {\Gamma_n}^- \leq \alpha_n^*(1+\bar{K}) \,, \\ Q_n(r) &= 1 - ru_n \geq 1 - \bar{r}_n \alpha_n^*(1+\bar{K}) \geq 1 - \bar{q} \,. \end{aligned}$$

Hence, with a) we get

$$\begin{aligned} &\alpha_n^*(1-\bar{q}) \leq u_n \leq \frac{u_n}{1-ru_n} = v_n(r) \leq \frac{\alpha_n^*(1+\bar{K})}{1-\bar{q}} \,, \\ &\left| \frac{\Gamma_n - \alpha_n^* u_n}{1-ru_n} \right| \leq \frac{|\,\Gamma_n\,| + \alpha_n^* u_n}{1-\bar{q}} \leq \frac{|\,\Gamma_n\,| + \alpha_n^{*2}(1+\bar{K})}{1-\bar{q}} \,. \end{aligned}$$

Applying now assumptions (Y10) and (Y11), the assertions in b), c) and d) follow.

*Example 6.15.* Consider $Q_n(r)$ given by

$$Q_n(r) = \exp(-r\alpha_n^* + r^2\Gamma_n) \,,\ r \in [0, \bar{r}_n] \tag{6.128a}$$

$$\bar{r}_n := \min\left\{ \bar{r} \,,\ \bar{q}\,\frac{\alpha_n^*}{{\Gamma_n}^+} \right\} . \tag{6.128b}$$

In (6.128b) $\bar{r} > 0$ and $\bar{q} \in (0,1)$ are arbitrary constants. The feasibility of this approach is guaranteed by the next lemma:

**Lemma 6.36.** *Let $Q_n(r)$ and $\bar{r}_n$ be given by (6.128a,b). Then $Q_n(r)$ can be represented by (6.114d), where for each $r \in [0, \bar{r}_n]$ we have*

a) $v_n(r) = \frac{1}{r}\left(\exp(r\alpha_n^* - r^2\Gamma_n) - 1\right) = \alpha_n^* + \Delta\alpha_n^*(r)\, r$ *with*

$$\Delta\alpha_n^*(r) := \frac{1}{r}(\alpha_n^* - r\Gamma_n)\left( \frac{\exp(r\alpha_n^* - r^2\Gamma_n) - 1}{r\alpha_n^* - r^2\Gamma_n} - 1 \right) - \Gamma_n \,,$$

b) $\underline{\vartheta}\,(1-\bar{q}) \leq v_n(r) \leq \bar{u}(1 + \frac{\bar{r}\,\bar{u}}{2}\exp(\bar{r}\,\bar{u}))$ *a.s., where* $\bar{u} := \bar{\vartheta} + \bar{r}\,\bar{\gamma}$,

c) $|\,\Delta\alpha_n^*(r)\,| \leq \frac{\bar{u}^2}{2}\exp(\bar{r}\,\bar{u}) + \bar{\gamma}$ *a.s.* ,

d) $\bar{r}_n \geq \bar{r}_0 := \min\left\{ \bar{r} \,,\ \bar{q}\,\frac{\underline{\vartheta}}{\bar{\gamma}} \right\} > 0$ *a.s.* .

*Proof.* Assertion a) follows by an elementary calculation. According to (6.128b) we have

$$\alpha_n^*(1-\bar{q}) \leq \alpha_n^* - \bar{r}_n {\Gamma_n}^+ \leq \alpha_n^* - r\Gamma_n =: u_n \leq \alpha_n^* + \bar{r}\,|\,\Gamma_n\,| \,. \tag{i}$$

Furthermore, for each $t > 0$ we may write

$$\exp(t) = 1 + t + \frac{t^2}{2}\exp(\tau(t)) \ \text{ with } \ \tau(t) \in [0, t] \,. \tag{ii}$$

Hence, with a) we get

$$v_n(r) = u_n \frac{\exp(ru_n) - 1}{ru_n} = u_n \left(1 + \frac{ru_n}{2} \exp\bigl(\tau(ru_n)\bigr)\right) .$$

Together with (i) we find

$$\alpha_n^*(1 - \bar{q}) \le v_n(r) \le u_n \left(1 + \frac{\bar{r}u_n}{2} \exp(\bar{r}u_n)\right) , \tag{iii}$$

and (i), (ii) yield also the relation

$$\begin{aligned} \mid \Delta\alpha_n^*(r) \mid &= \left| \frac{u_n}{r} \left( \frac{\exp(ru_n) - 1}{ru_n} - 1 \right) - \Gamma_n \right| \\ &= \left| \frac{u_n}{r} \frac{ru_n}{2} \exp\bigl(\tau(ru_n)\bigr) - \Gamma_n \right| \le \frac{u_n^2}{2} \exp(\bar{r}u_n) + \mid \Gamma_n \mid . \end{aligned} \tag{iv}$$

The assertion in b), c) and d) follow now from (iii), (iv) and the assumptions (Y10), (Y11).

## Choice of $(\alpha_n^*)$

In the above examples for $Q_n(r)$ condition (Y6) holds, and

$$v_n(0) = \alpha_n^* \text{ a.s. , for all } n > 1. \tag{6.129}$$

Hence, assymption (Y7) is equivalent to

$$\lim_{n \to \infty} \alpha_n^* = \widehat{\alpha} \quad \text{a.s. .} \tag{Y12}$$

Note that (Y12) and (Y9) are exactly the conditions which are required for $(\alpha_n^*)$ in Section 6.7.2 (see (X8) and (X9)).

Theorem 6.15 guarantees also in the present case that $X_n \longrightarrow x^*$ and $H_n^{(0)} \longrightarrow H^*$ a.s. . Thus, sequence $(\alpha_n^*)$ can be chosen also according to (6.106). Then (Y10) holds because of (X13), and due to Lemma 6.27 also conditions (Y12) and (Y9) are fulfilled.

## Selection of sequence $(\Gamma_n)$

The $\mathcal{A}_n$-measurable factors $\Gamma_n$ in the above examples for $Q_n(r)$ can be chosen arbitrarily, provided that condition (Y11) holds. According to Lemma 6.32, Theorem 6.16 and 6.17, $(\Gamma_n)$ does not influence the limit of $(nr_n)$ and the asymptotic variance of $(\sqrt{n}(X_n - x^*))$. Of course, the most simple choice is

$$\Gamma_n \equiv 0, \; n > 1 . \tag{6.130}$$

A positive $\Gamma_n$ means that the step size $r_n$ has a slower convergence to zero than in the standard case (6.130).

Example 6.13 suggest the following choice for $(\Gamma_n)$:

$$\Gamma_n := \begin{cases} \beta_n^2 & , \text{ if } \beta_n^2 \leq \bar{\gamma} \\ \gamma_0 & , \text{ else} \end{cases} \quad \text{for } n = 2, 3, \ldots , \tag{6.131}$$

where $\beta_n$ is an $\mathcal{A}_n$-measurable estimate of the maximum eigenvalue $\beta^*$ of the matrix $\frac{1}{2}\left(H(x^*) + H(x^*)^T\right)$, and $\gamma_0$ is a constant such that $\mid \gamma_0 \mid \leq \bar{\gamma}$ .

Since

$$H_n^{(0)} \longrightarrow H^* := H(x^*) \quad \text{a.s.} , \tag{6.132}$$

see Theorem 6.15 b), the following approximates $\beta_n$ of $\beta^*$ can be used:

$$\text{a)} \ \beta_n = \psi^{(1)}(H_n^{(0)}) \ \text{ with } \ \psi^{(1)}(H) := tr(H) , \tag{6.133}$$

$$\text{b)} \ \beta_n = \psi^{(2)}(H_n^{(0)}) \ \text{ with } \ \psi^{(2)}(H) := \max_{i \leq \nu} h_{ii} , \tag{6.134}$$

where $h_{ii}$ denotes the $i$-th diagonal element of $H$.

Because of (6.132) we have

$$\beta_n \longrightarrow \psi^{(i)}(H^*) \quad \text{a.s.}, \tag{6.135}$$

where $i = 1$ for (6.133) and $i = 2$ for (6.134). The bound $\bar{\gamma}$ in (6.131) is chosen such that

$$\left(\psi^{(i)}(H^*)\right)^2 \leq \bar{\gamma} . \tag{6.136}$$

Instead of (6.131) we can also take

$$\Gamma_n := \alpha_n^{*2}, n > 1. \tag{6.137}$$

In this case very simple expressions are obtained for the step size bound $\bar{r}_n$ in Examples 6.12-6.15.

### 6.8.3 Optimal Sequence $(w_n)$

The factors $w_n$ in (6.114a-d) are defined in the literature often by

$$w_n \equiv 1, \ n \geq 1, \tag{6.138}$$

cf. Examples 6.11, 6.13. However, this approach does not yield the minimum asymptotic variance bound for $\sqrt{n}(X_n - x^*)$, see Theorem 6.16 b).

According to Section 6.7.2, the $\mathcal{A}_n$-measurable random variables $\widehat{w}_n^{(1)}, \ldots,$ $\widehat{w}_n^{(\kappa)}$, defined by (6.103), fulfill for $k = 1, \ldots, \kappa$ the following conditions:

$$\underline{w} \leq \widehat{w}_n^{(k)} \leq \bar{w} , \tag{6.139}$$

$$\lim_{n \in \mathbb{N}^{(k)}} \widehat{w}_n^{(k)} = \lim_{n \in \mathbb{N}^{(k)}} E_n \|Z_n - E_n Z_n\|^2 \ \text{ a.s.} . \tag{6.140}$$

Since the limits $w^{(1)}, \ldots, w^{(\kappa)}$ of $(w_n)$ from (Y8) should fulfill in the best case equation (6.123), because of (6.140) we prefer the following choice of $(w_n)$:

For an integer $n \geq 1$ belonging to the set $\mathbb{N}^{(k)}$ with $k \in \{1, \ldots, \kappa\}$, put

$$w_n := \widehat{w}_n^{(k)} . \tag{6.141}$$

Then, because of (6.139), (6.140) also conditions (Y3) and (Y8) hold.

### 6.8.4 Sequence $(K_n)$

According to Lemma 6.32, Theorem 6.16, 6.17, the positive $\mathcal{A}_n$-measurable factors $K_n$ in the step size rule (6.114b) play no role in the formulas for the limit of $(nr_n)$ and the asymptotic variance of $(\sqrt{n}(X_n - x^*))$.

Selecting $(K_n)$, we have only to take into account the convergence condition (Y4). This requires

$$K_n \cdot K_{n+1} \cdot \ldots \approx 1 \text{ for "large" } n.$$

Hence, the factors $K_n$ are responsible only for the *initial behavior* of the step size $r_n$ and the iterates $X_n$.

By the factor $K_n$ (as well as by $w_n$) the step size can be enlarged at $r_{n-1} \to r_n$. This is necessary e.g. in case of a too small starting step size $r_1$.

An obvious definition of $K_n$ reads

$$K_n \equiv 1, \quad n > 1 . \tag{6.142}$$

However, it is preferable to take $K_n > 1$ as long as algorithm (6.2a,b) shows an "almost deterministic" behavior. This holds as long as the influence of the estimation error $Z_{n-1}$ to the step direction $Y_{n-1} = G_{n-1} + Z_{n-1}$ is small.

Based on the above considerations, some definitions for $(K_n)$ are presented:

#### Adaptive selection of $(K_n)$

Suppose

$$K_n = \exp(s_n^2 \, k_n), \quad n > 1 , \tag{6.143}$$

where $(s_n)$ is a deterministic sequence and $k_n$ denotes an $\mathcal{A}_n$-measurable random number such that

$$\sum_n s_n^2 < \infty , \tag{Y13}$$

$$\sup_n E \mid k_n \mid < \infty . \tag{Y14}$$

Consequently,

$$E\Big(\sum_n s_n^2 \mid k_n \mid\Big) < \infty \quad \text{and especially} \quad \sum_n s_n^2 \mid k_n \mid < \infty \text{ a.s. .}$$

Hence, due to (6.143), sequence $(K_n)$ fulfills condition (Y4).

E.g., $(s_n)$ can be chosen as follows:

$$s_n = \frac{s_1}{n}, \quad n > 1, \quad \text{with} \quad s_1 > 0\,. \tag{6.144}$$

In the following some examples are given for sequence $(k_n)$.

*Example 6.16.* According to assumption a) in Section 6.3, the norm of the $\mathcal{A}_n$-measurable difference $\Delta X_n := X_n - X_{n-1}$ is bounded by $\delta_0$. Hence,

$$\text{a)}\;\; k_n := \langle \Delta X_n\,, \Delta X_{n-1}\rangle\,, \tag{6.145}$$

$$\text{b)}\;\; k_n := \langle \Delta X_n\,, \Delta X_{n-1}\rangle^+\,, \tag{6.146}$$

$$\text{c)}\;\; k_n := \cos\,(\Delta X_n\,, \Delta X_{n-1}) := \frac{\langle \Delta X_n\,, \Delta X_{n-1}\rangle}{\|\Delta X_n\|\|\Delta X_{n-1}\|}\,, \tag{6.147}$$

$$\text{d)}\;\; k_n := \cos\,(\Delta X_n\,, \Delta X_{n-1})^+ \tag{6.148}$$

fulfill condition (Y14) and are therefore appropriate definitions for $(k_n)$.

*Example 6.17.* The factors $k_n$ can be chosen also as functions of the approximates $Y_{n-1}^{(0)}, \ldots, Y_{n-1}^{(2l_n-1)}$ for the step direction $Y_{n-1}$, cf. (W1a-e) in Section 6.6. Thereby we need the following result:

**Lemma 6.37.** *For arbitrary index $n \in \mathbb{N}$ and $j, k \in \{0, \ldots, 2l_n\}$ we have*

$$E\|Y_n^{(j)}\|\|Y_n^{(k)}\| \le L^2 := 4\big(\|G^*\|^2 + (\beta^2 + t_1^2)\,\delta_0^2 + t_0^2 + L_1^2\big) < \infty\,.$$

*Here, $\delta_0$ denotes the diameter of $D$, and $\beta, t_0, t_1$ and $L_1$ are given by (6.3), (6.71), (6.73) and (W2a).*

*Proof.* Corresponding to the proof of Lemma 6.8 we find

$$\begin{aligned}\|Y_n^{(j)}\|^2 &= \big\|G^* + (G_n^{(j)} - G^*) + (Z_n^{(j)} - E_n Z_n^{(j)}) + E_n Z_n^{(j)}\big\|^2 \\ &\le 4\big(\|G^*\|^2 + \|G_n^{(j)} - G^*\|^2 + \|Z_n^{(j)} - E_n Z_n^{(j)}\|^2 + \|E_n Z_n^{(j)}\|^2\big)\,.\end{aligned}$$

By (6.4), (6.70), (6.71) and (W2a) we have

$$\begin{aligned}\|G_n^{(j)} - G^*\| &\le \beta\|X_n^{(j)} - x^*\|\,, \\ E\|Z_n^{(j)} - E_n Z_n^{(j)}\|^2 &\le t_0^2 + t_1^2 E\|\Delta_n\|^2\,, \\ \|E_n Z_n^{(j)}\| &\le L_1\,.\end{aligned}$$

Hence, due to $\|X_n^{(j)} - x^*\|$ , $\|\Delta_n\| \le \delta_0$ and the above inequalities we get

$$E\|Y_n^{(j)}\|^2 \le L^2 \quad \text{as well as} \quad E\|Y_n^{(k)}\|^2 \le L^2\,.$$

Applying now Schwarz' inequality, the assertion follows.

Having this lemma, we find that condition (Y14) holds in the following cases:

$$\text{a)}\quad k_n := \frac{1}{l_{n-1}} \sum_{i=1}^{l_{n-1}} \langle Y_{n-1}^{(i)} , Y_{n-1}^{(l_{n-1}+i)} \rangle , \tag{6.149}$$

$$\text{b)}\quad k_n := \frac{1}{l_{n-1}} \sum_{i=1}^{l_{n-1}} \langle Y_{n-1}^{(i)} , Y_{n-1}^{(l_{n-1}+i)} \rangle^+ , \tag{6.150}$$

$$\text{c)}\quad k_n := \|Y_{n-1}\|^2 I_{\{\min\limits_{i=1,\dots,l_{n-1}} \langle Y_{n-1}^{(i)} , Y_{n-1}^{(l_{n-1}+i)} \rangle > 0\}} , \tag{6.151}$$

$$\text{d)}\quad k_n := \frac{1}{2l_{n-1}} \sum_{j=1}^{2l_{n-1}} \langle Y_{n-1}^{(0)} , Y_{n-1}^{(j)} \rangle^+ . \tag{6.152}$$

Similar examples may be found easily.

# Part V

# Technical Applications

# 7

# Approximation of the Probability of Failure/Survival in Plastic Structural Analysis and Optimal Plastic Design

## 7.1 Introduction

According to Section 2.6 and Theorem 2.3 for the safety of structures made of elastoplastic materials we have the following criterion (2.54a,b):

$$s^* = s^*(R, P) \begin{cases} < 0 : \text{the structure is in a safe stress state} \\ \geq 0 : \text{a safe stress state cannot be guaranteed.} \end{cases} \tag{7.1}$$

Here, the state function (limit state function or performance function) $s^*$ is defined, see (2.53a-c), by the minimum value of the optimization problem

$$\min s$$

s.t.

$$C\sigma = P$$
$$\pi(R_{id}^{-1}\sigma_i|K_i) - 1 \leq s, i = 1, \ldots, n_G.$$

The total vector $R$ of material resistances

$$R = R(\sigma_y, x) = R\Big(\sigma_y(\omega), x\Big) \tag{7.2a}$$

depends on the random vector $\sigma_y = \sigma_y(\omega)$ of yield stresses and an $r$–vector $x$ of design variables.

Moreover, the external load vector

$$P = P\Big(p(\omega), x\Big) \tag{7.2b}$$

is a function of a vector $p = p(\omega)$ of random load parameters and some load factors $\mu$ included in the design vector $x$. We may assume that the

random $\nu_\sigma$–vector $\sigma_y = \sigma_y(\omega)$ and the random
$\nu_p$–vector $p = p(\omega)$ are stochastically independent. (7.2c)

Let $\nu := \nu_\sigma + \nu_p$, and define the total $\nu$–random parameter vector

$$y = y(\omega) := \begin{pmatrix} \sigma_y(\omega) \\ p(\omega) \end{pmatrix}. \tag{7.2d}$$

In the following, $x \in D$ denotes an arbitrary, but fixed vector of design variables and/or load factors. By means of (7.2a-d), the state function $s^*$ can be represented

$$\tilde{s}^*(y, x) = \tilde{s}^*\left(\begin{pmatrix} \sigma_y \\ p \end{pmatrix}, x\right) := s^*\Big(R(\sigma_y, x), P(p, x)\Big) \tag{7.3}$$

as a function of the parameter vector $y$ and the design vector $x$.

## 7.2 Probability of Survival/Failure $p_s, p_f$

According to (7.1) and (7.3), for a given vector $x$, the so–called safe domain $B_{s,x}$ is defined by

$$B_{s,x} := \{y \in \mathbb{R}^\nu : \tilde{s}^*(y, x) < 0\}; \tag{7.4a}$$

moreover, the unsafe or failure domain $B_{f,x}$ is given by

$$B_{f,x} := \{y \in \mathbb{R}^\nu : \tilde{s}^*(y, x) \geq 0\}. \tag{7.4b}$$

Correspondingly, the probability of survival $p_s$ and the probability of failure $p_f$ is given by

$$\begin{aligned} p_s &:= \mathcal{P}\ (\text{the structure is in a safe stress state}) \\ &= \mathcal{P}\Big(y(\omega) \in B_{s,x}\Big) = \mathcal{P}\Big(\tilde{s}^*\big(y(\omega), x\big) < 0\Big) \\ &= \mathcal{P}\left(s^*\Big(R\big(\sigma_y(\omega), x\big), P\big(p(\omega), x\big)\Big) < 0\right), \end{aligned} \tag{7.5a}$$

and

$$\begin{aligned} p_f &:= \mathcal{P}\ (\text{a safe stress state cannot be guaranteed}) \\ &= \mathcal{P}\big(y(\omega) \in B_{f,x}\big) = \mathcal{P}\Big(\tilde{s}^*\big(y(\omega), x\big) \geq 0\Big) \\ &= \mathcal{P}\left(s^*\Big(R\big(\sigma_y(\omega), x\big), P\big(p(\omega), x\big)\Big) \geq 0\right). \end{aligned} \tag{7.5b}$$

In the following we assume that

$$\bar{y} = Ey(\omega) = \begin{pmatrix} E\sigma_y(\omega) \\ Ep(\omega) \end{pmatrix} = \begin{pmatrix} \bar{\sigma}_y \\ \bar{p} \end{pmatrix} \in B_{s,x}, \tag{7.6}$$

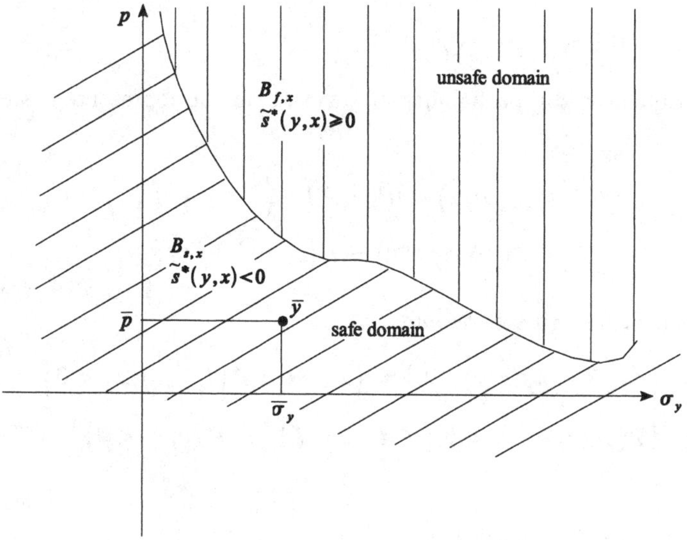


**Fig. 7.1.** Safe/unsafe domain in the $y$-variables

i.e. the expectation $\bar{y}$ of the parameter vector $y(\omega)$ lies in the safe domain $B_{s,x}$.

For a large class of probability distributions $\mathcal{P}_{(\cdot)}$, e.g. for continuous ones, the random vector

$$y = y(\omega) = \begin{pmatrix} \sigma_y(\omega) \\ p(\omega) \end{pmatrix}$$

can be represented [94] by

$$y(\omega) = T^{-1}\Big(z(\omega)\Big) \text{ or } z(\omega) = T\Big(y(\omega)\Big) \tag{7.7a}$$

where $T : \mathbb{R}^\nu \to \mathbb{R}^\nu$ is a certain 1-1–transformation, and $z = z(\omega)$ is a normal distributed random $\nu$–vector with

$$Ez(\omega) = 0 \text{ and } \operatorname{cov}\big(z(\cdot)\big) = I. \tag{7.7b}$$

Because of (7.2c), hence, the stochastic independence of $\sigma_y(\omega)$ and $p(\omega)$, the transformation $T^{-1} = T^{-1}(z)$ can be selected such that

$$y = \begin{pmatrix} \sigma_y \\ p \end{pmatrix} = T^{-1}(z) = \begin{pmatrix} T_\sigma^{-1}(z_\sigma) \\ T_p^{-1}(z_p) \end{pmatrix}, z = \begin{pmatrix} z_\sigma \\ z_p \end{pmatrix}. \tag{7.7c}$$

Inserting then (7.7a-c) into (7.3), the state function $\tilde{s}^* = \tilde{s}^*(y,x)$ is transformed into

$$\begin{aligned} \tilde{s}_T^* = \tilde{s}_T^*(z,x) &:= \tilde{s}^*\left(T^{-1}(z), x\right) && (7.8a) \\ &= s^*\Big(R\big(T_\sigma^{-1}(z_\sigma), x\big), P\big(T_p^{-1}(z_p), x\big)\Big), \end{aligned}$$

hence,

$$\tilde{s}^*_T(\cdot, x) = \tilde{s}^*(\cdot, x) \circ T^{-1}. \tag{7.8b}$$

Consequently, the probability of survival can be represented, see (7.5a), by

$$\begin{aligned} p_s &= \mathcal{P}\Big(\tilde{s}^*\big(y(\omega), x\big) < 0\Big) = \mathcal{P}\left(\tilde{s}^*\Big(T^{-1}\big(z(\omega)\big), x\Big) < 0\right) \\ &= \mathcal{P}\Big(\tilde{s}^*_T\big(z(\omega), x\big) < 0\Big) = \mathcal{P}\big(z(\omega) \in B^T_{s,x}\big) \end{aligned} \tag{7.9a}$$

with the transformed safe domain

$$\begin{aligned} B^T_{s,x} &:= \left\{z \in \mathbb{R}^\nu : \tilde{s}^*_T(z, x) < 0\right\} = \left\{z \in \mathbb{R}^\nu : \tilde{s}^*\left(T^{-1}(z), x\right) < 0\right\} \\ &= \left\{T(y) : \tilde{s}^*(y, x) < 0, y \in \mathbb{R}^\nu\right\} = T\Big(\left\{y : \tilde{s}^*(y, x) < 0\right\}\Big) = T(B_{s,x}), \end{aligned} \tag{7.9b}$$

see (7.4a).

In the same way we find, see (7.5b), (7.4b),

$$\begin{aligned} p_f &= \mathcal{P}\Big(\tilde{s}^*\big(y(\omega), x\big) \geq 0\Big) = \mathcal{P}\Big(\tilde{s}^*_T\big(z(\omega), x\big) \geq 0\Big) \\ &= \mathcal{P}\big(z(\omega) \in B^T_{f,x}\big), \end{aligned} \tag{7.10a}$$

where

$$B^T_{f,x} = T(B_{f,x}) = \left\{z \in \mathbb{R}^\nu : \tilde{s}^*_T(z, x) \geq 0\right\} \tag{7.10b}$$

Corresponding to the assumption (7.6), i.e.

$$\bar{y} \in B_{s,x},$$

it is supposed in reliability analysis that

$$0 \in B^T_{s,x} = T(B_{s,x}) \text{ or } y_0 := T^{-1}(0) \in B_{s,x}. \tag{7.11a}$$

In many cases we have

$$\bar{y} = T^{-1}(0) \text{ or } T(\bar{y}) = 0. \tag{7.11b}$$

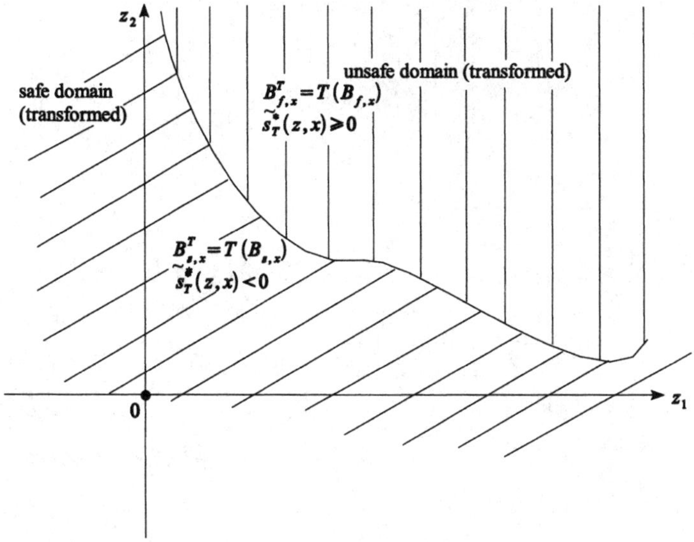


**Fig. 7.2.** Transformed safe/unsafe domain

In any case, relation (7.6), thus $\bar{y} \in B_{s,x}$, yields

$$z_0 := T(\bar{y}) \in T(B_{s,x}) = B^T_{s,x}. \tag{7.11c}$$

## 7.3 Approximation of $p_s, p_f$ by Linearization of the Transformed Limit State Function

### 7.3.1 The Origin Lies in the Transformed Safe Domain

In order to approximate $p_s$, first the so–called $\beta$–point $z^*$, i.e. the projection of the origin 0 in the $z$–domain onto the failure domain $B^T_{f,x}$ is determined:

$$\min \|z\|^2 \tag{7.12a}$$

s.t.

$$\tilde{s}^*_T(z,x) \geq 0. \tag{7.12b}$$

Obviously, (7.12b) is equivalent to

$$-\tilde{s}^*_T(z,x) \leq 0. \tag{7.12b'}$$

The Lagrangian $L = L(z, \lambda)$ of (7.12a), (7.12b') reads

$$L(z,\lambda) := \|z\|^2 + \lambda\big(-\tilde{s}^*_T(z,x)\big). \tag{7.13}$$

Therefore, one has the following necessary optimality conditions for $z = z^*$:

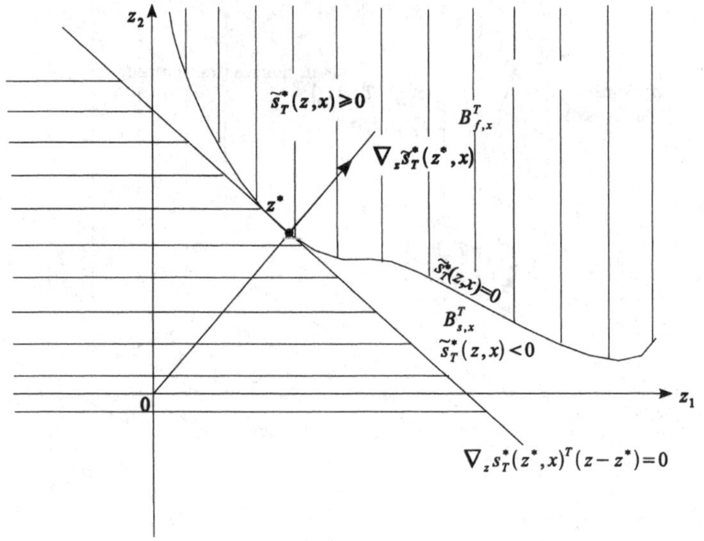


**Fig. 7.3.** Projection and linearization *I*

$$\nabla_z L(z,\lambda) = 2z + \lambda\big(-\nabla_z \tilde{s}_T^*(z,x)\big) = 0 \tag{7.14a}$$

$$\frac{\partial L}{\partial \lambda}(z,\lambda) = -\tilde{s}_T^*(z,x) \qquad \leq 0 \tag{7.14b}$$

$$\lambda \frac{\partial L}{\partial \lambda}(z,\lambda) = \lambda\big(-\tilde{s}_T^*(z,x)\big) \qquad = 0 \tag{7.14c}$$

$$\lambda \quad \geq 0. \tag{7.14d}$$

**Theorem 7.1 (Necessary optimality conditions).** *Let $z^*$ be an optimal solution of the projection problem (7.12a,b'). If $z^*$ is regular, i.e. if*

$$\tilde{s}_T^*(z^*,x) > 0 \text{ or} \tag{7.15a}$$

$$\nabla_z \tilde{s}_T^*(z^*,x) \neq 0 \text{ in case } \tilde{s}_T^*(z^*,x) = 0, \tag{7.15b}$$

*then there exists $\lambda^* \in \mathbb{R}$ such that $(z^*,\lambda^*)$ fulfills the Kuhn–Tucker conditions (7.14a-d).*

From the above theorem we obtain the following consequences:

a) In the first case (7.15a) we get $\lambda^* = 0$ and therefore $z^* = 0$ with $\tilde{s}_T^*(0,x) > 0$. Hence, $0 \in B_{f,x}^T$ which contradicts to the assumption that $0 \in B_{s,x}^T$. Thus, under this assumption, case (7.15a) is not possible.
b) In the second case it is $\tilde{s}_T^*(z^*,x) = 0$ and

$$z^* = \frac{\lambda^*}{2}\nabla_z \tilde{s}_T^*(z^*,x) \text{ with } \lambda^* > 0. \tag{7.16}$$

Condition $\lambda^* > 0$ holds, since otherwise we get $z^* = 0$ and therefore again $\tilde{s}_T^*(0,x) = 0$ or $0 \in B_{f,x}^T$ which contradicts our assumption in Section 7.3.1.

Having $z^*$, the state function $\tilde{s}_T^* = \tilde{s}_T^*(z,x)$ is now linearized at $z = z^*$:

$$\begin{aligned} \tilde{s}_T^*(z,x) &= \tilde{s}_T^*(z^*,x) + \nabla_z \tilde{s}_T^*(z^*,x)^T(z - z^*) + \dots \\ &= \nabla_z \tilde{s}_T^*(z^*,x)^T(z - z^*) + \dots \end{aligned} \tag{7.17}$$

According to the definition (7.9a), the probability $p_s$ of survival can be approximated by

$$\begin{aligned} p_s = P\Big(\tilde{s}_T^*\big(z(\omega),x\big) < 0\Big) &\approx P\Big(\nabla_z \tilde{s}_T^*(z^*,x)^T\big(z(\omega) - z^*\big) < 0\Big) \\ &= P\big(\nabla_z \tilde{s}_T^*(z^*,x)^T z(\omega) < \nabla_z \tilde{s}_T^*(z^*,x)^T z^*\big). \end{aligned} \tag{7.18}$$

Due to our construction, $z = z(\omega)$ is a standard normal distributed random vector, see (7.7b). Hence, the random variable

$$\xi(\omega) = \nabla_z \tilde{s}_T^*(z^*,x)^T z(\omega) \text{ is } N\Big(0, \big\|\nabla_z \tilde{s}_T^*(z^*,x)\big\|^2\Big) - \text{distributed,} \tag{7.19}$$

and (7.18) yields therefore

$$p_s \approx \Phi\left(\frac{\nabla_z \tilde{s}_T^*(z^*,x)^T z^* - 0}{\big\|\nabla_z \tilde{s}_T^*(z^*,x)\big\|}\right) = \Phi\left(\frac{\nabla_z \tilde{s}_T^*(z^*,x)^T z^*}{\big\|\nabla_z \tilde{s}_T^*(z^*,x)\big\|}\right). \tag{7.20}$$

Inserting (7.16) into (7.20) yields

$$\begin{aligned} p_s \approx \Phi\left(\frac{\nabla_z \tilde{s}_T^*(z^*,x)^T \frac{\lambda^*}{2} \nabla_z \tilde{s}_T^*(z^*,x)}{\big\|\nabla_z \tilde{s}_T^*(z^*,x)\big\|}\right) &= \Phi\left(\frac{\lambda^*}{2}\big\|\nabla_z \tilde{s}_T^*(z^*,x)\big\|\right) \\ &= \Phi\big(\|z^*\|\big). \end{aligned}$$

Consequently, we have this result:

**Theorem 7.2.** *The probability of survival $p_s$ can be approximated by*

$$p_s \approx \Phi\big(\|z^*\|\big), \tag{7.21}$$

*where $\|z^*\|$ is the length of the regular projection $z^*$ of the origin 0 onto the transformed failure domain $B_{f,x}^T$.*

*Having (7.21), the probabilistic constrains*

$$p_s \geq \alpha_s \tag{7.22a}$$

*with given $\alpha_s$, e.g. $\alpha_s = 0.99$, can be approximated by the condition*

$$\|z^*\| \geq \Phi^{-1}(\alpha_s). \tag{7.22b}$$

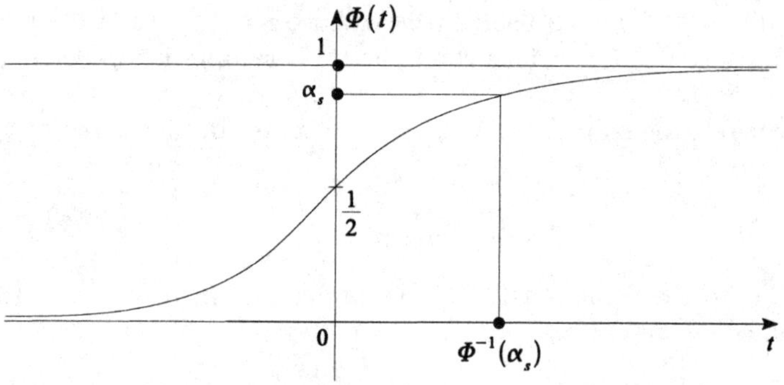


**Fig. 7.4.** Distribution function $N(0,1)$

### 7.3.2 The Origin Lies in the Transformed Failure Domain

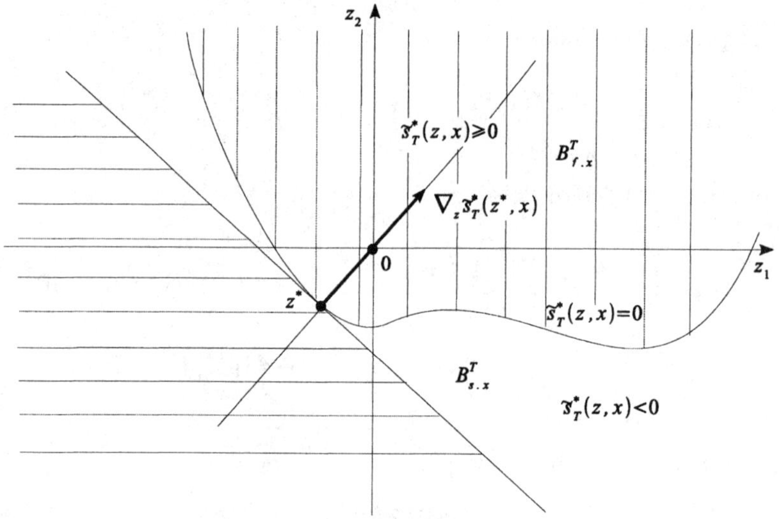


**Fig. 7.5.** Projection and linearization II

Considering here the projection of the origin 0 onto the closure of the safe domain $B_{s,x}^T$, we have the minimization problem

$$\min \|z\|^2 \tag{7.23a}$$

s.t.

$$\tilde{s}_T^*(z,x) \leq 0. \tag{7.23b}$$

In this case we obtain the Lagrangian

$$L(z, \lambda) = \|z\|^2 + \lambda \tilde{s}_T^*(z, x), \tag{7.24}$$

and the Kuhn–Tucker conditions for a projection $z^*$ of the origin 0 onto $B_{s,x}^T$ read, cf. (7.14a-d),

$$\nabla_z L(z, \lambda) = 2z + \lambda \nabla_z \tilde{s}_T^*(z, x) = 0 \tag{7.25a}$$

$$\frac{\partial L}{\partial \lambda}(z, \lambda) = \tilde{s}_T^*(z, x) \leq 0 \tag{7.25b}$$

$$\lambda \frac{\partial L}{\partial \lambda}(z, \lambda) = 0 \tag{7.25c}$$

$$\lambda \geq 0 \tag{7.25d}$$

The necessary optimality conditions can then be discussed as before:

a) If $z = z^*$ is a regular point such that $\tilde{s}_T^*(z^*, x) < 0$, then $\lambda^* = 0$, and therefore $z^* = 0$. However, this means that $\tilde{s}_T^*(0, x) < 0$, hence $0 \in B_{s,x}^T$, which contradicts our assumption in Section 7.3.2. Thus, this case is not possible.

b) Let $z = z^*$ be a regular point with $\tilde{s}_T^*(z^*, x) = 0$ and $\nabla \tilde{s}_T^*(z^*, x) \neq 0$. Then

$$z^* = -\frac{\lambda^*}{2} \nabla_z \tilde{s}_T^*(z^*, x) \text{ with } \lambda^* \geq 0, \tag{7.26}$$

see (7.16). Now, the state function $\tilde{s}_T^* = \tilde{s}_T^*(z, x)$, is linearized at the $\beta$–point $z^*$, thus

$$\begin{aligned} \tilde{s}_T^*(z, x) &= \tilde{s}_T^*(z^*, x) + \nabla_z \tilde{s}_T^*(z^*, x)^T (z - z^*) + \ldots \\ &= \nabla_z \tilde{s}_T^*(z^*, x)^T (z - z^*) + \ldots . \end{aligned} \tag{7.27}$$

As before, the probability of survival $p_s$ is approximated by

$$\begin{aligned} p_s &= P\Big(\tilde{s}_T^*(z(\omega), x) < 0\Big) \approx P\Big(\nabla_z \tilde{s}_T^*(z^*, x)^T (z(\omega) - z^*) < 0\Big) \\ &= \Phi\left(\frac{\nabla_z \tilde{s}^*(z^*, x)^T z^*}{\|\nabla_z \tilde{s}_T^*(z^*, x)\|}\right). \end{aligned} \tag{7.28}$$

Inserting (7.26) into (7.28), we find

$$\begin{aligned} p_s &\approx \Phi\left(\frac{\nabla_z \tilde{s}^*(z^*, x)^T \left(-\frac{\lambda^*}{2} \nabla_z \tilde{s}^*(z^*, x)\right)}{\|\nabla_z \tilde{s}_T^*(z^*, x)\|}\right) \\ &= \Phi\left(-\frac{\lambda^*}{2} \|\nabla_z \tilde{s}_T^*(z^*, x)\|\right) = \Phi\left(-\left\|\frac{\lambda^*}{2} \nabla_z \tilde{s}_T^*(z^*, x)\right\|\right) \\ &= \Phi(-\|z^*\|). \end{aligned}$$

In addition to Theorem 7.2, here we have this result:

**Theorem 7.3.** *If* $0 \in B^T_{f,x}$, *then*

$$p_s \approx \Phi(-\|z^*\|), \tag{7.29}$$

*where* $\|z^*\|$ *is the length of the regular projection* $z^*$ *of the origin* $0$ *onto the closure of the transformed safe domain* $B^T_{s,x}$.

*Remark 7.1.* Since $\Phi(-\|z^*\|) \leq \frac{1}{2}$, the case $0 \in B^T_{f,x}$ or $0 \notin B^T_{s,x}$ is of minor interest in practice. Hence, in the following we consider the case $0 \in B^T_{s,x}$ only.

## 7.4 Computation of the $\beta$–Point $z^*$

According to Section 7.3, for a given vector $x$, a $\beta$–point $z^* = z^*_x$ is defined by the projection problem (7.12a,b), i.e.

$$\min \|z\|^2$$

s.t.

$$\tilde{s}^*_T(z,x) \geq 0.$$

In the present case of elastoplastic materials, according to (2.53a-c) in Section 2.6.1, the transformed state function $\tilde{s}^*_T = \tilde{s}^*_T(z,x)$ is the minimum value of the convex minimization problem:

$$\min s \tag{7.30a}$$

s.t.

$$C\sigma = P\big(T_p^{-1}(z_p), x\big) \tag{7.30b}$$

$$\pi\Big(R_{id}\big(T_\sigma^{-1}(z_\sigma), x\big)^{-1}\sigma_i | K_i\Big) - 1 \leq s, i = 1, \ldots, n_G. \tag{7.30c}$$

Since (7.30a-c) is a convex minimization problem in the variables $(s, \sigma)$, and the problem fulfills the Slater condition, an optimal solution $(\tilde{s}^*_T, \sigma^*_T) = \big(\tilde{s}^*_T(z,x), \sigma^*_T(z,x)\big)$ can be characterized by the necessary and sufficient Kuhn–Tucker conditions [42]. Hence, let

$$L = s + \lambda^T(C\sigma - P) + \sum_{i=1}^{n_G} \mu_i\big(\pi(R_{id}^{-1}\sigma_i|K_i) - (1+s)\big) \tag{7.31a}$$

denote the Lagrange function of (7.30a-c), where

$$P = P\big(T_p^{-1}(z_p), x\big) = \tilde{P}(z_p, x) \tag{7.31b}$$

$$R_i = R_i\big(T_\sigma^{-1}(z_\sigma), x\big) = \tilde{R}_i(z_\sigma, x), i = 1, 2, \ldots, n_G. \tag{7.31c}$$

Then, $(\tilde{s}_T^*, \sigma_T^*)$ is an optimal solution of (7.30a-c) if and only if there exist Lagrange multipliers $\lambda^* = \lambda^*(z,x), \mu^* = \mu^*(z,x)$ such that

$$\frac{\partial L}{\partial s} = 1 - \sum_{i=1}^{n_G} \mu_i^* = 0 \tag{7.32a}$$

$$\frac{\partial L}{\partial \sigma_i} = C_{(i)}^T \lambda^* + \mu_i^* \tilde{R}_{id}(z,x)^{-1} \nabla_z \pi\left(\tilde{R}_{id}(z,x)^{-1} \sigma_i^* | K_i\right) = 0,$$
$$i = 1, \ldots, n_G \tag{7.32b}$$

$$\frac{\partial L}{\partial \lambda} = C\sigma^* - \tilde{P}(z,x) = 0 \tag{7.32c}$$

$$\frac{\partial L}{\partial \mu_i} = \pi\left(\tilde{R}_{id}(z,x)^{-1} \sigma_i^* | K_i\right) - (1 + \tilde{s}_T^*) \le 0, i = 1, \ldots, n_G \tag{7.32d}$$

$$\mu_i \frac{\partial L}{\partial \mu_i} = \mu_i \left(\pi\left(\tilde{R}_{id}(z,x)^{-1} \sigma_i^* | K_i\right) - (1 + \tilde{s}_T^*)\right) = 0,$$
$$i = 1, \ldots, n_G \tag{7.32e}$$

$$\mu_i \ge 0, i = 1, \ldots, n_G, \tag{7.32f}$$

where $C_{(i)}$ is the submatrix of $C$ related to $\sigma_i$, see also Section 2.6.

Thus, the projection problem (7.12a,b) can be represented in the following *explicit* form:

**Lemma 7.1.** *Let $x$ be a given $r$-vector. The projection problem (7.12a,b) for the computation of a $\beta$–point $z^* = z_x^*$ and the minimum distance $\beta_x = \|z_x^*\|$ between the origin 0 in $z$–space and the failure domain $B_{f,x}^T$ reads:*

$$\min \|z\|^2 \tag{7.33a}$$

*s.t.*

$$1 - \sum_{i=1}^{n_G} \mu_i = 0 \tag{7.33b}$$

$$C_{(i)}^T \lambda + \mu_i \tilde{R}_{id}(z,x)^{-1} \nabla_z \pi\left(\tilde{R}_{id}(z,x)^{-1} \sigma_i | K_i\right) = 0, i = 1, \ldots, n_G \tag{7.33c}$$

$$C\sigma - \tilde{P}(z,x) = 0 \tag{7.33d}$$

$$\pi\left(\tilde{R}_{id}(z,x)^{-1} \sigma_i | K_i\right) - (1 + s) \le 0, i = 1, \ldots, n_G \tag{7.33e}$$

$$\mu_i \left(\pi\left(\tilde{R}_{id}(z,x)^{-1} \sigma_i | K_i\right) - (1 + s)\right) = 0, li = 1, \ldots, n_G \tag{7.33f}$$

$$\mu_i \ge 0, i = 1, \ldots, n_G \tag{7.33g}$$

$$s \ge 0. \tag{7.33h}$$

A further representation can be obtained if, by piecewise linearization, the convex feasible domains $K_i, i = 1, \ldots, n_G$, are replaced by convex polyhedrons $\tilde{K}_i$ defined by

$$\tilde{K}_i = \{z \in \mathbb{R}^{n_0} : \tilde{N}_i z \le 1\}, i = 1, \ldots, n_G, \tag{7.34}$$

with given $\tilde{m}_y \times n_0$ matrices $\tilde{N}_i, i = 1, \ldots, n_G$, and the vector $1 = 1_{\tilde{m}_y} = (1, \ldots, 1)^T \in \mathbb{R}^{\tilde{m}_y}$, see [89].

Then, the state function $\tilde{s}_T^* = \tilde{s}_T^*(z, x)$ is the minimum value function of the linear program

$$\min s \tag{7.35a}$$

s.t.

$$C\sigma = \tilde{P}(z, x) \tag{7.35b}$$

$$\tilde{N}_i \tilde{R}_{id}(z, x)^{-1} \sigma_i \leq (1 + s) 1_{\tilde{m}_y}, i = 1, \ldots, n_G. \tag{7.35c}$$

Introducing fixed reference material strength values $R_{ij}^{\circ} > 0, j = 1, \ldots, n_0, i = 1, \ldots, n_G$, a sufficient condition for (7.35c) reads, cf. [89],

$$\tilde{N}_i R_{id}^{0-1} \sigma_i \leq \left(\tilde{\rho}_{i\,\min}(z, x) + s\right) 1_{\tilde{m}_y}, \tag{7.35c'}$$

where $R_i^0 = (R_{ij}^0)_{i \leq j \leq n_0}$, and

$$\tilde{\rho}_{i\,\min}(z, x) := \min_{1 \leq j \leq n_0} \frac{R_{ij}\left(T_\sigma^{-1}(z_\sigma), x\right)}{R_{ij}^0}. \tag{7.36}$$

Consequently, the state function $\tilde{s}_T^* = \tilde{s}_T^*(z, x)$ can be represented then also by the maximum value function of the dual program to (7.35a-c):

$$\max \tilde{P}(z, x)^T u - \tilde{F}_0(z, x)^T \tilde{u} \tag{7.37a}$$

s.t.

$$C^T u - \hat{R}(z, x)_d^{-1} N^T \tilde{u} = 0 \tag{7.37b}$$

$$1_{\nu_G}^T \tilde{u} = 1 \tag{7.37c}$$

$$\tilde{u} \geq 0, \tag{7.37d}$$

where, with a generalized unit matrix $U$ and $\tilde{\varrho}_{\min} := (\tilde{\varrho}_{i\,\min})_{1 \leq i \leq n_G}$,

$$\hat{R}(z, x) := \tilde{R}(z, x), \hat{R}(z, x) = R^0, \text{ resp.}, \tag{7.37e}$$

$$\tilde{F}_0(z, x) = U 1_{n_G}, \tilde{F}_0(z, x) = U \tilde{\varrho}_{\min}(z, x), \text{ resp.}, \tag{7.37f}$$

corresponding to the cases (7.35c), (7.35c'), cf. [89].

Thus, due to this duality relation, the constraint $\tilde{s}_T^*(z, x) \geq 0$ holds if and only if there is a vector $(u, \tilde{u})$ of dual variables fulfilling the relations

$$\tilde{P}(z, x)^T u - \tilde{F}_0(z, x)^T \tilde{u} \geq 0 \tag{7.38a}$$

$$C^T u - \hat{R}(z, x)_d^{-1} N^T \tilde{u} = 0 \tag{7.38b}$$

$$1_{\nu_G}^T \tilde{u} = 1 \tag{7.38c}$$

$$\tilde{u} \geq 0. \tag{7.38d}$$

Here, corresponding to Lemma 7.1, we find this characterization of a $\beta$–point:

**Lemma 7.2.** *Let $x$ be a given $r$-vector. Replace the feasible domains $K_i$ by convex polyhedrons $\tilde{K}_i, i = 1, \dots, n_G$. Then the projection problem (7.12a,b) for the computation of a $\beta$–point $z_x^*$ has the following explicit form:*

$$\min \|z\|^2 \tag{7.39a}$$

*s.t.*

$$\tilde{P}(z,x)^T u - \tilde{F}_0(z,x)^T \tilde{u} \geq 0 \tag{7.39b}$$
$$C^T u - \hat{R}(z,x)_d^{-1} N^T \tilde{u} = 0 \tag{7.39c}$$
$$1_{\nu_G}^T \tilde{u} = 1 \tag{7.39d}$$
$$\tilde{u} \geq 0. \tag{7.39e}$$

## 7.5 Trusses

For trusses [89, 117, 133] the state function $s^* = s^*(R,P) = s^*(F_0,P)$ can be represented, cf. (2.53a-c), by the linear program:

$$\min s \tag{7.40a}$$

s.t.

$$CF = P(x) \tag{7.40b}$$
$$HF - F_0(\sigma_y, x) \leq s1, \tag{7.40c}$$

where $F$ denotes the $n$–vector of member (axial) forces, $C$ is the $m \times n$ equilibrium matrix, $P = P(x)$ is the $m$–vector of external loads, and $x$ is a given vector of design variables or load factors. Moreover, the matrices and vectors in (7.40c) are given by

$$H = \begin{pmatrix} I \\ -I \end{pmatrix}, F_0(\sigma_y, x) = \begin{pmatrix} \sigma_d^U A(x) \\ -\sigma_d^L A(x) \end{pmatrix}, \sigma_y = \begin{pmatrix} \sigma^U \\ \sigma^L \end{pmatrix}, A(x) = \big(A_k(x)\big).$$

Here, $I$ is the $m \times n$ identy matrix, $\sigma^U, \sigma^L$ denotes the $n$–vector of yield stresses $\sigma_j^U, \sigma_j^L, j = 1, \dots, n$, in tension, compression, resp., and $A$ is the $n$–vector of cross sectional areas $A_k, k = 1, \dots, n$. The dual maximization problem of LP (7.40a-c) has then the following form, cf. (7.37a-f):

$$\max P(x)^T u - F_0(\sigma_y, x)^T \tilde{u} \tag{7.41a}$$

s.t.

$$C^T u - H^T \tilde{u} = 0 \tag{7.41b}$$
$$1^T \tilde{u} = 1 \tag{7.41c}$$
$$\tilde{u} \geq 0. \tag{7.41d}$$

Consequently, due to Lemma 7.2, (7.2a-d) and (7.7a-c), the $\beta$–point $z^* = z_x^*$ is determined by the minimization problem

$$\min \|z\|^2 \tag{7.42a}$$

s.t.

$$P\big(T_p^{-1}(z_p), x\big)^T u - F_0\big(T_\sigma^{-1}(z_\sigma), x\big)^T \tilde{u} \geq 0 \tag{7.42b}$$

$$C^T u - H^T \tilde{u} = 0 \tag{7.42c}$$

$$1^T \tilde{u} = 1 \tag{7.42d}$$

$$\tilde{u} \geq 0. \tag{7.42e}$$

Let

$$U := \big\{(u, \tilde{u}) : C^T u - H^T \tilde{u} = 0, 1^T \tilde{u} = 1, \tilde{u} \geq 0\big\} \tag{7.43}$$

denote the set of feasible dual vectors $(u, \tilde{u})$ in (7.42a-e). Then, for a given vector $(u, \tilde{u}) \in U$ of dual variables, define now the subproblem:

$$\min \|z\|^2 \tag{7.44a}$$

s.t.

$$P\Big(T_p^{-1}(z_p), x\Big)^T u - F_0\Big(T_\sigma^{-1}(z_\sigma), x\Big)^T \tilde{u} \geq 0. \tag{7.44b}$$

Furthermore, let denote $z_x^* = z_x^*(u, \tilde{u})$ an optimal solution of the above projection problem (7.44a,b), hence, a $\beta$–point with respect to the failure domain indicated by the dual variables (failure mode) $(u, \tilde{u})$.

Then, for a $\beta$–point $z_x^*$, i.e. an optimal solution of (7.42a-e), we have the following representation

$$\begin{aligned} \|z_x^*\|^2 &= \inf\left\{\Big\|z_x^*(u, \tilde{u})\Big\|^2 : (u, \tilde{u}) \in U\right\} \\ &= \inf\left\{\Big\|z_x^*(u, \tilde{u})\Big\|^2 : C^T u - H^T \tilde{u} = 0, 1^T \tilde{u} = 1, \tilde{u} \geq 0\right\}. \end{aligned} \tag{7.45}$$

Note that $\|z_x^*(u, \tilde{u})\|$ is the distance between the origin 0 and the surface in $z$–space given by

$$P\Big(T_p^{-1}(z_p), x\Big)^T u - F_0\Big(F_0^{-1}(z_\sigma), x\Big)^T \tilde{u} \geq 0. \tag{7.44b'}$$

*Remark 7.2.* Inserting a nonnegative slack variable $w \geq 0$ into (7.44b), we also have

$$\|z_x^*\|^2 = \inf\left\{\Big\|z_x^*(u, \tilde{u}, w)\Big\|^2 : C^T u - H^T \tilde{u} = 0, 1^T \tilde{u} = 1, \tilde{u} \geq 0, w \geq 0\right\}, \tag{7.45'}$$

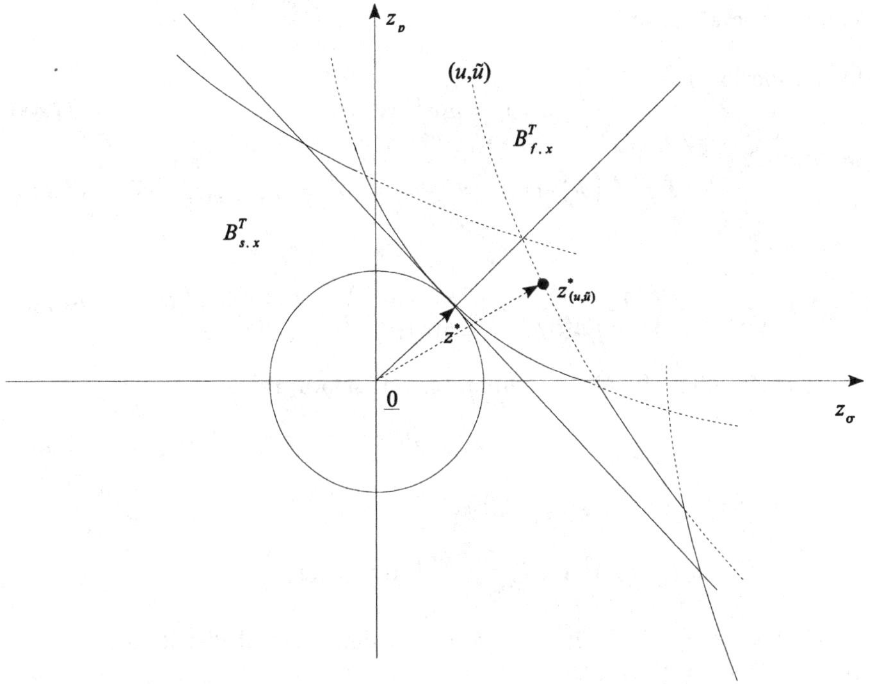


**Fig. 7.6.** $\beta$–points in case of trusses

where $z_x^*(u, \tilde{u}, w)$ is an optimal solution of

$$\min \|z\|^2 \tag{7.46a}$$

s.t.

$$P\Big(T_p^{-1}(z_p), x\Big)^T u - F_0\Big(T_\sigma^{-1}(z_\sigma), x\Big)^T \tilde{u} + w = 0. \tag{7.46b}$$

As an important consequence of the minimum representation (7.45), (7.45′) we find that the reliability constraint (7.22b), hence,

$$\|z_x^*\|^2 \geq \beta_s^2 \text{ with } \beta_s := \Phi^{-1}(\alpha_s)$$

holds if and only if

$$\Big\|z_x^*(u, \tilde{u})\Big\|^2 \geq \beta_s^2 \text{ for all } (u, \tilde{u}) \in U, \tag{7.47a}$$

where again

$$U := \Big\{(u, \tilde{u}) : C^T u - H^T \tilde{u} = 0, 1^T \tilde{u} = 1, \tilde{u} \geq 0\Big\}. \tag{7.47b}$$

### 7.5.1 Special Case

In the special case

$$P(p) = p, \sigma^L = -\sigma^U \tag{7.48}$$

we have

$$P = P\Big(T_p^{-1}(z_p)\Big) = T_p^{-1}(z_p), \sigma^U = T_\sigma^{-1}(z_\sigma) \tag{7.49a}$$

and

$$F_0(\sigma_y, x) = \begin{pmatrix} \sigma_d^U A(x) \\ -\sigma_d^L A(x) \end{pmatrix} = \begin{pmatrix} A(x)_d \sigma^U \\ A(x)_d \sigma^U \end{pmatrix} = \begin{pmatrix} A(x)_d \\ A(x)_d \end{pmatrix} \sigma^U. \tag{7.49b}$$

Here, for given $(u, \tilde{u}) \in U$, subproblem (7.44a,b) reads

$$\min \|z\|^2 \tag{7.50a}$$

s.t.

$$T_p^{-1}(z_p)^T u - T_\sigma^{-1}(z_\sigma)^T \Big(A(x)_d \vdots A(x)_d\Big) \tilde{u} \geq 0. \tag{7.50b}$$

If $P = P(\omega), \sigma^U = \sigma^U(\omega)$ are independent normal distributed random vectors, then, cf. (7.7a-c),

$$P(\omega) = T_p^{-1}\Big(z_p(\omega)\Big) = \bar{P} + \Gamma_p z_p(\omega) \tag{7.51a}$$

$$\sigma^U(\omega) = T_\sigma^{-1}\Big(z_\sigma(\omega)\Big) = \bar{\sigma}^U + \Gamma_\sigma z_\sigma(\omega), \tag{7.51b}$$

where $\overline{P}, \overline{\sigma}^U$ denote the expectations of $P(\omega), \sigma^U(\omega)$, and $\Gamma_p, \Gamma_\sigma$ are given regular matrices. Furthermore, $z_p = z_p(\omega), z_\sigma = z_\sigma(\omega)$ are standard normal distributed random vectors with mean zero and identity covariance matrix.

Inserting (7.51a,b) into (7.50b), problem (7.50a,b) for the computation of the projections $z_x^*(u, \tilde{u})$ reads

$$\min \|z_p\|^2 + \|z_\sigma\|^2 \tag{7.52a}$$

s.t.

$$(\bar{P} + \Gamma_p z_p)^T u - (\bar{\sigma}^U + \Gamma_\sigma z_\sigma)^T \Big(A(x)_d \vdots A(x)_d\Big) \tilde{u} \geq 0. \tag{7.52b}$$

Furthermore, using subproblem (7.46a,b), in (7.52a,b) we have to replace the constraint (7.52b) by

$$(\bar{P}+\Gamma_p z_p)^T u - (\bar{\sigma}^U + \Gamma_\sigma z_p)^T \Big(A(x) \vdots A(x)\Big) \tilde{u} + w = 0. \tag{7.52b'}$$

## 7.6 Reliability–Based Design Optimization (RBDO)

According to Section 2.1.1 , we have to solve problems of the type

$$\min f(x) \tag{7.53a}$$

s.t.

$$p_s(x) \geq \alpha_s \tag{7.53b}$$
$$x \in D, \tag{7.53c}$$

where $p_s = p_s(x)$ is the probability that the external load is carried safely by the structure, $\alpha_s$ is the minimum reliability, $f = f(x)$ is a cost function, (an expected cost function), as e.g. the volume, weight of the structure or the negtive load factor, and $D$ is the deterministic feasible domain for the design $r$–vector $x$. Using the first order approximate (7.22b), i.e. the condition

$$||z_x^*||^2 \geq \beta_s^2 \text{ with } \beta_s := \Phi^{-1}(\alpha_s),$$

problem (7.53a-c) can be approximated, see Lemma 7.2, by

$$\min f(x) \tag{7.54a}$$

s.t.

$$||z||^2 \geq \beta_s^2 \tag{7.54b}$$
$$z \in \text{ argmin}(7.39\text{a-e}) \mid_x \tag{7.54c}$$
$$x \in D. \tag{7.54d}$$

A related problem is obtained if we want to maximize the survival probability $p_s = p_s(x)$ subject to a cost constraint $f(x) \leq c^{\max}, x \in D$. According to (7.21) we have

$$\max ||z||^2 \tag{7.55a}$$

s.t.

$$z \in \text{ argmin}(7.39\text{a-e})|_x \tag{7.55b}$$
$$f(x) \leq c^{\max} \tag{7.55c}$$
$$x \in D. \tag{7.55d}$$

Solving (7.54a-d) or (7.55a-d), the main difficulty is the solution of the internal optimization problem, i.e. the projection problem (7.39a-e) depending on a design vector $x$. Hence, problem (7.54a-d), (7.55a-d), resp., with the complicated constraint (7.54c), (7.55b) must be solved by applying some approximative strategies.

### 7.6.1 Necessary Optimality Conditions for the $\beta$–Point

Assume that the necessary optimality conditions for a $\beta$–point $z_x^*$, i.e. an optimal solution of (7.39a-e) is given by the relation

$$(z_x^*, \lambda^*) \in \mathcal{R}_x, \tag{7.56}$$

where $\lambda$ is a certain vector of Lagrange multipliers. Replacing (7.54c) by the necessary optimality condition (7.56), problem (7.54a-d) can be approximated by

$$\min f(x) \tag{7.57a}$$

s.t.

$$\|z\|^2 \geq \beta_s^2 \tag{7.57b}$$
$$(z, \lambda) \in \mathcal{R}_x \tag{7.57c}$$
$$x \in D. \tag{7.57d}$$

Of course, (7.55a-d) can be approximated in the same way. Problem (7.57a-d) is now an ordinary parameter optimization problem to be solved by numerical optimization procedures.

In order to study the relationship between (7.54a-d) and (7.57a-d) we suppose that $(x, z)$ is a feasible point in (7.54a-d). Thus, $x \in D$ and $z = z_x^*$ is an optimal solution of (7.39a-e) with $\|z\|^2 \geq \beta_s^2$. According to the above assumption, there is a vector $\lambda$ such that (7.56) is fulfilled. Hence, the following relations hold:

$$\|z\|^2 \geq \beta_s^2$$
$$(z, \lambda) \in \mathcal{R}_x$$
$$x \in D.$$

However, this shows that the tuple $(z, \lambda, x)$ is a feasible point in (7.57a-d). As a consequence, we find this result:

**Lemma 7.3.** *Suppose that the necessary optimality conditions of (7.39a-e) are given by relation (7.56). Then*

$$\inf \text{(7.54a-d)} \geq \inf \text{(7.57a-d)}\,. \tag{7.58}$$

### 7.6.2 Duality Relations

Suppose that for given $x \in D$ there is a dual maximization problem related to (7.39a-e) which can be represented by

$$\max g(v, x) \tag{7.59a}$$

s.t.

$$H_I(v,x) \leq 0 \tag{7.59b}$$
$$H_{II}(v,x) = 0 \tag{7.59c}$$
$$v \in V. \tag{7.59d}$$

Moreover, assume that

$$\min\,(7.39\text{a-e}) \ \geq \max\,(7.59\text{a-d}) \text{ for each } x \in D. \tag{7.60}$$

Consequently, from

$$\|z_x^*\|^2 = \inf\,(7.39\text{a-e}) \ \geq \max\,(7.59\text{a-d}) \ = g(v^*, x)$$

for a feasible point $v^*$ of (7.59a-d), the next lemma follows:

**Lemma 7.4.** *Suppose that (7.60) holds. Then (7.54a-d) may be approximated by*

$$\min f(x) \tag{7.61a}$$

*s.t.*

$$g(v,x) \geq \beta_s^2 \tag{7.61b}$$
$$H_I(v,x) \leq 0 \tag{7.61c}$$
$$H_{II}(v,x) = 0 \tag{7.61d}$$
$$v \in V, x \in D. \tag{7.61e}$$

*If (7.60) holds with "=", then (7.54a-d) is equivalent to the above program. If "≥" holds, then (7.61a-e) yields at least an upper cost bound for (7.54a-d).*

# Part VI

# Appendix

# A

# Sequences, Series and Products

In this appendix let $\mathbb{N} = \mathbb{N}^{(1)} \cup \ldots \cup \mathbb{N}^{(\kappa)}$ denote a partition of $\mathbb{N}$ fulfilling condition (U6) from Chapter 6.3.

## A.1 Mean Value Theorems for Deterministic Sequences

Here, often used statements about generalized mean values of scalar sequences are given.

In this context the following double sequences are of importance. We use the following notations:

$$\begin{aligned}
\mathcal{F} &:= \{(a_n) \;:\; a_n \in \mathbb{R} \quad \text{for} \quad n \in \mathbb{N}\} \,, \\
\mathcal{D} &:= \{(a_{m,n})_{m,n} \;:\; a_{m,n} \in \mathbb{R} \quad \text{for} \quad 0 \le m \le n\} \,, \\
\mathcal{D}_+ &:= \{(a_{m,n})_{m,n} \in \mathcal{D} \;:\; \text{it exists} \quad n_0 \in \mathbb{N} \quad \text{with} \\
&\qquad\quad a_{m,n} > 0 \quad \text{for} \quad n_0 \le m \le n\} \,, \\
\mathcal{D}_0 &:= \{(a_{m,n})_{m,n} \in \mathcal{D}_+ \;:\; \lim_{n\to\infty} a_{m,n} = 0 \quad \text{for} \quad m = 0,1,2,\ldots\} \,.
\end{aligned}$$

Two sequences $(a_{m,n})$ and $(b_{m,n})$ in $\mathcal{D}_+$ can be compared if we define

a) $(a_{m,n})$ is *not greater* than $(b_{m,n})$ , denoted as $a_{m,n} \ll b_{m,n}$ , if to any number $\varepsilon > 0$ there exists $n_0 \in \mathbb{N}$ with

$$0 < a_{m,n} \le (1+\varepsilon) b_{m,n} \quad \text{for all} \quad n_0 \le m \le n \,.$$

b) $(a_{m,n})$ and $(b_{m,n})$ are *equivalent*, denoted as $a_{m,n} \sim b_{m,n}$ , if

$$a_{m,n} \ll b_{m,n} \quad \text{and} \quad b_{m,n} \ll a_{m,n} \,.$$

Hence, "$\ll$" is an order, and "$\sim$" is an equivalence relation on $\mathcal{D}_+$ having following properties:

**Lemma A.1.** *Let* $(a_{m,n}), (b_{m,n}), (\alpha_{m,n})$ *and* $(\beta_{m,n})$ *denote sequences from* $\mathcal{D}_+$.

*a) If* $a_{m,n} \ll \alpha_{m,n}$ *and* $b_{m,n} \ll \beta_{m,n}$, *then*

$$a_{m,n} + b_{m,n} \ll \alpha_{m,n} + \beta_{m,n}\,, \quad a_{m,n}b_{m,n} \ll \alpha_{m,n}\beta_{m,n}\,,$$
$$\frac{a_{m,n}}{\beta_{m,n}} \ll \frac{\alpha_{m,n}}{b_{m,n}}\,, \quad {a_{m,n}}^s \ll {\alpha_{m,n}}^s \quad \textit{for} \quad s > 1\,.$$

*b) If* $a_{m,n} \sim \alpha_{m,n}$ *and* $b_{m,n} \sim \beta_{m,n}$, *then*

$$a_{m,n} + b_{m,n} \sim \alpha_{m,n} + \beta_{m,n}\,, \quad a_{m,n}b_{m,n} \sim \alpha_{m,n}\beta_{m,n}\,,$$
$$\frac{a_{m,n}}{b_{m,n}} \sim \frac{\alpha_{m,n}}{\beta_{m,n}}\,, \quad {a_{m,n}}^s \sim {b_{m,n}}^s \quad \textit{for} \quad s \in \mathbb{R}\,.$$

*Proof.* In case of a) there exists to any $\varepsilon > 0$ an index $n_0 \in \mathbb{N}$ with:

$$0 < a_{m,n} \le (1+\varepsilon)\alpha_{m,n},$$
$$0 < b_{m,n} \le (1+\varepsilon)\beta_{m,n} \quad \text{for} \quad n_0 \le m \le n\,.$$

Hence, for $n_0 \le m \le n$ we get

$$0 < a_{m,n} + b_{m,n} \le (1+\varepsilon)(\alpha_{m,n} + \beta_{m,n}),$$
$$0 < a_{m,n}b_{m,n} \le (1+\varepsilon)^2(\alpha_{m,n}\beta_{m,n}),$$
$$0 < \frac{a_{m,n}}{\beta_{m,n}} \le (1+\varepsilon)^2\frac{\alpha_{m,n}}{b_{m,n}},$$
$$0 < {a_{m,n}}^s \le (1+\varepsilon)^s{\alpha_{m,n}}^s \quad \text{for} \quad s \ge 1.$$

Thus, assertion a) follows and therfore also assertion b).

The following mean value theorem is a modification of a theorem by O. Toeplitz (cf. [57, Section 8, Theorem 5]).

**Lemma A.2.** *Let* $(a_{m,n}), (\alpha_{m,n}) \in \mathcal{D}_0$ *and* $(\beta_m) \in \mathcal{F}$, *and assume*

$$\lim_{n\to\infty} \sum_{m=1}^{n} a_{m,n} =: a \in (0,\infty).$$

*Defining* $\underline{\beta} := \liminf_n \beta_n$ *and* $\bar{\beta} := \limsup_n \beta_n$, *we get for each index* $k \in \mathbb{N}$:

*a) If* $\alpha_{m,n} \ll a_{m,n}$ *and* $\bar{\beta} \ge 0$, *then* $\limsup_n \sum_{m=k}^{n} \alpha_{m,n}\beta_m \le a\bar{\beta}$.

*b) If* $a_{m,n} \ll \alpha_{m,n}$ *and* $\underline{\beta} > 0$, *then* $a\underline{\beta} \le \liminf_n \sum_{m=k}^{n} \alpha_{m,n}\beta_m$.

*c) If* $\alpha_{m,n} \sim a_{m,n}$, *then* $\operatorname{limextr}_n \sum_{m=k}^{n} \alpha_{m,n}\beta_m \in [a\underline{\beta}\,, a\bar{\beta}]$.

In Lemma A.2 *limextr* denotes lim sup or lim inf.

*Proof.* Due to $(a_{m,n}), (\alpha_{m,n}) \in \mathcal{D}_0$ , we have for any $k \in \mathbb{N}$

$$\lim_n \sum_{m=1}^{k-1} a_{m,n} = 0 \quad \text{and} \quad \lim_n \sum_{m=1}^{k-1} \alpha_{m,n}\beta_m = 0 \,. \tag{i}$$

Hence, all assertions have only to be shown for $k = 1$.

For the proof of a) assume $\bar{\beta} < \infty$. Hence, for any $\varepsilon > 0$ there exists an index $n_0$ such that

$$\beta_m \le \bar{\beta} + \varepsilon > 0 \quad \text{and} \quad 0 < \alpha_{m,n} \le (1+\varepsilon)a_{m,n}$$

for all indices $m, n$ with $n_0 \le m \le n$ . Hence, for $n \ge n_0$ we find

$$\begin{aligned}\sum_{m=n_0}^{n} \alpha_{m,n}\beta_m &\le \sum_{m=n_0}^{n} \alpha_{m,n}(\bar{\beta}+\varepsilon) \le (1+\varepsilon)(\bar{\beta}+\varepsilon) \sum_{m=n_0}^{n} a_{m,n} \\ &= (1+\varepsilon)(\bar{\beta}+\varepsilon)\Big(\sum_{m=1}^{n} a_{m,n} - \sum_{m=1}^{n_0-1} a_{m,n}\Big).\end{aligned}$$

Now, by means of (i) we get $\limsup_n \sum_{m=1}^{n} \alpha_{m,n}\beta_m \le (1+\varepsilon)(\bar{\beta}+\varepsilon)a$. Since $\varepsilon > 0$ was chosen arbitrarily, this shows assertion a).

Using the assumptions of b), there exists to any $\varepsilon \in (0, \underline{\beta}]$ an index $n_0$ with

$$0 \le \underline{\beta} - \varepsilon \le \beta_m \quad \text{and} \quad 0 \le a_{m,n} \le (1+\varepsilon)\alpha_{m,n}$$

for $n_0 \le m \le n$. Hence, for $n \ge n_0$

$$\sum_{m=n_0}^{n} \alpha_{m,n}\beta_m \ge \frac{\underline{\beta}-\varepsilon}{1+\varepsilon}\Big(\sum_{m=1}^{n} a_{m,n} - \sum_{m=1}^{n_0-1} a_{m,n}\Big)$$

holds. Therefore again by means of (i) we have $\liminf_n \sum_{m=1}^{n} \alpha_{m,n}\beta_m \ge (\underline{\beta} - \varepsilon)(1+\varepsilon)^{-1}a$ . Thus, we obtain now statement b).

In the proof of c) we consider the following three possible cases:

Case 1: $\underline{\beta} > 0$. The relation in c) is fulfilled due to a) and b). Case 2: $\bar{\beta} < 0$. This case is reduced to the previous case by means of the transition from $\beta_n$ to $-\beta_n$. Case 3: $\underline{\beta} \le 0 \le \bar{\beta}$. Here, from a) we get

$$\limsup_n \sum_{m=k}^{n} \alpha_{m,n}\beta_m \le a\bar{\beta} \quad \text{and} \quad \limsup_n \sum_{m=k}^{n} \alpha_{m,n}(-\beta_m) \le a(-\underline{\beta}) \,.$$

Hence, also statement c) holds.

Given a sequence $(a_n) \in \mathcal{F}$, for $m, n \in \mathbb{N}_0, m \leq n$, we define now

$$b_{m,n} := \begin{cases} \prod_{j=m+1}^{n} (1 - a_j) & , \text{ for } \quad m < n \\ 1 & , \text{ for } \quad m = n \, . \end{cases}$$

About this sequence the following statement can be made:

**Lemma A.3.** *a) If*

*α) there exists an index $n_0$ with $a_n < 1$ for $n \geq n_0$ and β) $\sum_n a_n = \infty$ ,*

*then $(b_{m,n}) \in \mathcal{D}_0$ and $\lim_{n\to\infty} \sum_{m=k}^{n} b_{m,n} a_m = 1$ for $k = 1, 2, \ldots$ .*

*b) In case $\sum_n a_n^2 < \infty$ it is $b_{m,n} \sim \exp(-(a_{m+1} + \ldots + a_n))$ .*

*c) If $\sum_n | a_n |< \infty$, then there exists an index $n_2$ with*

$$\operatorname*{limextr}_n b_{m,n} \in (0, \infty) \quad \text{for} \quad m \geq n_2 \, .$$

*Proof.* In all cases a), b) and c) there exists an index $n_0$ with $a_m < 1$ for $m \geq n_0$ . Hence, it holds

$$0 < b_{m,n} \leq \exp(-(a_{m+1} + \ldots + a_n)) \quad \text{for} \quad n_0 \leq m < n. \tag{i}$$

Consequently, in case $\sum_n a_n = \infty$ we have $\lim_{n\to\infty} b_{m,n} = 0$ for $m = 0, 1, 2, \ldots$ and therefore $(b_{m,n}) \in \mathcal{D}_0$. Hence, for any index $k$

$$\sum_{m=k}^{n} b_{m,n} a_m = \sum_{m=k}^{n} b_{m,n}(1 - (1 - a_m)) = 1 - b_{k-1,n} \longrightarrow 1 \quad \text{for} \quad n \longrightarrow \infty.$$

In case $\sum_n a_n^2 < \infty$ exists to each $\varepsilon > 0$ an index $n_1 \geq n_0$ having

$$0 < \exp(-2a_j) \leq 1 - a_j^2 \quad \text{for} \quad j \geq n_1,$$

$$\exp\left(2 \sum_{j\geq n_1} a_j^2\right) \leq 1 + \varepsilon.$$

Hence, for $n_1 \leq m < n$ we find

$$b_{m,n} \exp(a_{m+1} + \ldots + a_n) \geq b_{m,n}(1 + a_{m+1}) \cdot \ldots \cdot (1 + a_n)$$
$$= (1 - a_{m+1}^2) \cdot \ldots \cdot (1 - a_n^2) \geq \exp\Big(-2(a_{m+1}^2 + \ldots + a_n^2)\Big) \geq \frac{1}{1+\varepsilon} ,$$

or $\exp(-(a_{m+1}+\ldots+a_n)) \leq (1+\varepsilon)b_{m,n}$, respectively. Therfore b) follows by means of (i).

Now, let $\sum\limits_n \mid a_n \mid < \infty$ . Then $(a_n)$ converges to zero and series $\sum\limits_n b_n$ converges, if $b_n, n = 1, 2, \ldots$, is defined by

$$b_n := \begin{cases} 2a_n & , \text{ for } \ a_n \geq 0 \\ \frac{1}{2}a_n & , \text{ for } \ a_n < 0 \,. \end{cases}$$

There exists an index $n_2 \geq n_0$ with

$$0 < \exp(-b_n) \leq 1 - a_n \quad \text{for} \quad n \geq n_2 \,.$$

Hence, using (i), assertion c) follows.

A simple application of Lemma A.2 und Lemma A.3 is given next:

**Lemma A.4.** *Let $T_1 \in \mathbb{R}$ and $(a_n)$ , $(b_n) \in \mathcal{F}$ with following properties: i) There exists an index $n_0$ with $a_n \in [0,1)$ for $n \geq n_0$, ii) $\sum\limits_n a_n = \infty$ . Then limextr $T_n \in [\liminf\limits_n b_n \ , \ \limsup\limits_n b_n]$ for sequence*

$$T_{n+1} := (1-a_n)T_n + a_n b_n, \quad n = 1, 2, \ldots \,.$$

*Proof.* By means of complete induction, formula

$$T_{n+1} = b_{0,n}T_1 + \sum_{m=1}^{n} (b_{m,n}a_m)b_m \quad \text{for} \quad n = 1, 2, \ldots$$

is obtained. Hence, the assertion follows from Lemma A.3 a) and Lemma A.2 c).

Now, consider to given number $a \in \mathbb{R}$ the following double sequence in $\mathcal{D}_+$:

$$\Phi_{m,n}(a) := \begin{cases} \prod\limits_{j=m+1}^{n} (1 - \frac{a}{j}) & , \text{ if } \ m < n \\ 1 & , \text{ if } \ m = n \,. \end{cases}$$

Moreover, let $\mathbb{M}$ denote a subset of $\mathbb{N}$ such that the limit

$$q := \lim_{n\to\infty} \frac{\mid \{1,\ldots,n\} \cap \mathbb{M} \mid}{n}$$

exists and is positive. Then we get the following result:

**Lemma A.5.** *For any $a \in \mathbb{R}$ holds:*

*a)* $\Phi_{m,n}(a) \sim (\frac{m}{n})^a$ ,

*b)* $\frac{m}{n} \sim \frac{|\mathbb{M}_{0,m}|}{|\mathbb{M}_{0,n}|}$ ,

*c)* $\Phi_{m,n}(a) \sim \Phi_{|\mathbb{M}_{0,m}|,|\mathbb{M}_{0,n}|}(a)$ ,

*d) If* $a > 0$, *then* $\lim_{n\to\infty} \sum_{m\in \mathbb{M}_{k,n}} \Phi_{m,n}(a)\frac{1}{m} = \frac{q}{a}$ *for each* $k \in \mathbb{N}_0$ .

*Proof.* Due to representation

$$\exp(-a(\frac{1}{m+1} + \ldots + \frac{1}{n})) = \exp(a(c_m - c_n))(\frac{m}{n})^a$$

with the convergent sequence (cf. [57])

$$c_k := \frac{1}{1} + \ldots + \frac{1}{k} - \ln(k), \quad k = 1, 2, \ldots ,$$

by means of Lemma A.3 b) we have

$$\left(\frac{m}{n}\right)^a \sim \exp\left(-a\left(\frac{1}{m+1} + \ldots + \frac{1}{n}\right)\right) \sim \Phi_{m,n}(a) .$$

Since the limit $q$ is positive, to any $\varepsilon > 0$ there exists an index $n_0$ such that

$$0 < \frac{q}{1+\varepsilon} \leq \frac{|\mathbb{M}_{0,k}|}{k} \leq (1+\varepsilon)q \quad \text{for} \quad k \geq n_0 .$$

Hence, for $n_0 \leq m \leq n$ we get

$$\frac{m}{(1+\varepsilon)^2\, n} \leq \frac{|\mathbb{M}_{0,m}|}{|\mathbb{M}_{0,n}|} \leq (1+\varepsilon)^2 \frac{m}{n} ,$$

and therefore b) holds. Due to a), b), and Lemma A.1 b) we have

$$\Phi_{m,n}(a) \sim \left(\frac{m}{n}\right)^a \sim \left(\frac{|\mathbb{M}_{0,m}|}{|\mathbb{M}_{0,n}|}\right)^a \sim \Phi_{|\mathbb{M}_{0,m}|,|\mathbb{M}_{0,n}|}(a). \tag{i}$$

For $a > 0$, due to Lemma A.3 a) we have

$$\lim_{n\to\infty} \sum_{m\in \mathbb{M}_{k,n}} \Phi_{|\mathbb{M}_{0,m}|,|\mathbb{M}_{0,n}|}(a)\frac{1}{|\mathbb{M}_{0,m}|}$$

$$= \frac{1}{a} \lim_{n\to\infty} \sum_{M=|\mathbb{M}_{0,k}|+1}^{|\mathbb{M}_{0,n}|} \Phi_{M,|\mathbb{M}_{0,n}|}(a)\frac{a}{M} = \frac{1}{a} . \tag{ii}$$

Note that the index transformation in (ii) is feasible. Indeed,

$$m \in \mathbb{M}_{k,n} := \{k+1,\ldots,n\} \cap \mathbb{M} \iff |\, \mathbb{M}_{0,m} \,| \in \{|\, \mathbb{M}_{0,k} \,| + 1, \ldots, |\, \mathbb{M}_{0,n} \,|\} .$$

Hence, by means of (i) and the definition of $q$ it is

$$\Phi_{|\mathbb{M}_{0,m}|,|\mathbb{M}_{0,n}|}(a) \frac{1}{|\, \mathbb{M}_{0,m} \,|} \sim \Phi_{m,n}(a) \frac{1}{qm} .$$

Therefore, (ii) and Lemma A.2 c) yields now assertion d).

Now let $(A_n) \in \mathcal{F}$ denote a given sequence with

$$\lim_{n \in \mathbb{N}^{(k)}} A_n = A^{(k)} \quad \text{for} \quad k = 1, \ldots, \kappa,$$
$$A := q_1 A^{(1)} + \ldots + q_\kappa A^{(\kappa)} > 0 .$$

Here again $q_1, \ldots, q_\kappa$ denote the limit parts of sets $\mathbb{N}^{(1)}, \ldots, \mathbb{N}^{(\kappa)}$ in $\mathbb{N}$ according to (U6). Then double sequence

$$b_{m,n} := \begin{cases} \prod_{j=m+1}^{n} (1 - \frac{A_j}{j}) & , \text{ for } \quad m < n \\ 1 & , \text{ for } \quad m = n \end{cases}$$

is in $\mathcal{D}_+$ and has the following properties:

**Lemma A.6.** *a) For any* $\delta \in (0, A)$ *we have* $\Phi_{m,n}(A + \delta) \ll b_{m,n} \ll \Phi_{m,n}(A - \delta)$ .

*b)* $\lim_{n \to \infty} b_{m,n} = 0$ *for* $m = 0, 1, 2, \ldots$ .

*c) For each* $k \in \mathbb{N}_0$ *it is* $\lim_{n \to \infty} \sum_{m \in \mathbb{M}_{k,n}} b_{m,n} \frac{1}{m} = \frac{q}{A}$ .

*Proof.* We have representation

$$b_{m,n} = \prod_{k=1}^{\kappa} \left( \prod_{l \in \mathbb{N}_{m,n}^{(k)}} \left( 1 - \frac{A_l \frac{|\mathbb{N}_{0,l}^{(k)}|}{l}}{|\, \mathbb{N}_{0,l}^{(k)} \,|} \right) \right) , \tag{i}$$

and for arbitrary numbers $b_1, \ldots, b_\kappa \in \mathbb{R}$ it is due to Lemma A.1 and Lemma A.5

$$\prod_{k=1}^{\kappa} \left( \prod_{l \in \mathbb{N}_{m,n}^{(k)}} \left( 1 - \frac{b_k}{|\, \mathbb{N}_{0,l}^{(k)} \,|} \right) \right) = \prod_{k=1}^{\kappa} \left( \prod_{L=|\mathbb{N}_{0,m}^{(k)}|+1}^{|\mathbb{N}_{0,n}^{(k)}|} \left( 1 - \frac{b_k}{L} \right) \right)$$
$$= \prod_{k=1}^{\kappa} \Phi_{|\mathbb{N}_{0,m}^{(k)}|,|\mathbb{N}_{0,n}^{(k)}|}(b_k) \sim \prod_{k=1}^{\kappa} \Phi_{m,n}(b_k)$$
$$\sim \prod_{k=1}^{\kappa} (\frac{m}{n})^{b_k} \sim \Phi_{m,n}(b_1 + \ldots + b_k) . \tag{ii}$$

For each $\delta \in (0, A)$ there exists an index $n_0$ with

$$\underline{b}_k := q_k A^{(k)} - \frac{\delta}{k} \le A_l \frac{|\,\mathbb{N}_{0,l}^{(k)}\,|}{l} \le q_k A^{(k)} + \frac{\delta}{k} =: \bar{b}_k$$

for all $n_0 \le l \in \mathbb{N}^{(k)}$ and $k = 1, \ldots, \kappa$ . Hence, due to (i) and (ii)

$$\Phi_{m,n}(A+\delta) \ll b_{m,n} \ll \Phi_{m,n}(A-\delta).$$

Since $A \pm \delta > 0$, Lemma A.5 a) yields

$$\lim_{n\to\infty} b_{m,n} = 0 \quad \text{for} \quad m = 0, 1, 2, \ldots,$$

and from Lemma A.5 d) and Lemma A.2 one obtains

$$\tfrac{q}{A+\delta} \le \operatorname*{limextr}_{n} \sum_{m \in \mathbf{M}_{k,n}} b_{m,n} \tfrac{1}{m} \le \tfrac{q}{A-\delta}.$$

This holds true for any $\delta \in (0, A)$ , and therefore c) holds.

Applying now Lemma A.6, the limit of sequence

$$T_{n+1} := (1 - \frac{A_n}{n}) T_n + \frac{f_n}{n} + g_n \quad \text{for} \quad n = 1, 2, \ldots$$

can be calculated. Here, $T_1 \in \mathbb{R}$ and $(f_n)$ , $(g_n) \in \mathcal{F}$ with

$$\lim_{n \in \mathbf{N}^{(k)}} f_n = f^{(k)} \quad \text{for} \quad k = 1, \ldots, \kappa\,, \quad \sum_n g_n \quad \text{convergent.}$$

**Lemma A.7.** *For sequence $(T_n)$ equation*

$$\lim_{n\to\infty} T_n = \sum_{k=1}^{\kappa} \frac{q_k f^{(k)}}{A}$$

*holds.*

*Proof.* With abbreviations $g_0 := 0$ and

$$F_n := f_n - A_n(g_0 + \ldots + g_{n-1})\,, \; S_n := T_n - (g_0 + \ldots + g_{n-1})$$

we get

$$\begin{aligned} S_{n+1} &= T_{n+1} - (g_0 + \ldots + g_n) \\ &= (T_n + \frac{1}{n}(f_n - A_n T_n) + g_n) - (g_0 + \ldots + g_n) \\ &= (T_n - (g_0 + \ldots + g_{n-1})) + \frac{1}{n}(F_n - A_n S_n) \\ &= \left(1 - \frac{A_n}{n}\right) S_n + \frac{1}{n} F_n \quad \text{for} \quad n = 1, 2, \ldots . \end{aligned}$$

Hence, by means of complete induction we get

$$S_{n+1} = b_{0,n}S_1 + \sum_{k=1}^{\kappa} \sum_{m\in \mathbf{N}_{0,n}^{(k)}} b_{m,n}\frac{1}{m}F_m \,.$$

Due to the assertion we have with $G := \sum_n g_n$

$$\lim_{m\in\mathbf{N}^{(k)}} F_m = f^{(k)} - A^{(k)}G \quad \text{for} \quad k = 1,\ldots,\kappa\,.$$

Now, from Lemma A.6 b), c) and Lemma A.2 c) follows

$$\lim_{n\to\infty} S_n = \sum_{k=1}^{\kappa} \frac{q_k}{A}(f^{(k)} - A^{(k)}G) = \sum_{k=1}^{\kappa} \frac{q_k f^{(k)}}{A} - G.$$

Finally, we get our assertion by means of the relation $T_n = S_n + (g_0 + \ldots + g_{n-1})$.

The following simple mean value theorem is used a few times:

**Lemma A.8.** *Let $(f_n)$ be selected as above. Then*

$$\frac{1}{n}(f_1 + \ldots + f_n) \longrightarrow \sum_{k=1}^{\kappa} q_k f^{(k)} \;, n\to\infty.$$

*Proof.* Due to representation

$$T_{n+1} := \frac{1}{n}(f_1 + \ldots + f_n) = \left(1 - \frac{1}{n}\right) T_n + \frac{f_n}{n},$$

the assertion follows from Lemma A.7.

## A.2 Iterative Solution of a Lyapunov Matrix Equation

In Section 6.5 (or more precisely in the proof of Lemma 6.13) the following recursive matrix equation is of interest: For $n = 1, 2, \ldots$ let

$$V_{n+1} = \left(I - \frac{1}{n}B_n\right) V_n \left(I - \frac{1}{n}B_n\right)^T + \frac{1}{n}C_n$$

where $V_1$ and $C_n$ denote symmetric and $B_n$ real $\nu\times\nu$-matrices with the following properties:

$$\lim_{n\in\mathbf{N}^{(k)}} C_n = C^{(k)}, \lim_{n\in\mathbf{N}^{(k)}} B_n = B^{(k)}, B^{(k)} + B^{(k)T} \geq 2u^{(k)}I$$

for $k = 1,\ldots,\kappa$ , where $u^{(1)},\ldots,u^{(\kappa)}$ denote reals with

$$\bar{u} := q_1u^{(1)} + \ldots + q_\kappa u^{(\kappa)} > 0\,.$$

Convergence of matrix sequence $(V_n)$ is guaranteed by the following result:

**Lemma A.9.** *The limit* $V := \lim_{n\to\infty} V_n$ *exists and is unique solution of* Lyapunov equation

$$BV + VB^T = C, \qquad (*)$$

*where we put*

$$B := q_1 B^{(1)} + \ldots + q_\kappa B^{(\kappa)} \quad \textit{and} \quad C := q_1 C^{(1)} + \ldots + q_\kappa C^{(\kappa)} .$$

For the proof of this theorem we need further auxiliary results. First of we state the following remark:

*Remark A.1.* If for $n \in \mathbb{N}$ holds

$$B_n + B_n^T \geq 2\widetilde{a}_n I \quad \text{for} \quad \|B_n\| \leq \widetilde{b}_n$$

with numbers $\widetilde{a}_n$ , $\widetilde{b}_n$ , then (cf. Lemma 6.6)

$$\left\| I - \frac{1}{n} B_n \right\|^2 \leq 1 - \frac{2u_n}{n} \quad \text{with} \quad u_n := \widetilde{a}_n - \frac{1}{2n}\widetilde{b}_n^2 .$$

Due to the assumptions about $(B_n)$, $(\widetilde{a}_n)$ and $(\widetilde{b}_n)$ can be chosen such that

$$\lim_{n\in\mathbb{N}^{(k)}} u_n = u^{(k)} \quad \text{for} \quad k = 1, \ldots, \kappa .$$

**Lemma A.10.** *For* $n \in \mathbb{N}$ *let* $T_1, E_n$ *and* $F_n$ *denote symmetric* $\nu \times \nu$*-matrices having*

$$\widehat{E}_{n+1} := \frac{1}{n}(E_1 + \ldots + E_n) \longrightarrow 0 \quad \textit{and} \quad \sum_n \|F_n\| < \infty .$$

*Then* $T_n \to 0, n \to \infty$, *if the sequence* $(T_n)$ *is defined by*

$$T_{n+1} := \left(I - \frac{1}{n}B_n\right) T_n \left(I - \frac{1}{n}B_n\right)^T + \frac{1}{n}E_n + F_n \ , \quad n = 1, 2, \ldots .$$

*Proof.* With matrices $T_1^{(1)} := T_1$ , $T_1^{(2)} := 0$ and

$$T_{n+1}^{(1)} := \left(I - \frac{1}{n}B_n\right) T_n^{(1)} \left(I - \frac{1}{n}B_n\right)^T + F_n,$$

$$T_{n+1}^{(2)} := \left(I - \frac{1}{n}B_n\right) T_n^{(2)} \left(I - \frac{1}{n}B_n\right)^T + \frac{1}{n}E_n$$

we can write $T_n = T_n^{(1)} + T_n^{(2)}$ for $n = 1, 2, \ldots$. Because of the above remark we get now

$$\|T_{n+1}^{(1)}\| \leq \left(1 - \frac{2}{n}u_n\right) \|T_n^{(1)}\| + \|F_n\|.$$

Hence, Lemma A.7 yields $T_n^{(1)} \longrightarrow 0$ .

Due to $\widehat{E}_{n+1} = \left(1 - \frac{1}{n}\right) \widehat{E}_n + \frac{1}{n} E_n$ it is

$$\begin{aligned} T_{n+1}^{(2)} - \widehat{E}_{n+1} &= (I - \frac{1}{n} B_n)((T_n^{(2)} - \widehat{E}_n) + \widehat{E}_n) \left(I - \frac{1}{n} B_n\right)^T + \left(\frac{1}{n} - 1\right) \widehat{E}_n \\ &= \left(I - \frac{1}{n} B_n\right) (T_n^{(2)} - \widehat{E}_n) \left(I - \frac{1}{n} B_n\right)^T \\ &\quad + \frac{1}{n} [\widehat{E}_n - (B_n \widehat{E}_n + \widehat{E}_n B_n^T) + \frac{1}{n} B_n \widehat{E}_n B_n^T]. \end{aligned}$$

Hence, the above remark gives us

$$\begin{aligned} \|T_{n+1}^{(2)} - \widehat{E}_{n+1}\| \leq \left(1 - \frac{2}{n} u_n\right) \|T_n^{(2)} - \widehat{E}_n\| \\ + \frac{\|\widehat{E}_n\|}{n} \left[1 + 2\|B_n\| + \frac{1}{n}\|B_n\|^2\right]. \end{aligned}$$

Again, using Lemma A.7, $T_n^{(2)} - \widehat{E}_n \longrightarrow 0$ follows. Finally, because of the assumptions about $E_n$ also $T_n^{(2)}$ converges to 0.

*Proof (of Lemma A.9).* Having a solution $V$ of equation $(*)$ it holds

$$\begin{aligned} V_{n+1} - V &= \left(I - \frac{1}{n} B_n\right) \left((V_n - V) + V\right) \left(I - \frac{1}{n} B_n\right)^T + \frac{1}{n} C_n - V \\ &= \left(I - \frac{1}{n} B_n\right) (V_n - V) \left(I - \frac{1}{n} B_n\right)^T \\ &\quad + \frac{1}{n} \left[C_n - (B_n V + V B_n^T)\right] + \frac{1}{n^2} B_n V B_n^T. \end{aligned}$$

Moreover, due to Lemma A.8 and the above assumptions we obtain

$$\frac{1}{n}(B_1 + \ldots + B_n) \longrightarrow \sum_{k=1}^{\kappa} q_k B^{(k)} =: B,$$

$$\frac{1}{n}(C_1 + \ldots + C_n) \longrightarrow \sum_{k=1}^{\kappa} q_k C^{(k)} =: C.$$

Hence, due to Lemma A.10 we have $V_n \longrightarrow V, n \to \infty$.

The following theorem guarantees the uniqueness of the solution $V$ of equation $(*)$.

**Lemma A.11 (Lyapunov).** *If* $-B$ is stable, *that is*

$$Re(\lambda) > 0 \textit{ for any eigenvalue } \lambda \textit{ of } B,$$

*then* $(*)$ *has an unique solution* $V$ .

This theorem can be found e.g. in [8]. Due to relations

$$Re(\lambda) \geq \frac{1}{2}\lambda_{\min}(B + B^T) \geq \bar{u} > 0 \text{ for all eigenvalues } \lambda \text{ of } B,$$

in our situation $-B$ is stable.

# B

# Convergence Theorems for Stochastic Sequences

The following theorems giving assertions about "almost sure convergence", "convergence in the mean" and "convergence in distribution" of random sequences, are of fundamental importance for Chapter 6.

Let, as already stated in Section 6.3, $(\Omega, \mathcal{A}, P)$ denote a probability space, $\mathcal{A}_1 \subset \mathcal{A}_2 \subset \ldots \subset \mathcal{A}$ $\sigma$-algebras, $E_n(\cdot) := E(\cdot \mid \mathcal{A}_n)$ the expectation with respect to $\mathcal{A}_n$ and $\mathbb{N} = \mathbb{N}^{(1)} \cup \ldots \cup \mathbb{N}^{(\kappa)}$ a disjunct partition of $\mathbb{N}$ fulfilling condition (U6).

## B.1 A Convergence Result of Robbins-Siegmund

The following lemma due to H. Robbins and D. Siegmund, cf. [115], gives an assertion about almost sure convergence of nonnegative random sequences $(T_n)$.

Let $T_n$, $\alpha_n$, $\alpha_n^{(1)}$ and $\alpha_n^{(2)}$ denote nonnegative $\mathcal{A}_n$-measurable random variables with

$$E_n T_{n+1} \leq (1 + \alpha_n^{(1)}) T_n - \alpha_n + \alpha_n^{(2)} \quad \text{a.s.}$$

for $n = 1, 2, \ldots$, where $\sum_n \alpha_n^{(i)} < \infty$ a.s. for $i = 1, 2$.

**Lemma B.1.** *The limit* $\lim_{n \to \infty} T_n < \infty$ *exists, and it is* $\sum_n \alpha_n < \infty$ *a.s.* .

### B.1.1 Consequences

In special case

$$\alpha_n = a_n T_n \quad \text{for} \quad n = 1, 2, \ldots$$

with nonnegative $\mathcal{A}_n$-measurable random variables $a_n$ , fulfilling $\sum_n a_n = \infty$ a.s., we get this result:

**Lemma B.2.** $T_n \longrightarrow 0, n \to \infty$ *a.s.* .

*Proof.* According to Lemma B.1 we know that $T := \lim_{n\to\infty} T_n$ exists a.s. and $\sum_n a_n T_n < \infty$ a.s.. Hence, due to $\sum_n a_n = \infty$ a.s. we obtain the proposition.

Now we abandon the nonnegativity condition of $a_n$.

To $n \in \mathbb{N}$, let $0 \leq S_n$ and $A_n$ denote $\mathcal{A}_n$-measurable random numbers with

$$E_n S_{n+1} \leq \left(1 - \frac{A_n}{n} + \alpha_n^{(1)}\right) S_n + \alpha_n^{(2)} \quad \text{a.s.}$$

for $n = 1, 2, \ldots$ , where we demand for $(A_n)$ :

a) $\lim_{n\in\mathbb{N}^{(k)}} A_n = A^{(k)}$ a.s. for $k = 1, \ldots, \kappa$ with numbers $A^{(1)}, \ldots, A^{(\kappa)} \in \mathbb{R}$ ,

b) $A := q_1 A^{(1)} + \ldots + q_\kappa A^{(\kappa)} > 0$ .

Since for large $n$ sequence $(A_n)$ is positive "in the mean", $(S_n)$ converges a.s. to zero.

**Lemma B.3.** $S_n \longrightarrow 0, n \to \infty$ *a.s.* .

*Proof.* To arbitrary $\varepsilon \in (0, A)$, let

$$B_n := q_k (A^{(k)} - \varepsilon) \frac{n}{|\mathbb{N}_{0,n}^{(k)}|}$$

for indices $n \in \mathbb{N}^{(k)}$, where $k \in \{1, \ldots, \kappa\}$ . Then, for all $k = 1, \ldots, \kappa$ and $n \in \mathbb{N}$ we get

$$\lim_{n\in\mathbb{N}^{(k)}} B_n = A^{(k)} - \varepsilon \ , \ A_n \geq B_n - \Delta B_n,$$

with the $\mathcal{A}_n$-measurable random variable $\Delta B_n := \max\{0, B_n - A_n\} \geq 0$ . Since $S_n$ is nonnegative, it follows

$$E_n S_{n+1} \leq \left(1 - \frac{B_n}{n} + \beta_n^{(1)}\right) S_n + \alpha_n^{(2)},$$

where $\beta_n^{(1)} := \alpha_n^{(1)} + \frac{\Delta B_n}{n}$ . For sufficiently large indices $n$ it is $\Delta B_n = 0$ a.s., and therfore $\sum_n \frac{\Delta B_n}{n} < \infty$ a.s.. Hence, it holds

$$0 \leq \beta_n^{(1)} \ , \ \sum_n \beta_n^{(1)} < \infty \quad \text{a.s.} \ .$$

Furthermore, there exists an index $n_0$ with

$$1 - \frac{A - \varepsilon}{n}, \ 1 - \frac{B_n}{n} \geq \frac{1}{2} \quad \text{for} \quad n \geq n_0 \ .$$

Hence, for $n \geq n_0$ exists the $\mathcal{A}_n$-measurable positive random variable

$$p_n := \frac{1-\frac{A-\varepsilon}{n}}{1-\frac{B_n}{n}+\beta_n^{(1)}} = \frac{1-\frac{A-\varepsilon}{n}}{1-\frac{B_n}{n}}\left(1-\frac{\beta_n^{(1)}}{1-\frac{B_n}{n}+\beta_n^{(1)}}\right).$$

Now, let $P_{n_0} := 1$ and for $n \geq n_0$

$$P_{n+1} := p_{n_0} \cdot \ldots \cdot p_n \ , \ \bar{S}_n := P_n S_n \ , \ \beta_n^{(2)} := P_{n+1}\alpha_n^{(2)} \ .$$

Then $P_{n+1}$ , $\bar{S}_n$ and $\beta_n^{(2)}$ are nonnegative $\mathcal{A}_n$-measurable random variables, where

$$E_n\bar{S}_{n+1} \leq \left(1-\frac{A-\varepsilon}{n}\right)\bar{S}_n + \beta_n^{(2)}, \quad n \geq n_0.$$

Now, only the following relations have to be verified:

$$\limsup_n P_n < \infty \quad , \quad \liminf_n P_n > 0 \quad \text{a.s.} \ .$$

Using the first inequality it follows

$$\sum_n \beta_n^{(2)} < \infty \text{ a.s.} \ ,$$

and therfore $(\bar{S}_n)$ converges to zero a.s. due to Lemma B.2. Hence, due to the second inequality and the definition also $(S_n)$ converges a.s. to 0.

For $n \geq n_0$ we have

$$\frac{1-\frac{A-\varepsilon}{n}}{1-\frac{B_n}{n}}(1-2\beta_n^{(1)}) \leq p_n \leq \frac{1-\frac{A-\varepsilon}{n}}{1-\frac{B_n}{n}}.$$

As in the proof of Lemma A.6 it can be shown that for $n_0 \leq m < n$ it holds

$$\prod_{j=m+1}^{n}\left(1-\frac{B_j}{j}\right) = \prod_{k=1}^{\kappa}\left(\prod_{l\in \mathbb{N}_{m,n}^{(k)}}\left(1-\frac{B_l\frac{|\mathbb{N}_{0,l}^{(k)}|}{l}}{|\mathbb{N}_{0,l}^{(k)}|}\right)\right)$$

$$= \prod_{k=1}^{\kappa}\left(\prod_{l\in \mathbb{N}_{m,n}^{(k)}}\left(1-\frac{q_k(A^{(k)}-\varepsilon)}{\mathbb{N}_{0,l}^{(k)}}\right)\right) \sim \Phi_{m,n}(A-\varepsilon) \ ,$$

or stated in another way

$$\prod_{j=m+1}^{n}\frac{1-\frac{A-\varepsilon}{j}}{1-\frac{B_j}{j}} \sim 1.$$

Moreover, according to Lemma A.3 c) for an index $n_2$ we have

$$\liminf_{n} \prod_{j=n_2+1}^{n} (1 - 2\beta_j^{(1)}) > 0 \quad \text{a.s.} \; .$$

The last two relations and the above inequality for $p_n$ finally yield

$$\underset{n}{limextr}\; P_n \in (0, \infty) \quad \text{a.s..}$$

## B.2 Convergence in the Mean

For real $\mathcal{A}_n$-measurable random variables $T_n$, $A_n$, $B_n$, $C_n$ and $v_n$ assume for any $n \in \mathbb{N}$ and $k = 1, \ldots, \kappa$

$$E_n T_{n+1} \leq (1 - \frac{A_n}{n}) T_n + \frac{1}{n}(B_n v_n + C_n) \text{ a.s.,}$$

as well as

$$\begin{aligned}
&T_n, B_n, v_n \geq 0, \\
&\liminf_{n \in \mathbb{N}^{(k)}} A_n \geq A^{(k)} \;, \; \limsup_{n \in \mathbb{N}^{(k)}} B_n \leq B^{(k)} \;, \; \limsup_{n \in \mathbb{N}^{(k)}} C_n \leq C^{(k)} \quad \text{a.s.} \;, \\
&\limsup_{n \in \mathbb{N}^{(k)}} E v_n \leq w^{(k)} \;, \; A := q_1 A^{(1)} + \ldots + q_\kappa A^{(\kappa)} > 0
\end{aligned}$$

with real numbers $A^{(1)}, \ldots, A^{(\kappa)}, B^{(1)}, \ldots, B^{(\kappa)}, C^{(1)}, \ldots, C^{(\kappa)}$ and $w^{(1)}, \ldots, w^{(\kappa)}$

Now, for sequence $(T_n)$ the following convergence theorem holds:

**Lemma B.4.** *To any $\varepsilon_0 > 0$ there exists a set $\Omega_0 \in \mathcal{A}$ such that*

*a)* $\mathcal{P}(\bar{\Omega}_0) < \varepsilon_0$,

*b)* $\displaystyle \limsup_{n \in \mathbb{N}} \; E T_n I_{\Omega_0} \; \leq \sum_{k=1}^{\kappa} \frac{q_k (B^{(k)} w^{(k)} + C^{(k)})}{A}.$

For the proof of this theorem the following theorem of Egorov (cf. e.g. [59] page 285) is needed:

**Lemma B.5.** *Let $U$, $U_1$, $U_2, \ldots$ denote real random variables having*

$$\limsup_{n} U_n \leq U \quad a.s. \; .$$

*Then to any positive number $\varepsilon_0$ there exists a set $\Omega_0 \in \mathcal{A}$ with following properties:*

*a)* $\mathcal{P}(\bar{\Omega}_0) < \varepsilon_0$ ,

*b)* $\limsup_n U_n \le U$ *is fulfilled uniformly on* $\Omega_0$, *that is: To any number* $\varepsilon > 0$ *there exists an index* $n_0 \in \mathbb{N}$ *such that*

$$U_n(\omega) \le U(\omega) + \varepsilon \quad \textit{for all } n \ge n_0 \textit{ and } \omega \in \Omega_0 \, .$$

*Proof (of Lemma B.4).* Let $\varepsilon_0 > 0$ be chosen arbitrarily. According to Lemma B.5 there exists a set $\Omega_0 \in \mathcal{A}$ such that inequalities

$$\liminf_{n \in \mathbb{N}^{(k)}} A_n \ge A^{(k)} \, , \; \limsup_{n \in \mathbb{N}^{(k)}} B_n \le B^{(k)} \, , \; \limsup_{n \in \mathbb{N}^{(k)}} C_n \le C^{(k)}$$

for all $k = 1, \ldots, \kappa$ hold uniformly on $\Omega_0$. Choose now an arbitrary number $\varepsilon \in (0, A)$. Then there exists an index $n_0$ such that for all $n \ge n_0$ and $n \in \mathbb{N}^{(k)}$ on $\Omega_0$ holds:

$$A_n \ge A^{(k)} - \varepsilon =: a_n \, , \; B_n \le B^{(k)} + \varepsilon =: b_n \, , \; C_n \le C^{(k)} + \varepsilon =: c_n \, .$$

Hence, for the $\mathcal{A}_n$-measurable sets

$$\Omega_n := \begin{cases} \{A_n \ge a_n\} \cap \{B_n \le b_n\} \cap \{C_n \le c_n\}, & \text{for } \; n \ge n_0 \\ \Omega \qquad\qquad\qquad\qquad\qquad\qquad , & \text{for } \; n < n_0 \end{cases}$$

we have

$$\Omega_0 \subset \Omega_1 \cap \ldots \cap \Omega_n =: \Omega^{(n)} \in \mathcal{A}_n \, .$$

Let now $S_1 := T_1$, and for $n \ge 1$ define

$$S_{n+1} := \left(1 - \frac{a_n}{n}\right) S_n + \left[T_{n+1} - \left(1 - \frac{a_n}{n}\right) T_n\right] I_{\Omega^{(n)}} .$$

Then, due to $\Omega^{(1)} \supset \Omega^{(2)} \supset \ldots \supset \Omega^{(n)}$ we get

$$S_{n+1} = \begin{cases} T_{n+1} & , \text{ on } \Omega^{(n)} \\ \left(1 - \frac{a_n}{n}\right) S_n & , \text{ otherwise} \, . \end{cases}$$

It is $\Omega_n \supset \Omega^{(n)} \in \mathcal{A}_n$ and therfore

$$\begin{aligned} E_n S_{n+1} &= \left(1 - \frac{a_n}{n}\right) S_n + \left[E_n T_{n+1} - \left(1 - \frac{u_n}{n}\right) T_n\right] I_{\Omega^{(n)}} \\ &\le \left(1 - \frac{a_n}{n}\right) S_n + \frac{1}{n}(B_n v_n + C_n) I_{\Omega^{(n)}} \le \left(1 - \frac{a_n}{n}\right) S_n + \frac{1}{n}(b_n v_n + c_n) \, . \end{aligned}$$

Taking expectations, for $n = 1, 2, \ldots$ we get

$$ES_{n+1} \le \left(1 - \frac{a_n}{n}\right) ES_n + \frac{1}{n}(b_n E v_n + c_n) \, .$$

Now, applying Lemma A.7 it follows

$$\limsup_n ES_n \le \sum_{k=1}^{\kappa} \frac{q_k \Big( (B^{(k)} + \varepsilon) w^{(k)} + C^{(k)} + \varepsilon \Big)}{A - \varepsilon} \, .$$

Moreover, because of the above statements, due to $S_n \geq 0$ and $\Omega_0 \subset \Omega^{(n)}$ we find

$$ES_{n+1} \geq ES_{n+1}I_{\Omega_0} = ET_{n+1}I_{\Omega_0} .$$

Since $\varepsilon \in (0, A)$ was chosen arbitrarily, last two relations yield

$$\limsup_n ET_n I_{\Omega_0} \leq \sum_{k=1}^{\kappa} \frac{q_k(B^{(k)}w^{(k)} + C^{(k)})}{A}.$$

## B.3 The Strong Law of Large Numbers for Dependent Matrix Sequences

For all $n \in \mathbb{N}$ let denote $a_n$ real random numbers and $T_n, C_n$ real random $\mu \times \nu$–matrices with the following properties:

$$T_1 \ \mathcal{A}_1\text{-measurable}, \ a_n \ \mathcal{A}_n\text{-measurable}, \ C_n \ \mathcal{A}_{n+1}\text{-measurable},$$
$$a_n \in [0,1) \ , \ \sum_n a_n = \infty \text{ a.s.} \ ,$$
$$\sum_n a_n^2 E_n \|C_n - E_n C_n\|^2 < \infty \ , \ E_n C_n \longrightarrow C, n \to \infty \text{ a.s.} \ ,$$

with a further matrix $C$. For the averaged sequence

$$T_{n+1} := (1 - a_n)T_n + a_n C_n, n = 1, 2, \ldots,$$

the following *strong law of large numbers* holds:

**Lemma B.6.** $T_n \longrightarrow C, n \to \infty$ *a.s.* .

*Proof.* With abbreviations $T_1^{(1)} := T_1$ , $T_1^{(2)} := 0$ and

$$T_{n+1}^{(1)} := (1 - a_n)T_n^{(1)} + a_n(C_n - E_n C_n) \ ,$$
$$T_{n+1}^{(2)} := (1 - a_n)T_n^{(2)} + a_n E_n C_n \ ,$$

we write $T_n = T_n^{(1)} + T_n^{(2)}$ for $n = 1, 2, \ldots$ . Hence, according to Lemma A.4 it is $\lim_n T_n^{(2)} = C$ a.s.. Let now $0 \neq h \in \mathbb{R}^\nu$ be chosen arbitrarily. With $S_n := \|T_n^{(1)}h\|^2$ we then get

$$\begin{aligned} S_{n+1} = (1 - a_n)^2 S_n + a_n^2 \left\| (C_n - E_n C_n)h \right\|^2 \\ + 2(1 - a_n)a_n \langle T_n^{(1)}h, (C_n - E_n C_n)h \rangle \ . \end{aligned}$$

Therefore, by taking expectations we find

$$\begin{aligned} E_n S_{n+1} &= (1 - a_n)^2 S_n + a_n^2 E_n \left\| (C_n - E_n C_n)h \right\|^2 \\ &\leq (1 - a_n)S_n + \|h\|^2 a_n^2 E_n \|C_n - E_n C_n\|^2. \end{aligned}$$

Now, due to Lemma B.2 we get $S_n \longrightarrow 0$ a.s.. Since $h$ is chosen arbitrarily, it follows $T_n^{(1)} \longrightarrow 0$ a.s. and therefore due to the above also $T_n \longrightarrow C$ a.s..

## B.4 A Central Limit Theorem for Dependent Vector Sequences

For indices $m \leq n$, let $A_{m,n}$ denote real deterministic $\mu \times \nu$-matrices, and let $U_n : \Omega \longrightarrow \mathbb{R}^\nu$ be $\mathcal{A}_{n+1}$ -measurable random vectors with the following properties:

a) For $n = 1, 2, \ldots$, assume

$$E_n U_n \equiv 0 \text{ a.s. .} \tag{K0}$$

b) It exists a deterministic matrix $V$ with

$$\sum_{m=1}^{n} A_{m,n} E_m (U_m U_m^T) A_{m,n}^T \overset{stoch}{\longrightarrow} V \, . \tag{KV}$$

c) For arbitrary $\epsilon > 0$ assume

$$\sum_{m=1}^{n} \|A_{m,n}\|^2 E_m \left( \|U_m\|^2 I_{\{\|A_{m,n}U_m\| > \epsilon\}} \right) \overset{stoch}{\longrightarrow} 0 \, . \tag{KL}$$

Then, sequence $(\sum_{m=1}^{n} A_{m,n} U_m)_n$ holds an asymptotic normal distribution:

**Lemma B.7.** *$(\sum_{m=1}^{n} A_{m,n} U_m)_n$ converges in distribution to a Gaussian distributed random vector $U$ with $EU = 0$ and $cov(U) = V$ .*

*Proof.* Let $U$ denote an arbitrary $N(0, V)$-distributed random vector, and let $0 \neq h \in \mathbb{R}^\mu$ be chosen arbitrarily. Now, for $m \leq n$ define

$$Y_{m,n} := h^T A_{m,n} U_m \, , \; Y_n := \sum_{m=1}^{n} Y_{m,n} \, , \; Y := h^T U \, .$$

Due to the "Cramér-Wold-Devise" (cf. e.g. [38], Theorem 8.7.6), only the convergence of $(Y_n)$ in distribution to $Y$ has to be shown.

Due to (K0), $(Y_{m,n}, \mathcal{A}_{m+1})_{m \leq n}$ is a martingale difference scheme. Therefore, according to the theorem of Brown (cf. [38], Theorem 9.2.3), the following conditions have to be verified:

$$\sum_{m=1}^{n} E_m Y_{m,n}^2 \overset{stoch}{\longrightarrow} h^T V h \, , \tag{KV1}$$

$$\sum_{m=1}^{n} E_m \left( Y_{m,n}^2 I_{\{|Y_{m,n}| > \delta\}} \right) \overset{stoch}{\longrightarrow} 0 \quad \text{for all} \quad \delta > 0 \, . \tag{KL1}$$

(KV1) follows from (KV), since $E_m Y_{m,n}^2 = h^T A_{m,n} E_m(U_m U_m^T) A_{m,n}^T h$ . Moreover, it is $|Y_{m,n}| \leq \|h\| \|A_{m,n} U_m\| \leq \|h\| \|A_{m,n}\| \|U_m\|$, and therefore for $\delta > 0$ we get

$$E_m(Y_{m,n}^2 I_{\{|Y_{m,n}|>\delta\}}) \\ \leq \|h\|^2 \|A_{m,n}\|^2 E_m\left(\|U_m\|^2 I_{\{\|A_{m,n} U_m\| > \frac{\delta}{\|h\|}\}}\right).$$

Hence, (KL) is sufficient for (KL1).

# C

# Tools from Matrix Calculus

We consider here real matrices of the size $\nu \times \nu$ .

## C.1 Miscellaneous

First we give inequalities about the error occuring in the inversion of a perturbed matrix, cf. [60].

**Lemma C.1.** *Perturbation lemma. Let $A$ and $\Delta A$ denote matrices such that*

$$A \text{ is regular and } \|A^{-1}\|_0 \cdot \|\Delta A\|_0 < 1 ,$$

*where $\|\cdot\|_0$ is an arbitrary matrix norm such that $\|I\|_0 = 1$.*

*Then, also $A + \Delta A$ is regular and*

$$a) \ \|(A+\Delta A)^{-1} - A^{-1}\|_0 \leq \frac{\|A^{-1}\|_0^2 \cdot \|\Delta A\|_0}{1 - \|A^{-1}\|_0 \cdot \|\Delta A\|_0} ,$$

$$b) \ \|(A+\Delta A)^{-1}\|_0 \leq \frac{\|A^{-1}\|_0}{1 - \|A^{-1}\|_0 \cdot \|\Delta A\|_0} .$$

A consequence of Lemma C.1 is given next.

**Lemma C.2.** *Let $(A_n)$ and $(B_n)$ denote two matrix sequences with the following properties:*

$$i) \ A_n \text{ regular, } \|A_n^{-1}\| \leq L < \infty \text{ for } n \in \mathbb{N} ,$$

$$ii) \ B_n - A_n \longrightarrow 0 .$$

*Then there is an index $n_0$ such that*

$$a) \ B_n \text{ is regular for } n \geq n_0 ,$$

$$b) \ B_n^{-1} - A_n^{-1} \longrightarrow 0 .$$

*Proof.* Due to assumption ii) there is an index $n_0$ such that

$$||B_n - A_n|| < \frac{1}{L}, n \geq n_0.$$

Hence, because of i) we get $||A_n^{-1}||||B_n - A_n|| < 1$. According to Lemma C.1, matrix $B_n$ is regular and

$$||B_n^{-1} - A_n^{-1}|| \leq \frac{L^2||B_n - A_n||}{1 - L||B_n - A_n||}.$$

Using ii), assertion b) follows.

If a rank-1-matrix is added to a regular matrix, then the determinant and the inverse of the new matrix can be computed easily. This is contained in the *Sherman-Morrison-lemma:*

**Lemma C.3.** *If* $e, d \in \mathbb{R}^\nu$ *and* $A$ *is a regular matrix, then*

a) $\det(A + ed^T) = \det(A)\alpha$, *where* $\alpha := 1 + d^T A^{-1} e$.

b) *If* $\alpha \neq 0$, *then* $(A + ed^T)^{-1} = A^{-1} - \frac{1}{\alpha}(A^{-1}e)(d^T A^{-1})$.

The proof of Lemma C.3 can be found e.g. in [60].

A lower bound for the minimum eigenvalue of a positiv definite matrix is given next, cf. [9], page 223:

**Lemma C.4.** *If* $A$ *is a symmetric, positiv definite* $\nu \times \nu$*-matrix, then*

$$\lambda_{\min}(A) > \det(A)(\frac{\nu - 1}{tr(A)})^{\nu-1}.$$

*Proof.* If $0 < \lambda_1 \leq \ldots \leq \lambda_\nu$ denotes all eigenvalues of $A$, then

$$tr(A) = \lambda_1 + \ldots + \lambda_n, \quad \det(A) = \lambda_1 \cdot \ldots \cdot \lambda_n.$$

Hence

$$\frac{\det(A)}{\lambda_1} = \lambda_2 \cdot \ldots \cdot \lambda_\nu \leq \left(\frac{\lambda_2 + \ldots + \lambda_\nu}{\nu - 1}\right)^{\nu-1} < \left(\frac{tr(A)}{\nu - 1}\right)^{\nu-1}.$$

## C.2 The v. Mises-Procedure in Case of Errors

For a symmetric, positiv semidefinite $\nu \times \nu$ matrix $A$, the maximum eigenvalue $\bar{\lambda} := \lambda_{\max}(A)$ of $A$ can be determined recursively by the v. Mises procedure, cf. [137], page 203:

Given a vector $0 \neq u_1 \in \mathbb{R}^\nu$, define recursively for $n = 1, 2, \ldots$

$$v_n := Au_n \quad \text{and} \quad u_{n+1} := \frac{v_n}{\|v_n\|} \,.$$

For a suitable start vector $u_1$, sequence $(u_n)$ converges to an eigenvector of the eigenvalue $\bar{\lambda}$ and $\langle u_n, v_n \rangle$ converges to $\bar{\lambda}$.

Suppose now that approximates $A_1$, $A_2$, ... of $A$ are given. Corresponding to the above procedure, for a start vector $u_1 \in \mathbb{R}^\nu$ with $\|u_1\| = 1$, the sequences $(u_n)$ and $(v_n)$ are determined by the following rule:

$$v_n := A_n u_n \quad \text{and} \quad u_{n+1} := \frac{v_n}{\|v_n\|} \quad \text{for} \quad n = 1, 2, \ldots \,. \tag{$*$}$$

We suppose that $A$ and $A_n$ fulfill the following properties:

a) $\nu > 1$, and $0 \neq A$ symmetric and positiv semidefinite,
b) $A_n \longrightarrow A$ .

Based on the following lemma, $\langle u_n, v_n \rangle$ converges to $\bar{\lambda}$, provided that an additional condition holds.

Let denote next $U := \{x \in \mathbb{R}^\nu : Ax = \bar{\lambda}x\}$ the eigenspace of the maximum eigenvalue $\bar{\lambda}$ of $A$. Each vector $z \in \mathbb{R}^\nu$ can then be decomposed uniquely as follows:

$$\begin{aligned} &z = z^{(1)} + z^{(2)}, \quad \text{where} \\ &z^{(1)} \in U \quad \text{and} \quad z^{(2)} \in U^\perp := \{y \in \mathbb{R}^\nu : \langle y, x \rangle = 0 \ \text{for all} \ x \in U\} \,. \end{aligned}$$

Now the already mentioned main result of this part of the appendix can be presented:

**Lemma C.5.** *Let $(u_n)$ and $(v_n)$ be determined according to $(*)$. Then the following assertions are equivalent:*

*a)* $\langle u_n, v_n \rangle \longrightarrow \bar{\lambda}$ .
*b) There is an index $n_0 \in \mathbb{N}$ such that* $\sup_{n \geq n_0} \|A_n - A\| < \frac{\bar{\lambda}-\mu}{3} \|u_{n_0}^{(1)}\|$.

*Here, $\mu := \lambda_{\nu-1}$ (cf. the proof of Lemma C.4) or $\mu := 0$ if $U = \mathbb{R}^\nu$ . The vector $u_{n_0}^{(1)}$ denotes the projection of $u_{n_0}$ onto $U$.*

In the standard case $A_n \equiv A$ the second condition in Lemma C.5 means that

$$\|u_{n_0}^{(1)}\| > 0 \quad \text{for an index } n_0 \in \mathbb{N} \,.$$

This is the known condition for the convergence of $\langle u_n, v_n \rangle$ to $\bar{\lambda}$ .

For the proof of Lemma C.5 we need several auxiliary results. Put $q := \frac{\mu}{\bar{\lambda}} \in [0, 1)$, and define for $n \in \mathbb{N}$

$$t_n := \begin{cases} \frac{\|u_n^{(2)}\|}{\|u_n^{(1)}\|} & , \quad \text{if } u_n^{(1)} \neq 0 \\ \infty & , \quad \text{else} , \end{cases}$$

$$d_n := \frac{1}{\bar{\lambda}} \|A_n - A\| \quad \text{and} \quad e_n := \sup_{k \geq n} d_k .$$

Then, $\|u_n^{(1)}\| = \frac{1}{\sqrt{1+t_n^2}}$, and the following lemma holds:

**Lemma C.6.** *If*

$$d_n \sqrt{1+t_n^2} < 1 \Longleftrightarrow \|A_n - A\| < \bar{\lambda} \|u_n^{(1)}\|,$$

*then*

$$t_{n+1} \leq \frac{q t_n + d_n \sqrt{1+t_n^2}}{1 - d_n \sqrt{1+t_n^2}} .$$

*Proof.* Because of $v_n = A_n u_n$, decomposition of $v_n$ yields:

$$v_n^{(1)} = P(A_n u_n) = P\big(A u_n + (A_n - A) u_n\big) = \bar{\lambda} u_n^{(1)} + P(A_n - A) u_n ,$$
$$v_n^{(2)} = Q(A_n u_n) = A u_n^{(2)} + Q(A_n - A) u_n .$$

Here, $P$ is the projection matrix describing the projection of $z \in \mathbb{R}^\nu$ onto $z^{(1)} \in U$, and $Q := I - P$. Taking the norm, one obtains

$$\|v_n^{(1)}\| \geq \bar{\lambda} \|u_n^{(1)}\| - \|A_n - A\| , \tag{i}$$
$$\|v_n^{(2)}\| \leq \mu \|u_n^{(2)}\| + \|A_n - A\| , \tag{ii}$$

if the following relations are taken into account:

$$\|A u_n^{(2)}\| \leq \mu \|u_n^{(2)}\| \quad , \quad \|P\|, \|Q\| \leq 1 \quad , \quad \|u_n\| = 1 .$$

From (i), (ii) and the assumption $\bar{\lambda}\|u_n^{(1)}\| - \|A_n - A\| > 0$ we finally get $v_n^{(1)} \neq 0$ and

$$t_{n+1} := \frac{\|u_{n+1}^{(2)}\|}{\|u_{n+1}^{(1)}\|} = \frac{\|v_n^{(2)}\|}{\|v_n^{(1)}\|} \leq \frac{\mu \|u_n^{(2)}\| + \|A_n - A\|}{\bar{\lambda} \|u_n^{(1)}\| - \|A_n - A\|}$$
$$= \frac{q t_n + d_n \sqrt{1+t_n^2}}{1 - d_n \sqrt{1+t_n^2}} .$$

In the next step conditions are given such that $(t_n)$ converges to zero:

**Lemma C.7.** *Assume that there is a number $T \in (0, \infty)$ and an index $n_0 \in \mathbb{N}$ with*

$$a) \; e_{n_0} \leq \epsilon(T) := \frac{(1-q)T}{(1+T)\sqrt{1+T^2}} ,$$

$$b) \; t_{n_0} \leq T .$$

*Then $t_n \to 0$, $n \to \infty$.*

*Proof.* By induction we show first that

$$t_n \le T, \quad n \ge n_0 . \tag{**}$$

Given $n \ge n_0$, suppose $t_n \le T$ . Because of a) we have

$$d_n\sqrt{1+t_n^2} \le e_{n_0}\sqrt{1+t_n^2} \le \frac{(1-q)T}{1+T}\sqrt{\frac{1+t_n^2}{1+T^2}} \le \frac{(1-q)T}{1+T} < 1 .$$

Hence, Lemma C.6 yields

$$t_{n+1} \le \frac{qt_n + d_n\sqrt{1+t_n^2}}{1-d_n\sqrt{1+t_n^2}} \le \frac{qT + d_n\sqrt{1+t_n^2}}{1-d_n\sqrt{1+t_n^2}} \le T$$

and (**) holds.

Since $(d_n)$ converges to zero, there is an index $n_1 \ge n_0$ with

$$1 - d_n\sqrt{1+T^2} \ge \frac{1+q}{2}, \quad n \ge n_1 .$$

By means of Lemma C.6 and (**), for $n \ge n_1$ we get

$$t_{n+1} \le \frac{qt_n + d_n\sqrt{1+T^2}}{\frac{1+q}{2}} = \frac{2q}{1+q}t_n + d_n\frac{2\sqrt{1+T^2}}{1+q} .$$

Because of $\frac{2q}{1+q} \in [0,1)$ and $d_n \longrightarrow 0$, we finally have $t_n \longrightarrow 0$.

The meaning of "$t_n \longrightarrow 0$" is shown in the next lemma:

**Lemma C.8.** *The following assertions are equivalent:*

*a)* $\langle u_n, v_n\rangle \longrightarrow \bar\lambda$ ,
*b)* $\|u_n^{(1)}\| \longrightarrow 1$ ,
*c)* $t_n \longrightarrow 0$ .

*Proof.* Because of $\|u_n^{(1)}\| = (1+t_n^2)^{-\frac{1}{2}}$ the equivalence of b) and c) is clear. For $n \in \mathbb{N}$ we have

$$\begin{aligned}\langle u_n, v_n\rangle &= u_n^T\big(A + (A_n - A)\big)u_n \\ &= (u_n^{(1)} + u_n^{(2)})^T A(u_n^{(1)} + u_n^{(2)}) + u_n^T(A_n - A)u_n \\ &= \bar\lambda\|u_n^{(1)}\|^2 + u_n^{(2)T}Au_n^{(2)} + u_n^T(A_n - A)u_n ,\end{aligned}$$

and

$$\left|u_n^T(A_n - A)u_n\right| \le \|A_n - A\| \quad , \quad 0 \le u_n^{(2)T}Au_n^{(2)} \le \mu\|u_n^{(2)}\|^2 .$$

Thus, for each $n \in \mathbb{N}$ we get

$$\|u_n^{(1)}\|^2 - d_n \leq \frac{\langle u_n, v_n\rangle}{\bar{\lambda}} \leq \|u_n^{(1)}\|^2 + q\|u_n^{(2)}\|^2 + d_n$$
$$= q + (1-q)\|u_n^{(1)}\|^2 + d_n \,.$$

Due to $d_n \longrightarrow 0$ and $q \in [0,1)$ we obtain now also the equivalence of a) and b).

Now Lemma C.5 can be proved.

*Proof (of Lemma C.5).* If

$$\langle u_n, v_n\rangle \longrightarrow \bar{\lambda}\,, \tag{i}$$

then because of $A_n \longrightarrow A$ and Lemma C.8 there is an index $n_0$ such that

$$\sup_{n \geq n_0} \|A_n - A\| < \frac{\bar{\lambda} - \mu}{3}\|u_{n_0}^{(1)}\| \iff e_{n_0} < \frac{1-q}{3}\|u_{n_0}^{(1)}\|\,. \tag{ii}$$

Conversely, suppose that this condition holds. Then there are two possibilities: Case 1: $e_{n_0} = 0$. Then $u_{n_0}^{(1)} \neq 0$, $t_{n_0} < \infty$, and also $e_{n_0} = 0 < \epsilon(t_{n_0})$ with function $\epsilon(t)$ from Lemma C.7. According to this Lemma we get then $t_n \longrightarrow 0$. Case 2: $e_{n_0} > 0$. Due to (ii) it is

$$0 < e_{n_0} \leq \frac{1-q}{3} < \frac{1}{2\sqrt{2}}(1-q) = \epsilon(1)\,.$$

Since $\epsilon(t)$ is continuous on $[1,\infty)$ and decreases monotonically to zero, there is a number $T \in [1,\infty)$ such that

$$e_{n_0} = \epsilon(T)\,. \tag{iii}$$

Because of $\epsilon(t)\cdot t \geq \frac{1-q}{3}$ for $t \geq 1$, with (ii) and (iii) we find

$$\frac{1-q}{3}\|u_{n_0}^{(2)}\| \leq \frac{1-q}{3} \leq \epsilon(T)T = e_{n_0}T < \frac{1-q}{3}\|u_{n_0}^{(1)}\|T\,,$$

and therefore

$$t_{n_0} := \frac{\|u_{n_0}^{(2)}\|}{\|u_{n_0}^{(1)}\|} \leq T\,. \tag{iv}$$

From (iii) and (iv) with Lemma C.7 we obtain again $t_n \longrightarrow 0$.

Due to Lemma C.8, condition " $\lim_n t_n = 0$ " implies (i) which concludes now the proof.

# References


1. Anbar D. (1978) A Stochastic Newton–Raphson Method. Journal of Statistical Planing and Inference **2**: 153–163
2. Augusti G, Baratta A., Casciati F. (1984) Probabilistic methods in structural engineering. Chapman and Hall, London, New York
3. Aurnhammer A. (2004) Optimale Stochastische Trajektorienplanung und Regelung von Industrierobotern. Fortschrittberichte VDI, Reihe 8, Nr. 1032. VDI-Verlag GmbH, Düsseldorf
4. Barner M., Flohr, F. (1996) Analysis II. Walter de Gruyter, Berlin, New York
5. Barthelemy J.–F., Haftka R. (1991) Recent Advances in Approximation Concepts for Optimum Structural Design. In: Rozvany, G. (ed.) Optimization of Large Structural Systems. Lecture Notes, Vol. 1: 96-108, NATO/DFG ASI, Berchtesgaden, Sept. 23-Oct. 4
6. Bastian G. et al (1996) Theory of robot control. Sringer Verlag, Berlin
7. Bauer H. (1996) Probability theory. De Gruyter, Berlin [etc.]
8. Bellman R. (1970) Introduction to Matrix Analysis. McGraw–Hill, New York
9. Beresin I.S.,Shidkow N.P. (1971) Numerische Methoden. 2. VEB Deutscher Verlag der Wissenschaften, Berlin
10. Berger J. (1985) Statistical Decision Theory and Bauyesian Analysis. Springer Verlag, New York [etc.]
11. Bertsekas D.P. (1973) Stochastic Optimization Problems with Nondifferentiable Cost Functionals. JOTA **12**: 218–231
12. Betro' B., de Biase L. (1978) A Newton–Like Method for Stochastic Optimization. Towards Global Optimization **2**, L.C.W. Dixon and G.P. Szegö (eds.), North–Holland Publishing Company, 269–289
13. Biles W.E., Swain J.J. (1979) Mathematical Programming and the Optimization of Computer Simulations. *Math. Programming Studies* 11: 189–207
14. Bleistein N., Handelsmann R. (1975) Asymptotic Expansions of Integrals. Holt, Rinehart and Winston, New York
15. Blum E., Oettli W. (1975) Mathematische Optimierung. Springer-Verlag, Berlin, Heidelberg, New York
16. Box G.E., Draper N.R. (1987) Empirical Model–Building and Response Surfaces. J. Wiley, New York, Chichester, Brisbane, Toronto, Singapore
17. Box G.E.P., Wilson K.G. (1951) On the experimental attainment of optimum conditions. *J. Royal Statistical Soc.* 13: 1-45

18. Breitung K., Hohenbichler M. (1989) Asymptotic approximations for multivariate integrals with an application to multinormal probabilities. Journal of Multivariate Analysis **30**: 80-97
19. Breitung K. (1984) Asymptotic Approximations for Multinormal Integrals. ASCE J. of the Engineering Mechanics Division 110, Nor. 1: 357–366
20. Breitung K. (1990) Parameter Sensitivity of Failure Probabilities. In: A. Der Kiureghian, P. Thoft–Christensen (eds.): Reliability and Optimization of Structural Systems '90. Lecture Notes in Engineering, Vol. 61, 43–51. Springer-Verlag, Berlin–Heidelberg–New York
21. Bronstein I.N., Semendjajew K.A. (1980) Taschenbuch der Mathematik. Verlag Harri Deutsch, Thun, Frankfurt/Main
22. Bucher C.G., Bourgund U. (1990) A Fast And Efficient Response Surface Approach For Structural Reliability Problems. Structural Safety **7**: 57–66
23. Chankong V., Haimes Y. Y. (1983) Multiobjective Decision Making. Oxford: North Holland, New York, Amsterdam
24. Chung K.L. (1954) On a Stochastic Approximation Method. Ann. Math. Statist. **25**: 463–483
25. Craig J.J. (1988) Adaptive Control of Mechanical Manipulators. Reading, Mass. (etc.). Addison-Wesley
26. Der Kiureghian A., Thoft-Christensen P. (eds.) (1990) Reliability and Optimization of Structural Systems '90. Lecture Notes in Engineering, Vol. 61. Springer-Verlag, Berlin, Heidelberg, New York
27. Dieudonné J. (1960) Foundations of modern analysis. Academic Press, New York (etc.)
28. Dolinski K. (1983) First–order second moment approximation in reliability of systems: critical review and alternative approach. Structural safety **1**: 211-213
29. Ermoliev Y. (1983) Stochastic Quasigradient Methods and their Application to System Optimization. Stochastics **9**: 1–36
30. Ermoliev Y., Gaivoronski A. (1983) Stochastic Quasigradient Methods and their Implementation. IIASA Workpaper, Laxenburg
31. Ermoliev Yu. (1988) Stochastic Quasigradient Methods. In: Ermoliev, Yu; Wets, R.J.–B. (eds.): Numerical Techniques for Stochastic Optimization, Springer Verlag, Berlin, Heidelberg, New York, 141–185
32. Ermoliev Yu., Wets R. (eds.) (1988) Numerical Techniques for Stochastic Optimization. Springer Series in Computational Mathematics Vol. 10. Springer-Verlag, Berlin, Heidelberg, New York
33. Eschenauer H.A., et al. (1991) Engineering Optimization of Design Processes. Lecture Notes in Engineering, Vol. 63. Springer-Verlag, Berlin
34. Fabian V. (1967) Stochastic approximation of minima with improved asymptotic speed. Ann. Math. Statist. **38**: 191–200
35. Fabian V. (1968) On Asymptotic Normality in Stochastic Approximation. Ann. Math. Statist. **39**: 1327–1332
36. Fabian V. (1971) Stochastic Approximation. In: J.S. Rustagi (ed.): Optimizing Methods in Statistics, Academic Press, New York–London, 439–470
37. Fleury C. (1989) First and second order approximation strategies in structural optimization. Structural Optimization **1**: 3–10
38. Gänssler P., Stute W. (1977) Wahrscheinlichkeitstheorie, Springer-Verlag, Berlin

39. Gaivoronski A. (1988) Stochastic Quasigradient Methods And Their Implementation. In: Yu. Ermolief; R.J.-B. Wets (eds.): Numerical Techniques for Stochastic Optimization, Springer-Verlag, Berlin, Heidelberg, New York, 313–351
40. Galambos J., Simonelli I. (1996) Bonferroni-type inequalities with applications. Springer-Verlag, New York [etc.]
41. Gates T.K., Wets R.J.-B., Grismer M.E. (1989) Stochastic Approximation Applied To Optimal Irrigation And Drainage Planning. Journal of Irrigation and Drainage Engineering, **115** No. 3: 488–502
42. Gill P. E., Murray W., Wright M. H. (1981) Practical Optimization. Academic Press, New York, London
43. Graf Fink von Finkenstein (1977) Einführung in die Numerische Mathematik, C. Hander Verlag, München
44. Han R., Reddy B.D. (1999) Plasticity-Mathematical Theory and Numerical Analysis. Springer-Verlag, New York (etc.)
45. Henrici P. (1964) Elements of Numerical Analysis. J. Wiley, New York
46. Hiriart-Urruty J.B. (1977) Contributions a la programmation mathematique: Cas deterministe et stochastique, Thesis, University of Clermont-Ferrand II
47. Hodge P.G. (1959) Plastic Analysis of Structurals. McGraw-Hill, New York
48. Hohenbichler M., Rackwitz R. et al. (1987) New light on first and second order reliability methods. Structural Safety **4**: 267-284
49. Jacobsen Sh.H., Schruben Lee W. (1989) Techniques For Simulation Response Optimization, Operations Research Letters **8**: 1-9
50. Johnson N.L., Kotz S. (1976) Distributions in Statistics: Continuous Multivariate Distributions. Wiley, New York-London
51. Kall P. (1976) Stochastic linear programming. Econometrics and Operations Research, Vol. XXI. Springer-Verlag, Berlin, Heidelberg, New York
52. Kall P., Wallace S.W. (1994) Stochastic Programming. J. Wiley, Chichester [etc.]
53. Kesten H. (1958) Accelerated stochastic approximation. Ann. Math. Statist. **29**: 41–59
54. Kiefer J., Wolfowitz J. (1952) Stochastic Estimation of the Maximum of a Regression Function. Ann. Math. Statist. **23**: 462–466
55. Kirsch U. (1993) Structural Optimization. Springer-Verlag, Berlin, Heidelberg, New York
56. Kleijnen J.P.C. (1987) Statistical tools for simulation practitioners. Marcel Decker, New York
57. Knopp K. (1964) Theorie und Anwendung der unendlichen Reihen. Springer-Verlag, Berlin
58. König J.A. (1987) Shakedown of elastic-Plastic Structures. Elsevier, Amsterdam [etc.]
59. Kolmogorof A.N., Fomin S.V. (1975) Reelle Funktionen und Funktionananalysis. VEB Deutscher Verlag der Wissenschaften, Berlin
60. Kosmol P. (1989) Methoden zur numerischen Behandlung nichtlinearer Gleichungen und Optimierungsaufgaben. B.G. Teubner, Stuttgart
61. Kushner H.J., Clark D.S. (1978) Stochastic Approximation Methods for Constrained and Unconstrained Systems. Springer-Verlag, Berlin, Heidelberg, New York
62. Lai T.L., Robbins, H. (1979) Adaptive Design and Stochastic Approximation. The Annals of Statistics Vol. 7, No. 6: 1196–1221

63. Ljung L. (1977) Analysis of Recursive Stochastic Algorithms. IEEE Transactions on Automatic Control, AC **22**: 551–575
64. Luenberger, D.G. (1969) Optimization by vector space methods. J. Wiley, New York (etc.)
65. Luenberger D.G. (1973) Introduction to linear and nonlinear programming. Addison-Wesley Publ. Comp., Reading, Ma (etc.)
66. Mangasarian O. L. (1969) Nonlinear Programming. McGraw-Hill, New York, London, Toronto
67. Marti K. (1977) Stochastische Dominanz und stochastische lineare Programme. Methods of Operations Research **23**: 141–160
68. Marti K. (1978) Stochastic linear programs with stable distributed random variables. In: J. Stoer (ed.): Optimization Techniques Part 2. *Lecture Notes in Control and Information Sciences* **7**: 76–86
69. Marti K. (1979) Approximationen Stochastischer Optimierungsprobleme. A. Hain, Königsstein/Ts.
70. Marti K. (1979) On solutions of stochastic programming problems by descent procedurees with stochastic and deterministic directions. Methods of Operations Research **33**: 281–293
71. Marti K. (1980) Solving stochastic linear programs by semi-stochastic approximation algorithms. In: P. Kall; A. Prekopa (eds.): Recent result in stochastic programming. *Lecture Notes in Economics and Mathematical Systems*, **179**: 191–213
72. Marti K. (1983) Descent Directions and Efficient Solutions in Discretely Distributed Stochastic Programs. LN in Economics and Mathematical Systems **299**. Springer-Verlag, Berlin, Heidelberg, New York
73. Marti K. (1984) On the construction of descent directions in a stochastic program having a discrete distribution. ZAMM **64**: T336–T338
74. Marti K. (1985) Computation of descent directions in stochastic optimization problems with invariant distributions. ZAMM **65**: 355–378
75. Marti K., Fuchs, E. (1986) Computation of descent directions and efficient points in stochastic optimization problems without using derivatives. Math. Prog. Study **28**: 132–156
76. Marti K., Fuchs E. (1986) Rates of Convergence of Semi-Stochastic Approximation Procedures for Solving Stochastic Optimization Problems. Optimization **17**: 243–265
77. Marti K. (1986) On accelerations of stochastic gradient methods by using more exact gradient estimations. Methods of Operations Research **53**: 327–336
78. Marti K. (1987) Solving Stochastic Optimization Problems By Semi–Stochastic Approximation Procedures. In: Borne/Tzafestas (eds.): Applied Modelling and Simulation of Technological Systems, 559–568. Elsevier Science Publ. B.V. (North Holland), Amsterdam, New York, Oxford, Tokyo
79. Marti K. (1990) Stochastic Optimization Methods in Structural Mechanics. ZAMM 70: T 742-T745
80. Marti K., Plöchinger E. (1990) Optimal Step Sizes in Semi-Stochastic Approximation Procedures, I and II, Optimization **21**, No. 1: 123–153 and No. 2: 281–312
81. Marti K. (1990) Stochastic Programming: Numerical Solution Techniques by Semi–Stochastic Approximation Methods. In: R. Slowinski, J. Teghem (eds.): Stochastic versus Fuzzy Approaches to Multiobjective Programming under Uncertainty, 23–43. Kluwer Academic Publ., Boston, Dordrecht, London

82. Marti K. (1990) Stochastic Optimization in Structural Design. ZAMM 72: T452-T464
83. Marti K. (1992) Approximations and Derivatives of Probabilities in Structural Design. ZAMM 72 **6**: T575-T578
84. Marti K. (1992) Computation of Efficient Solutions of Discretely Distributed Stochastic Optimization Problems. ZOR **36**: 259–296
85. Marti K. (1994) Approximations and derivatives of probability functions. In: C. Anastassiou, S.T. Rachev (eds.): Approximation, probability and related fields, 367–377. Plenum Publ. Corp, New York
86. Marti K. (1995) Differentiation of Probability Functions: The Transformation Method. Computers Math. Appl. 30, No. 3–6: 361–382
87. Marti K. (1995) Computation of Probability Functions and its Derivatives by means of Orthogonal Function Series expansions. In: K. Marti, P. Kall (eds.): Stochastic Programming: Numerical Techniques and Engineering Applications, 22–53, LNEMS 423. Springer-Verlag, Berlin, Heidelberg, New York
88. Marti K. (1999) Path Planning for Robots under Stochastic Uncertainty. Optimization **45**: 163–195
89. Marti K. (2003) Stochastic optimization methods in optimal engineering design under stochastic uncertainty. ZAMM 83, No. 11: 1–18
90. Marti K. (2003) Plastic Structural Analysis Under Stochastic Uncertainty. Math. and Comp. Mod. of Dyn. Syst. (MCMDS) **9**, No. 2 : 303–325
91. Marti K., Ermoliev Y., Pflug G. (eds.) (2004) Dynamic Stochastic Optimization. LNEMS 532. Springer-Verlag, Berlin, Heidelberg
92. Mauro C.A. (1984) On the performance of two-stage group screening experiments. Technometrics **26** (3): 255–264
93. McGuire W., Gallagher R.H. (1979) Matrix structural analysis. J. Wiley, New York, London
94. Melchers R.R. (1999) Structural reliability: Analysis and prediction. Second edition. Ellis Horwood Ltd., Chichester (England)
95. Montgomery C.A. (1984) Design and Analysis of Experiments. J. Wiley, New York
96. Myers R.H. (1971) Response surface methodology. Allyn and Bacon, Boston
97. Nevel'son M.B., Khas'minskii R.Z. (1973) An Adaptive Robbins-Monro Procedure. Automation and Remote Control **34**: 1594–1607
98. Pantel M. (1979) Adaptive Verfahren der Stochastischen Approximation. Dissertation, Universität Essen-Gesamthochschule, FB 6-Mathematik
99. Park S.H. (1996) Robust design and analysis for quality engineering. Chapman and Hall, London
100. Pfeiffer F., Johanni R. (1987) A concept for manipulator trajectory planning. IEEE J. Robotics and Automation RA-3: 115–123
101. Pflug G. Ch. (1988) Step Size Rules, Stopping Times And Their Implementation In Stochastic Quasigradient Algorithms. In: Yu. Ermoliev; R.J.-B. Wets (eds.): Numerical Techniques for Stochastic Optimization, Springer-Verlag, Berlin, Heidelberg, New York, 353–372
102. Phadke M.S. (1989) Quality engineering using robust design. P.T.R. Prentice Hall, Englewood Cliffs, N.J.
103. Pahl G., Beitz W. (2003) Engineering design: A systematic approach. Springer-Verlag, London (etc.)

104. Plöchinger E. (1992) Realisierung von adaptiven Schrittweiten für stochastische Approximationsverfahren bei unterschiedlichem Varianzverhalten des Schätzfehlers. Dissertation, Universität der Bundeswehr München
105. Polyak B.T., Tsypkin, Y.Z. (1980) Robust pseudogradient adaption algorithms. Automation and Remote Control **41**: 1404–1410
106. Polyak B.G., Tsypkin Y.Z. (1980) Optimal Pseudogradient Adaption Algorithms. Automatika i Telemekhanika **8**: 74-84
107. Prekopa A.; Szantai T. (1978) Flood control reservoir system design using stochastic programming. Math. Prog. Stuy **9**: 138-151
108. Press S.J. (1972) Applied Multivariate Analysis. Holt, Rinehart and Winston, New York-London
109. Rackwitz R., Cuntze R. (1987) Formulations of reliability–oriented optimization. Engineering Optimization **11**: 69-76
110. Reinhart J. (1997) Stochastische Optimierung von Faserkunststoffverbundplatten. Fortschritt–Berichte VDI, Reihe 5, Nr. 463. VDI Verlag GmbH, Düsseldorf
111. Reinhart J. (1998) Implementation of the Response Surface Method (RSM) for Stochastic Structural Optimization Problems. In: K. Marti, P. Kall (Eds.): Stochastic Programming Methods and Technical Applications. Lecture Notes in Economics and Mathematical Systems, 458: 394-409
112. Review of Economic Design, ISSN: 1434–4742/50. Springer-Verlag, Heidelberg
113. Richter H. (1966) Wahrscheinlichkeitstheorie. Springer-Verlag, Berlin
114. Robbins H., Monro, S. (1951) A Stochastic Approximation Method. Ann. Math. Statist. **22**: 400–407
115. Robbins H., Siegmund D. (1971) A Convergence Theorem for Non Negative Almost Supermaringales and Some Applications. Optimizing Methods in Statistics. Academic Press, New York, 233–257
116. Rommelfanger H. (1977) Differenzen– und Differentialgleichungen. Bibliographisches Institut Mannheim
117. Rozvany G.I.N. (1997) Topology Optimization in Structural Mechanics. Springer-Verlag, Wien, New York
118. Ruppert D. (1985) A Newton–Raphson Version of the Multivariate Robbins–Monro Procedure. The Annals of Statistics Vol. 13, No. 1: 236–245
119. Sacks J. (1958) Asymptotic Distributions of Stochastic Approximation Procedures. Ann. Math. Stat. **29**: 373–405
120. Sansone G. (1959) Orthogonal Functions. Interscience Publishers Inc., New York, London
121. Sawaragi Y., Nakayama, H., Tanino, T. (1985) Theory of Multiobjective Optimization. Academic Press, New York, London, Tokyo
122. Schmetterer L. (1960) Stochastic approximation. Proceedings 4th Berkeley Symposium on Mathematical Statistics and Probability **1**: 587–609
123. Schoofs A.J.G. (1987) Experimental Design and Structural Optimization. Doctoral Thesis, Technical University of Eindhoven
124. Schuëller G.I. (1987) A critical appraisal of methods to determine failure probabilities. J. structural Safety **4**: 293-309
125. Schuëller G.I., Gasser M. (1998) Some Baxic Principles of Reliability-Based Optimization (RBO) of Structure and Mechanical Components. In: K. Marti, P. Kall (eds.): Stochastic Programming-Numerical Techniques and Engineering. LNEMS Vol. 458: 80–103. Springer-Verlag, Berlin, Heidelberg

126. Schüler L. (1974) Schätzung von Dichten und Verteilungsfunktionen mehrdimensionaler Zufallsvariabler auf der Basis orthogonaler Reihen. Dissertation Naturwissenschaftliche Fakultät der TU Braunschweig, Braunschweig
127. Schüler L., Wolff H. (1976) Schätzungen mehrdimensionaler Dichten mit Hermiteschen Polynomen und lineare Verfahren zur Trendverfolgung. Forschungsbericht BMVg-FBWT 76–23, TU Braunschweig, Institut für Angwandte Mathematik, Braunschweig
128. Schwartz S.C. (1967) Estimation of Probability Density by an Orthogonal Series. Ann. Math. Statist. **38**: 1261–1265
129. Schwarz H.R. (1986) Numerische Mathematik. B.G. Teubner, Stuttgart
130. Sciavicco L., Siciliano B. (2000) Modeling and Control of Robot Manipulators. Springer-Verlag, London [etc.]
131. Slotine J.-J., Li W. (1991) Applied Nonlinear Control. Prentice-Hall Int. Inc., Englewood Chiffs, N.J.
132. Sobieszczanski–Sobieski J.G., Dovi A. (1983) Structural Optimization by Multi-Level Decomposition. AIAA J. **23**: 1775–1782
133. Spillers W.R. (1972) Automatd Structural Analysis: An Introduction. Pergamon, New York, Toronto, Oxford
134. Stengel R.F. (1986) Stochastic optimal control: theory and application. J. Wiley, New York [etc.]
135. Stöckl G. (2003) Optimaler Entwurf elastoplastischer mechanischer Strukturen unter stochastischer Unsicherheit. Fortschritt-Berichte VDI, Reihe 18, Nr. 278. VDI-Verlag GmbH, Düsseldorf
136. Streeter V.L., Wylie E.B. (1979) Fluid mechanics. McGraw–Hill, New York
137. Stummel F., Hainer K. (1982) Praktische Mathematik. B.B. Teubner, Stuttgart
138. Syski W. (1987) Stochastic Approximation Method with Subgradient Filtering and On-Line Stepsize Rules for Nonsmooth, Nonconvex and Unconstrained Problems. Control and Cybernetics Vl 16, **1**: 59–76
139. Szantai T. (1986) Evaluation of a special multivariate gamma distribution function. Math.Prog. Study **27**: 1-16
140. Thoft–Christensen P., Baker M.J. (1982) Structural Reliability Theory and its Applications. Springer-Verlag, Berlin
141. Tong Y.L. (1980) Probability Inequalities in Multivariate Distributions. Academic Press, New York-London
142. Toropov V.V. (1989) Simulation Approach to Structural Optimization. Structural Optimization **1** (1): 37–46
143. Tricomi F.G. (1970) Vorlesungen über Orthogonalreihen. Springer-Verlag, Berlin, Heidelberg, New York
144. Triebel H. (1981) Analysis und mathematische Physik. BSB B.B. Teubner Verlagsgesellschaft, Leipzig
145. Uryas'ev St. (1989) A Differentiation Formula for integrals over sets given by inclusion. Numer. Funct. Anal. and Optimiz. **10**: 827-841
146. Venter Y.H. (1966) On Dvoretzky Stochastic Aproximation Theorems. Ann. Math. Statist. **37**: 1534–1544
147. Venter Y.H. (1967) An Extention of the Robbings–Monro Procedure. Ann. Math. Statist. **38**: 181–190
148. Walk H. (1977) An Invariance Principle for the Robbings–Monro Process in a Hilbert Space. Z. Wahrscheinlichkeitstheorie verw. Gebiete **39**: 135–150
149. Wasan M.T. (1969) Stochastic Approximation. University Press, Cambridge

150. Wets R. (1983) Stochastic Programming: Solution Techniques and Approximation Schemes. In: Stochastic Programming A. Bachem; M. Grötschel; B. Korte (eds.): Mathematical Programming: The State of the Art. Springer-Verlag, Berlin, Heidelberg, New York, 566–603

# Index

Printing: Strauss GmbH, Mörlenbach
Binding: Schäffer, Grünstadt